Ecosystem Services for Well-Being in Deltas

Robert J. Nicholls
Craig W. Hutton
W. Neil Adger • Susan E. Hanson
Md. Munsur Rahman
Mashfiqus Salehin
Editors

Ecosystem Services for Well-Being in Deltas

Integrated Assessment for Policy Analysis

palgrave
macmillan

Editors
Robert J. Nicholls
Faculty of Engineering and the Environment
and Tyndall Centre for Climate Change
Research, University of Southampton
Southampton, UK

Craig W. Hutton
Geodata Institute
Geography and Environment
University of Southampton
Southampton, UK

W. Neil Adger
Geography, College of Life and
Environmental Sciences
University of Exeter
Exeter, UK

Susan E. Hanson
Faculty of Engineering and the
Environment and Tyndall Centre for
Climate Change Research
University of Southampton
Southampton, UK

Md. Munsur Rahman
Institute of Water and Flood Management
Bangladesh University of Engineering and
Technology
Dhaka, Bangladesh

Mashfiqus Salehin
Institute of Water and Flood Management
Bangladesh University of Engineering and
Technology
Dhaka, Bangladesh

espa
ecosystem services
for poverty alleviation

ISBN 978-3-319-71092-1   ISBN 978-3-319-71093-8   (eBook)
https://doi.org/10.1007/978-3-319-71093-8

Library of Congress Control Number: 2018941812

This Palgrave Macmillan imprint is published by the registered company Springer Nature Switzerland AG
The registered company address is: Gewerbestrasse 11, 6330 Cham, Switzerland

Dedication: To Dr. Nazmul Haq, University of Southampton, aka 'Uncle',
for his facilitation of the research herein and his ongoing commitment to the
people of Bangladesh.

Foreword

Deltas, Ecosystem Services and the Sustainable Well-Being of Humans and the Rest of Nature

Humanity is finally rediscovering an important relationship—the inter-dependent relationship between humans and the rest of nature. The industrial revolution and some religious traditions have emphasised the distinctions between humans and 'nature'—that humans are somehow above, apart from or fundamentally different from the rest of nature. In fact, the more we learn about the way the world and its complex inter-connected systems function, the more we recognise that *homo sapiens* is, and has always been, an integral component of the ecosystems it is embedded within. Humans are not apart from nature, but are a part of the natural world, and their health and well-being cannot be understood or managed separate from that complex and evolving context.

The concept of *ecosystem services* makes this interdependence with the rest of nature more apparent and quantitative. It does this by analysing, modelling, quantifying and valuing the degree to which humans are connected with and benefit from the ecosystems that enclose them. Ecosystems provide a range of services that are of fundamental impor-tance to human well-being, health, livelihoods and survival (Costanza et al. 1997; Daily 1997; MEA 2005; de Groot et al. 2014).

The idea that preserving the environment *as an asset*, rather than an impediment to economic and social development, is both very old and very new. For most of human history, at least until the start of the Industrial Revolution, the benefits humans derived from the rest of nature were well recognised and embedded in various cultural rules and norms. Parts of forests, lakes, wetlands or mountains were often deemed sacred and off limits. But it is no coincidence that these sacred natural assets also supplied essential life-support services for the communities involved. This is in stark contrast to the post-industrial view in much of the Western world that nature is merely a pretty picture—nice to enjoy if you can afford it but not essential to the more important business of 'growing the economy'. Too often, when the issue of conservation of the environment has entered public or political discussions, it has been purported to come at a cost, and the discussion has been framed as 'the environment versus the economy'.

Probably the most important contribution of the widespread recognition of *ecosystem services* is that it reframes the relationship between humans and the rest of nature to be more consistent with what we know. A better understanding of the role of ecosystem services emphasises our natural assets as critical ingredients to inclusive wealth, well-being and sustainability. Sustaining and enhancing human well-being requires a balance of all of our assets—individual people, society, the built economy and ecosystems. This reframing of the way we look at 'nature' is essential to solving the problem of how to build a sustainable and desirable future for humanity—a goal that we all share.

The ecosystem services concept makes it abundantly clear that the choice of 'the environment versus the economy' is a false choice. If the environment contributes significantly to human well-being, then it is a major contributor to the *real* economy and the choice becomes how to manage *all* our assets, including natural and human-made capital, more effectively and sustainably (Costanza et al. 2000).

Interest in ecosystem services in both the research and policy communities has grown rapidly (Braat and de Groot 2012). As of this writing, over 18,000 journal articles have been published on this topic, according to SCOPUS, and the number is growing exponentially. The most highly cited of these (with over 7,000 citations in SCOPUS as of this writing) is one that I and 12 co-authors published in *Nature* in 1997 that estimated the value of global ecosystem services to be in excess of US$33 trillion per year, a figure larger than global gross domestic product (GDP) at the time (Costanza et al. 1997). This admittedly crude underestimate, and a few other early studies, stimulated a huge surge in interest in this topic. In 2005, the concept of ecosystem services gained broader attention when the United Nations published its Millennium Ecosystem Assessment (MEA 2005). The MEA was a four-year, 1,300-scientist study for policymakers. In 2008, a second international initiative was undertaken by the UN Environment Programme, called The Economics of Ecosystems and Biodiversity (TEEB 2010). The TEEB report was picked up extensively by the mass media, bringing ecosystem services to a broader audience. Hundreds of projects and groups are currently working towards better understanding, modelling, valuation and management of ecosystem services and natural capital. In 2012 the Intergovernmental Science-Policy Platform on Biodiversity and Ecosystem Services (IPBES) was established. IPBES is an intergovernmental body (similar to the IPCC) which provides information on the state of biodiversity and ecosystem services for decision-making purposes. Its current membership includes 126 national governments. Emerging global, national and regional networks like the Ecosystem Services Partnership (www.es-partnership.org) have also emerged. Ecosystem services are now poised to provide real solutions to the problem of how to sustainably manage our critical natural capital assets.

From the perspective of ecosystem services, wetlands are among the most important and valuable ecosystems in the world (de Groot et al. 2012). The recognition of this value is a far cry from the situation not that long ago (and still prevalent in some places) when wetlands were considered to be 'wastelands' and every effort was made to drain, fill and convert them to other land uses.

Coastal wetlands, and in particular large river deltas, are especially important and valuable. River deltas contain the majority of coastal wetlands. However, they are also among the most impacted by human activities and 1 in 14 people globally live in deltaic regions (Day et al. 2016). The world's most populated delta is the Ganges–Brahmaputra–Meghna in Bangladesh.

This book is a compendium of some of the latest work on understanding, valuing and managing ecosystem services in this, one of the most important and vulnerable delta ecosystems in the world. It takes a much needed 'whole systems' approach to understanding the current status and trends in this complex system and focuses on the fundamental relationships between the biophysical system and the welfare of the diverse human communities that rely on it.

If we are to build the sustainable and desirable future we all want, we need to be able to understand, model and value complex social-ecological systems in the comprehensive way this book exemplifies. It is truly a model that needs to be broadly emulated.

Robert Costanza
Professor and VC's Chair in Public Policy
Crawford School of Public Policy
The Australian National University

References

Braat, L., and R. de Groot. 2012. The ecosystem services agenda: Bridging the worlds of natural science and economics, conservation and development, and public and private policy. *Ecosystem Services* 1: 4–15.

Costanza, R., R. d'Arge, R. de Groot, S. Farber, M. Grasso, B. Hannon, K. Limburg, S. Naeem, R.V. Oneill, J. Paruelo, R. G. Raskin, P. Sutton, and M. van den Belt. 1997. The value of the world's ecosystem services and natural capital. *Nature* 387 (6630): 253–260.

Costanza, R., M. Daly, C. Folke, P. Hawken, C. S. Holling, A.J. McMichael, D. Pimentel, and D. Rapport. 2000. Managing our environmental portfolio. *Bioscience* 50 (2): 149–155.

Daily, G.C. (1997). Nature's Services: Societal dependence on natural ecosystems. Washington, DC: Island Press.

Day, J.W., J. Agboola, Z. Chen, C. D'Elia, D.L. Forbes, L. Giosan, P. Kemp, C. Kuenzer, R.R. Lane, R. Ramachandran, J. Syvitski, and A. Yañez-Arancibia. 2016. Approaches to defining deltaic sustainability in the 21st century. *Estuarine, Coastal and Shelf Science* 183: 275–291.

de Groot, R., L. Brander, S, van der Ploeg, R. Costanza, F. Bernard, L. Braat, M. Christie, N. Crossman, A. Ghermandi, L. Hein, S. Hussain, P. Kumar, A. McVittie, R. Portela, L.C. Rodriguez, P. ten Brink, and P. van Beukering. 2012. Global estimates of the value of ecosystems and their services in monetary units. *Ecosystem Services* 1: 50–61.

Millennium Ecosystem Assessment (MEA). 2005. Ecosystems and human well-being: Synthesis. Washington DC: Island Press.

TEEB. 2010. *Mainstreaming the economics of nature: A synthesis of the approach, conclusions and recommendations of TEEB*. London/Washington: Earthscan.

Participants from the Following Organisations Attended Workshops Within the Project

Ministries of the Government of the People's Republic of Bangladesh

Department of Agricultural Extension (DAE)
Department of Fisheries (DoF)
Different Wings/Divisions of Planning Commission
General Economics Division of Planning Commission
Ministry of Environment and Forest (MoEF)
Ministry of Finance (MoF)
Ministry of Water Resources (MoWR)

Bangladesh Organisations

Bangladesh Agricultural Development Corporation (BADC)
Bangladesh Agricultural Research Institute (BARI)
Bangladesh Bureau of Statistics (BBS)
Bangladesh Institute of Development Studies (BIDS)
Bangladesh Rice Research Institute (BRRI)
Bangladesh Meteorological Department (BMD)
Bangladesh Water Development Board (BWDB)
Comprehensive Disaster Management Program (CDMP)

Dhaka University (DU)
Institute of Water Modeling (IWM)
River Research Institute (RRI)
Soil Resources Development Institute (SRDI)
Space Research and Remote Sensing Organization (SPARRSO)
Water Resources Planning Organisation (WARPO)
WildTeam

Local Organisations

Cyclone Preparedness Programme
Local community-based organisations
Local Disaster Risk Reduction (DDR) Volunteers
Local Government Engineering Department (LGED)
Local media
Local mosque committee
Local non-governmental organisations
Local schools
Shelter Management Committee
Small entrepreneurship
Switch gate committee
Union Parishad
Upazila Administration
Upazila Parishad
Water Management Committee

International Organisations

Asian Development Bank
CARE
Delta Plan Consultants (Deltares, Bandudeltas)
Food and Agriculture Organization of the United Nations (FAO)
German Development Cooperation (GIZ)
Global Water Partnership

International Organization for Migration (IOM)
International Union for Conservation of Nature (IUCN)
United Nations Development Programme (UNDP)
World Bank
World Food Program (WFP)
World Health Organization (WHO)

Preface

Deltas provide diverse ecosystem services and benefits for their large populations. At the same time, deltas are also recognised as one of the most vulnerable coastal environments, with a range of drivers operating at multiple scales, from global climate change and sea-level rise to delta-scale subsidence and land cover change. Lastly, many delta populations experience significant poverty. Hence when the Ecosystem Services for Poverty Alleviation (ESPA) programme was announced, we rapidly focussed on deltas as an issue for study. The focus of the book is the world's most populated delta, the Ganges–Brahmaputra–Meghna Delta, and more particularly within coastal Bangladesh west of the Lower Meghna River.

In our first visit to Dhaka, Bangladesh, in 2010, we recognised the complexity and challenges of understanding rural livelihoods in a dynamic delta. We held an intensive multidisciplinary workshop of UK and Bangladeshi scientists, followed by an inspiring visit to the Sundarbans. The resulting debates and conclusions, supported by acres of white board conceptual maps, formed the foundation that became the ESPA Deltas (Assessing Health, Livelihoods, Ecosystem Services and Poverty Alleviation in Populous Deltas) international consortium project. This involves more than 120 individuals and 21 institutions across Bangladesh, India and the UK. The collective thinking and experience of this team is distilled into this book, which examines the present and future of ecosystem services and livelihoods in coastal Bangladesh. It reflects the strong commitment to integration and a transdisciplinary approach, embracing disciplines as diverse as physical oceanography, sediment dynamics, agriculture, demographics and poverty. Input of policy experts and a substantial array of stakeholders are also fundamental. This study provided opportunities for substantial learning across standard discipline boundaries, providing co-produced policy relevant outputs and insights. It also fostered a family of researchers who developed a shared understanding that could be applied to this difficult and challenging problem. This included effective sharing of knowledge and learning to question and contribute effectively outside an individual's specialist field.

Integration is core to what has been accomplished here bringing together natural and social sciences in ways that are distinct and groundbreaking. Such integration needs to start as the research is initiated and is an ongoing process. Integration needs to be core to the project with key questions and themes that are properly resourced. To be policy relevant, the research must be guided by the perspectives, needs and expertise encapsulated by local stakeholders, especially the decision-making processes and governance context of the deltas in question. Stakeholders from civil society, the non-government sector and of course agencies of government are all involved in policy formulation. Indeed, one of the outstanding successes of the ESPA Deltas collaboration, which is reflected in this book, has been the engagement with and the impact on the policy context of Bangladesh. The research has raised, for the first time, consideration of ecosystem services, their links to poverty and livelihoods and their influence in the national policy and planning process across a range of government agencies including the Government of Bangladesh, Planning Commission and other government partners such as the Water Resources Planning Organization (WARPO). Indeed the Government of Bangladesh has requested continued engagement and further development of some of the modelling tools in the context of the Bangladesh Delta Plan 2100, which is a new national planning approach. Engaging with policy was always a main aspiration of the research and is perhaps the aspect of which we are most proud.

The research provides both the foundation and analyses which has led to some of its most innovative approaches and significant insights. This book offers an overarching and integrated framework to analyse changing ecosystem services in deltas and the implications for human well-being, focussing in particular on the provisioning ecosystem services of agriculture, inland and offshore capture fisheries, aquaculture and mangroves that directly support livelihoods. Each chapter contributes to the wider integrated assessment. Indeed, throughout the book there are reflections on the process of integrating information on the different environmental, social and economic dimensions of coastal management. The more detailed work supports significant conclusions that challenge elements of the perceived wisdom concerning human–environment relations and progress for the future, under the Sustainable Development Goals. We highlight, for example, that while ecosystem services support all populations in deltas, they act as a more critical safety net for the poorest and most marginalised delta populations. We show that while climate change has a real and tangible impact on the coastal zone, demographic, social dynamics and policy changes are likely to be more significant until at least 2050.

This book is not intended as a tool kit or specific guide to conducting integrated research in deltas or major coastal systems throughout the world. It offers, rather, a detailed account of a major integrative assessment relevant to development dilemmas in major ecosystems where biophysical, ecological and social dimensions are strongly coupled. This approach can be generalised beyond tropical deltas and even coastal zones; it addresses fundamental questions regarding the relationship of ecosystem services to the welfare of diverse rural communities that are important in every corner of the world today.

University of Southampton   Robert J. Nicholls
Southampton, UK   Craig W. Hutton
                                              Susan E. Hanson
University of Exeter   W. Neil Adger
Exeter, UK
Bangladesh University of Engineering   Md. Munsur Rahman
and Technology   Mashfiqus Salehin
Dhaka, Bangladesh

Acknowledgements

The book is a reflection on a diverse series of activities across biophysical, ecological and social science research on delta systems. The research is the outcome, firstly, of a major consortium project funded in the UK from 2012 to 2016. *The research 'ESPA Deltas (Assessing Health, Livelihoods, Ecosystem Services and Poverty Alleviation in Populous Deltas)' NE-J002755-1 was funded with support from the Ecosystem Services for Poverty Alleviation (ESPA) programme. The ESPA programme was funded by the Department for International Development (DFID), the Economic and Social Research Council (ESRC) and the Natural Environment Research Council (NERC).* We thank them for this funding and their support, including making this book an open access publication.

The ESPA Delta partnership involves researchers in Bangladesh, India and the UK and across a wide range of relevant disciplines. In the UK these are Universities of Southampton, Exeter, Dundee, Oxford, Bath, Plymouth Marine Laboratory, the National Oceanography Centre Liverpool and the Hadley Centre of the UK Met Office. The Bangladeshi partners are the Institute of Water and Flood Management at the Bangladesh University of Engineering and Technology, Bangladesh Agricultural Research Institute, Technological Assistance for Rural Advancement, Ashroy Foundation, Bangladesh Agricultural University, Bangladesh Institute of Development Studies, Center for Environmental and Geographical Information Services, Institute of Livelihood Studies, International Union for Conservation of

Nature, Water Resources Planning Organization and International Centre for Diarrhoeal Disease Research, Bangladesh. The project's Indian partners are the University of Jadavpur and the Indian Institute of Technology Kanpur.

A project with this large scope and duration, and strong stakeholder engagement, depends on the support of many individuals and organisations. They are too numerous to list comprehensively, but we acknowledge their critical contribution, such as participating in the numerous workshops and survey work that we organised.

In particular, the project team would also like to extend their thanks to the following individuals and their organisations who provided essential support during the research and greatly facilitated our engagement with the national policy process:

- Professor Shamsul Alam, Member (Senior Secretary) General Economics Division, Planning Commission, Government of the People's Republic of Bangladesh
- Dr. Sultan Ahmed, Director, Natural Resources Management, Department of Environment (DoE), Government of the People's Republic of Bangladesh
- Dr. Taibur Rahman, Senior Assistant Chief, General Economics Division, Planning Commission, Government of the People's Republic of Bangladesh
- Mr. Md. Mafidul Islam, Joint Chief, General Economics Division, Planning Commission, Government of the People's Republic of Bangladesh

Last but not least, we also express our gratitude to the people of Bangladesh who, during our research, provided hospitality and were ever willing to give their time and enthusiasm to ensure that our research remained relevant and successfully achieved its aim.

Contents

Notes on Contributors

Helen Adams Department of Geography, King's College London, London, UK

Munir Ahmed TARA, Dhaka, Bangladesh

Mahin Al Nahian Climate Change and Health, International Center for Diarrheal Disease Research, Bangladesh (icddr,b), Dhaka, Bangladesh

W. Neil Adger Geography, College of Life and Environmental Sciences, University of Exeter, Exeter, UK

Sate Ahmad Faculty of Agricultural and Environmental Sciences, University of Rostock, Rostock, Germany

Ali Ahmed Climate Change and Health, International Center for Diarrheal Disease Research, Bangladesh (icddr,b), Dhaka, Bangladesh

Abdur Razzaque Akanda Irrigation and Water Management Institute, Bangladesh Agricultural Research Institute, Gazipur, Bangladesh

Andrew Allan School of Law, University of Dundee, Dundee, UK

Manuel Barange Food and Agriculture Organization of the United Nations, Rome, Italy

Plymouth Marine Laboratory, Plymouth, UK

Emily J. Barbour School of Geography and the Environment, University of Oxford, Oxford, UK

Dilruba Begum Climate Change and Health, International Center for Diarrheal Disease Research, Bangladesh (icddr,b), Dhaka, Bangladesh

Sujit Kumar Biswas Irrigation and Water Management Institute, Bangladesh Agricultural Research Institute, Gazipur, Bangladesh

Lucy Bricheno National Oceanographic Centre, Liverpool, UK

Sally Brown Faculty of Engineering and the Environment and Tyndall Centre for Climate Change Research, University of Southampton, Southampton, UK

John Caesar Met Office Hadley Centre for Climate Science and Services, Exeter, Devon, UK

Mark Chan Department of Geography, King's College London, London, UK

Abhra Chanda School of Oceanographic Studies, Jadavpur University, Kolkata, India

Alexander Chapman Faculty of Engineering and the Environment and Tyndall Centre for Climate Change Research, University of Southampton, Southampton, UK

Md. Mahabub Arefin Chowdhury Institute of Water and Flood Management, Bangladesh University of Engineering and Technology, Dhaka, Bangladesh

Shahad Mahabub Chowdhury International Union for Conservation of Nature, Dhaka, Bangladesh

Derek Clarke Faculty of Engineering and the Environment and Tyndall Centre for Climate Change Research, University of Southampton, Southampton, UK

Stephen E. Darby Geography and Environment, Faculty of Social, Human and Mathematical Sciences, University of Southampton, Southampton, UK

John A. Dearing Geography and Environment, Faculty of Social, Human and Mathematical Sciences, University of Southampton, Southampton, UK

Jose A. Fernandes Plymouth Marine Laboratory, Plymouth, UK

Tuhin Ghosh School of Oceanographic Studies, Jadavpur University, Kolkata, India

Susan E. Hanson Faculty of Engineering and the Environment and Tyndall Centre for Climate Change Research, University of Southampton, Southampton, UK

Asadul Haque Department of Soil Science, Patuakhali Science and Technology University, Patuakhali, Bangladesh

Anisul Haque Institute of Water and Flood Management, Bangladesh University of Engineering and Technology, Dhaka, Bangladesh

Sugata Hazra School of Oceanographic Studies, Jadavpur University, Kolkata, India

Duncan D. Hornby Geodata Institute, Geography and Environment, University of Southampton, Southampton, UK

Mostafa A. R. Hossain Department of Fisheries Biology and Genetics, Bangladesh Agricultural University, Mymensingh, Bangladesh

Md. Sarwar Hossain Institute of Geography, University of Bern, Bern, Switzerland

Alistair Hunt Department of Economics, University of Bath, Bath, UK

Hamidul Huq Institute of Livelihood Studies, Bangladesh University of Engineering and Technology, Dhaka, Bangladesh

Craig W. Hutton Geodata Institute, Geography and Environment, University of Southampton, Southampton, UK

Mohammad Jahiruddin Department of Soil Science, Bangladesh Agricultural University, Mymensingh, Bangladesh

Fiifi Amoako Johnson Social Statistics and Demography, Faculty of Social, Human and Mathematical Sciences, University of Southampton, Southampton, UK

Tamara Janes Met Office Hadley Centre for Climate Science and Services, Exeter, Devon, UK

Rezaul Karim Institute of Water and Flood Management, Bangladesh University of Engineering and Technology, Dhaka, Bangladesh

Susan Kay Plymouth Marine Laboratory, Plymouth, UK

Valentina Lauria Plymouth Marine Laboratory, Plymouth, UK

Attila N. Lázár Faculty of Engineering and the Environment and Tyndall Centre for Climate Change Research, University of Southampton, Southampton, UK

Michelle Lim Adelaide Law School, University of Adelaide, Adelaide, SA, Australia

Zoe Matthews Social Statistics and Demography, Faculty of Social, Human and Mathematical Sciences, University of Southampton, Southampton, UK

Shahjahan Mondal Institute of Water and Flood Management, Bangladesh University of Engineering and Technology, Dhaka, Bangladesh

Anirban Mukhopadhyay School of Oceanographic Studies, Jadavpur University, Kolkata, India

Robert J. Nicholls Faculty of Engineering and the Environment and Tyndall Centre for Climate Change Research, University of Southampton, Southampton, UK

Sara Nowreen Institute of Water and Flood Management, Bangladesh University of Engineering and Technology, Dhaka, Bangladesh

Andres Payo British Geological Survey, Keyworth, Nottingham, UK

Mashrekur Rahman Institute of Water and Flood Management, Bangladesh University of Engineering and Technology, Dhaka, Bangladesh

Md. Munsur Rahman Institute of Water and Flood Management, Bangladesh University of Engineering and Technology, Dhaka, Bangladesh

Mohammed Mofizur Rahman International Center for Diarrheal Disease Research, Bangladesh (icddr,b), Dhaka, Bangladesh

Rezaur Rahman Institute of Water and Flood Management, Bangladesh University of Engineering and Technology, Dhaka, Bangladesh

Abul Fazal M. Saleh Institute of Water and Flood Management, Bangladesh University of Engineering and Technology, Dhaka, Bangladesh

Mashfiqus Salehin Institute of Water and Flood Management, Bangladesh University of Engineering and Technology, Dhaka, Bangladesh

Peter Kim Streatfield Formerly of the International Center for Diarrheal Disease Research, Bangladesh (icddr,b), Dhaka, Bangladesh

Sylvia Szabo Department of Development and Sustainability, Asian Institute of Technology, Bangkok, Thailand

Paul G. Whitehead School of Geography and the Environment, University of Oxford, Oxford, UK

Judith Wolf National Oceanographic Centre, Liverpool, UK

List of Figures

List of Tables

Part 1

Research Highlights and Framework

1

Ecosystem Services, Well-Being and Deltas: Current Knowledge and Understanding

W. Neil Adger, Helen Adams, Susan Kay,
Robert J. Nicholls, Craig W. Hutton, Susan E. Hanson,
Md. Munsur Rahman, and Mashfiqus Salehin

1.1 Introduction

Deltas are often attractive places to live and work, with more than 500 million people living in these environments worldwide (Ericson et al. 2006). Many large deltas have high population densities in productive rural areas as well as significant coastal cities. Such density of use and population is a legacy of their highly fertile soils, productive aquatic and coastal ecosystems, diverse landscapes and ease of navigation. In short, delta environments provide for and enhance the well-being of their human populations.

W. Neil Adger (✉)
Geography, College of Life and Environmental Sciences,
University of Exeter, Exeter, UK

H. Adams
Department of Geography, King's College London, London, UK

S. Kay
Plymouth Marine Laboratory, Plymouth, UK

© The Author(s) 2018
R. J. Nicholls et al. (eds.), *Ecosystem Services for Well-Being in Deltas*,
https://doi.org/10.1007/978-3-319-71093-8_1

In the past decade, the description of such positive relations has been increasingly expressed in terms of ecosystem services: the benefits to society from nature. At the most fundamental level, the survival and flourishing of human populations in deltas are, of course, entirely dependent on biotic and abiotic earth systems and how these systems interact with social-economic and governance structures.

In this introductory chapter, we describe the state of knowledge in the emerging interdisciplinary science of ecosystem services and how that science is applied to delta environments in order to highlight key research questions and issues that are explored in this book. We review the science of ecosystem services and describe relevant processes in deltas. We then examine the nature of well-being derived from these services and the mechanisms in delta environments that constrain and modify the distribution of those benefits, before outlining the contribution of the new underpinning research on the Ganges-Brahmaputra-Meghna (GBM) delta to this area of knowledge.

1.2   Ecosystem Services: Current Understanding

It is well established that ecosystem services support human well-being. The study of ecosystem services is a distinct field at the interface of natural and social sciences. It initially emerged at the boundaries of ecology, conservation biology and ecological economics, and highlighted the

R. J. Nicholls • S. E. Hanson
Faculty of Engineering and the Environment and Tyndall Centre for Climate Change Research, University of Southampton, Southampton, UK

C. W. Hutton
Geodata Institute, Geography and Environment, University of Southampton, Southampton, UK

Md. Munsur Rahman • M. Salehin
Institute of Water and Flood Management, Bangladesh University of Engineering and Technology, Dhaka, Bangladesh

benefits to humans of ecosystem processes. The Millennium Ecosystem Assessment (MEA 2005) popularised the term and drew on findings across hydrology, systems modelling, development economics, resilience theory and others to advance the concept and its applications. Much of the effort in ecosystem services has been in delineating and classifying ecosystem service types and processes, and in mapping and measuring ecosystem services and their benefits (Nicholls et al. 2016; Naidoo et al. 2008). In addition, the Millennium Ecosystem Assessment sought to characterise long-term future trends and sustainability through global and sub-global scenarios.

The principal tenets of ecosystem service science are firstly that the observed decline in ecosystem services at all scales is widely caused by human action. The drivers of decline include the scale of economic activity, not least in extraction of renewable resources such as forests and fisheries, unsustainable pollution loading and the trade-off between the selective enhancements of some ecosystem services at the detriment of others (MEA 2005; Bennett et al. 2009).

In the field of ecological economics, initial aggregation of the benefits of nature sought to make them commensurable, and to compare aggregate ecosystem service benefits to the scale of economic activity. Costanza et al. (1997) strongly argued that ecosystems provide unaccounted-for benefits to society that are of greater magnitude than the whole of the global economy by standard economic metrics. Assessments of interventions to preserve natural areas or conserve specific ecosystem functions show that such actions generate benefits that are often orders of magnitude greater than their costs (Balmford et al. 2002).

Yet many elements of the relationship between ecosystems and the environment remain poorly established, not least in coastal and marine environments, such as deltas. What constitutes an ecosystem service, for example, comes into sharp relief in delta environments. In practice, classification systems for ecosystem services are based on both the characteristics of the ecosystem of interest and the decision-making context. Ecosystems provide direct provisioning for humans, through processes of cultivation or extraction of food and fibre. Ecosystems regulate the environment through absorbing and processing pollutants or acting as shelter, barrier or other elements of human habitat. Ecosystems also have meaning beyond these direct benefits that mean their loss is keenly felt and ecosystems are highly valued. Hence there is a common distinction, used throughout this book, between provisioning, regulating and cultural ecosystem services.

There is also a distinction between ecosystem processes or functions on the one hand and the specific services that they provide on the other. Ecosystem services in aggregate are the benefits that humans derive from ecosystems, which means that they can either be defined as the end point (e.g. food production), rather than the intermediate process (agriculture), or they can be defined to include processes from which humans benefit indirectly (e.g. purification of water). Ecosystem processes can also be detrimental to well-being: the so-called dis-services such as agricultural diseases and pests (Zhang et al. 2007).

Fisher et al. (2009) suggest that ecosystem services are the aspects of ecosystems used actively or passively to generate human well-being and do not necessarily relate to specific ecosystem functions. For example, the benefits of ecosystem services in helping people to feel attached to their places of residence and work are highly contextualised and spatially variable, relating more to ideas of landscape. Indeed these benefits of nature to well-being may not necessarily relate to specific functions of the ecosystems. There is, therefore, a complex set of relationships between intermediate services, final services and benefits flowing from all ecosystems, including delta ecosystems. This uncertainty is reflected in the diversity of classification systems for ecosystem services.

Despite impressive estimates of the economic value of ecosystem services to society, net ecosystem services are in decline. When Costanza et al. (2014) re-estimated the economic value of nature, they showed a reduction of land-related ecosystem services of up to $20 trillion (US$ 2007 values) in the decade since 2005. Such declines reflect increased levels of pollution, human efforts to enhance specific services (e.g. agriculture) often at the expense of others, habitat loss, species decline, and loss of underlying environmental processes (Bennett et al. 2009). The decline in ecosystem services is charted in ecological as well as monetary metrics. The large-scale global assessment of ecosystem status under the Millennium Ecosystem Assessment in 2005 similarly documented the decline in ecological functions and benefits provided to humans, including in coastal areas (Agardy et al. 2005). Common delta habitats such as coastal marshes and mangroves are globally in decline with at least 11 mangrove species threatened with extinction (Polidoro et al. 2010), while the seagrass habitat has declined by approximately 30 per cent in the past century (Waycott et al. 2009). Many deltas are also losing relative elevation due to strong subsidence and a lack of sedimentation, reflecting the presence of polders and increasingly engineered landscapes (Syvitski et al. 2009).

A second area of research highlights the trade-offs between elements of ecosystem services and other societal goals. This area is more contested. It is clear that many regions of the world have increased their standard of living and well-being through exploitation of natural capital, turning this into other forms of built and human capital. Exploitation of forest and resources, for example, generated significant growth in countries such as Indonesia, but at the cost of declining natural capital stocks (Neumayer 2003). It is noted that economies sustain themselves and aggregate income and well-being have not collapsed in places where ecosystem services are in decline. Raudsepp-Hearne et al. (2010) explain this so-called environmentalist paradox in two potential ways. First, economies may not yet have reached key tipping points and the continued loss of ecosystem services may yet crash ecosystems, especially those that provide food and fibre, with catastrophic consequences, a contention supported by the concept of safe operating spaces (Rockström et al. 2009). Second, the paradox may be explained by the fact that standard measures do not account for the real, but hidden, costs of ecosystem degradation.

Other evidence highlights the positive benefits of conserving ecosystem services, but identifies the trade-offs involved in this process. Hence there is a continuing uncertainty on whether natural resources can be managed to optimise well-being, conservation of services and development processes (Barrett and Constas 2014). If there are trade-offs, then these may involve temporal questions: the short-term enhancement of human welfare compared to longer-term sustainability. And there are other issues such as distributional effects and the trade-offs between different users (Daw et al. 2011).

The ecosystem service approach also makes a strong argument that mapping and measuring ecosystem services and, in particular, the creation of economic incentives or markets for services, leads to greater recognition and balance in the optimisation processes (Daily et al. 2000). Efforts to incorporate ecosystem services into planning for conservation have been advocated and appraised as having enabled new insights and greater realisation of sustainability. For example, Arkema et al. (2015) facilitated new coastal planning that brought multiple benefits from coastal ecosystem services into a structured plan for coastal protection in Belize.

The monetisation of ecosystem services enables market solutions to their conservation, but there is less certainty about the benefits of such interventions. There are, for example, multiple examples of markets and interventions designed to promote conservation of ecosystem services, ranging from the carbon sequestration services of terrestrial ecosystems such as forests and wetlands, through to regulating services for clean water. While these have often been deemed to promote maintenance of specific ecosystem services, the distribution of the benefits and hence the legitimacy of the market mechanism are often in doubt (Pascual et al. 2014).

1.3   Ecosystems and Well-Being: The Current Debate

It is well established that ecosystem services provide benefits to society in terms of well-being, but less clear how those benefits are distributed in society. A key issue, not least in delta settings, is whether ecosystem services have greater importance for populations with low levels of well-being, and whether ecosystem services and their provision can represent a pathway out of poverty. The Millennium Ecosystem Assessment proposed that ecosystems bring benefits through maintaining and enhancing the health of people, through life-sustaining goods and services, and through options for use that represent opportunities for development in material and non-material ways. Beyond direct economic use of ecosystem services, benefits include impacts on health through pathways such as nutrition, clean air and clean water, but also psychological well-being. Further elements of well-being can include positive associations with place and identity in constructing meaning and purpose in life and how ecosystems ameliorate risk, providing a safety net and refuge.

The absence of well-being is most often regarded as poverty. Given that multiple dimensions of well-being are derived from nature, the absence of ecosystem services is likely to be manifest in multiple dimensions of deprivation and poverty. Hence poverty is the absence of material well-being for basic needs (food, water and shelter), along with the absence of positive health and nutrition, and an inability to participate fully in society. In essence, there are alternative ways of conceptualising poverty that focus not only on the command over commodities and income but also more broadly the capability to live a dignified life, which incorporates both the material dimensions and relational dimensions of an individual's place in society (Alkire 2007).

Many elements of poverty can be assessed and measured through objective indicators associated with income, expenditure, assets, educational attainment and objective health outcomes (Bourguignon and Chakravarty 2003). Yet the perceptions of exclusion and subjective elements of well-being are often hidden and are only revealed through social science approaches and methods, including direct measurement through social survey and participatory appraisals. Indeed, many elements of social well-being relate to the ability of individuals to perform their social roles and to participate meaningfully in society (Diener et al. 1985; Larson 1993).

How do ecosystem services relate to the distribution of well-being and poverty? At the most basic level, ecosystem services can provide well-being that lifts individuals above a poverty threshold or maintains levels of well-being above such thresholds. In other words, ecosystem services can serve to alleviate poverty in the short term and prevent poverty in the longer term (Daw et al. 2011). This distinction is important, as poverty is persistent and the ways to alleviate it are not clear. Direct benefits from ecosystems, including provisioning ecosystem services, have been argued to be more important for poverty prevention than for alleviation (Fisher et al. 2014). Thus direct consumption of food, materials for shelter and disaster mitigation have been shown to be critical as strategies for populations seeking to avoid poverty in various contexts (Daw et al. 2011), but not necessarily in improving welfare.

There are diverse interventions to try to alleviate poverty and then help people accumulate skills, capital and assets necessary to maintain raised well-being, preventing a return to poverty over time (e.g. Banerjee et al. 2015). Many provisioning services provide resources that are traded rather than consumed. Ecosystem services that provide a source of cash income can be significant in the alleviation of poverty but could potentially involve short-term sacrifice of ecosystem quality for long-term sustainability. For example, populations can raise their income levels by over-exploiting fisheries in the short term before switching to other forms of higher income employment. However, with time, this raises the potential of driving an ecosystem service system over a threshold from which an irreversible decline in service is inevitable (Hossain et al. 2017).

Poorer sections of populations are more directly dependent on ecosystem services than the general population in virtually all environments. This includes delta regions where there is high dependence on agriculture and fisheries, and undiversified rural economies mean that income and subsistence are often insecure and variable. Provisioning ecosystem services from collectively owned or open access resources such as mangrove forest areas and fisheries represent a disproportionate share of income for poorer populations (Dasgupta et al. 2016). The implications of this high dependence on ecosystem services are magnified by long- and short-term threats to the environment.

Hence if ecosystem services are threatened by long-term environmental change, then poor populations are likely to suffer disproportionately. This is a strong conclusion from global studies of disasters by the World Bank (Hallegatte et al. 2017) and those focused on coastal environments. Barbier (2015), for example, shows that 90 per cent of the poor populations in coastal zones globally reside in 15 countries (with Bangladesh being number two in the rankings) and that these countries are susceptible to significant climate and sea-level rise impacts that threaten the livelihood of those poor populations. Whilst it is clear that ecosystem services play a central role in the dynamic processes of welfare within rural poor of coastal zones, anthropogenic impacts of policy and interventions such as infrastructure development and market access can be a significant or even dominant role. For example, Amoako Johnson et al. (2016) identify a lack of road networks as a primary association with asset-based poverty west of the Lower Meghna River in coastal Bangladesh. Of course, the access that such a network provides can be thought of as facilitating the utilisation of ecosystem services and as such the two are not unrelated.

In summary, ecosystems play roles in poverty and well-being through direct provision of goods and services, by acting as a safety net for the poor, and potentially as a route out of poverty, sometimes in a non-sustainable manner. In the simplest sense, ecosystem services are effective at preventing downward movement into greater poverty, but less effective at actively elevating people out of poverty. All of the ways ecosystem services affect well-being are mediated by issues of access and control. Analysis in this book therefore considers the distribution of the benefits from these delta ecosystems accessed by different sets of people and the integrity and future of the functuioning and management of ecosystems themselves. Key questions include identifying how the presence of so-called provisioning and regulating ecosystem services make a particular difference for different sections of delta populations.

1.4 Ecosystem Processes and Services in Deltas

The Millennium Ecosystem Assessment showed in detail how many ecosystem services across the world are under stress and in decline. These trends are apparent in delta regions under stress (Nicholls et al. 2016). In the GBM delta, for example, there are high levels of soil and water salinity (Amoako Johnson et al. 2016). Globally, mangrove areas have been in significant decline, and natural habitats for aquatic species have shown stress due to over-exploitation and pollution (Polidoro et al. 2010).

Deltas are distinct in terms of the concentration of freshwater, nutrients and especially sediment inputs to a small concentrated area of the coastal zone, creating conditions ideal for fertile ecosystems, dense population and high economic activity (Bianchi 2016). Associated ecosystem services are high in number and include benefits such as productive agriculture and aquaculture, water provision and physical protection from the periodic impacts of extreme events such as coastal storms and cyclones. Ecosystem services can act as a safety net for poor populations. Akter and Mallick (2013) show in the GBM region that those populations with access to Sundarbans forest resources were more resilient following Cyclone Aila in 2009.

Importantly, virtually all services are directly affected by water and its flow through delta systems (Costanza et al. 1995), and there are multiple stresses which can significantly change their natural dynamics.

The processes supporting ecosystem services in deltas have developed over thousands of years (Woodroffe et al. 2006; Syvitski 2008; Wilson and Goodbred 2015), and historical development provides an important context to their present and future status. The nature of the link between the delta and its river catchment (Fig. 1.1) means processes occurring in one place within the deltaic system can lead to benefits or losses elsewhere. In particular, many human processes, such as river catchment management and land claim, modify water flows and the natural sedimentary

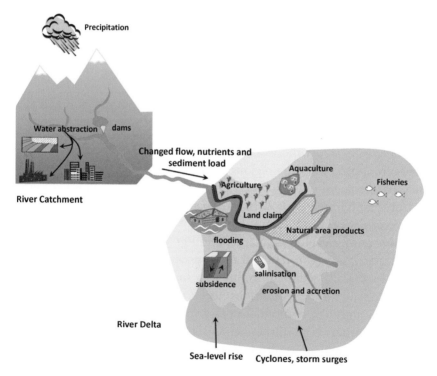

Fig. 1.1 Interventions and processes related to ecosystem service provision in delta environments

processes that maintain deltas (Syvitski et al. 2009; Day et al. 2016). This, combined with climate-driven processes such as precipitation, sea-level rise and storm intensity and frequency, means that delta areas are subject to changes such as periodic and permanent submergence (Ericson et al. 2006), erosion and accretion and salinisation.

More specifically, Fig. 1.2 outlines the principal ecological and physical processes of delta systems, which include both the ecosystems themselves and abiotic processes such as sediment transport that are definitive of delta environments.

The ecosystem service consequences of these process changes have been less considered and this is one of the key topics considered in this book. Erosion and submergence processes bring negative consequences to economic activity and health in terms of human well-being. Increases in water salinity are also important and have significant negative consequences for agricultural productivity and health. For all deltas, ecosystem services are highly diverse in terms of temporal and seasonal variation.

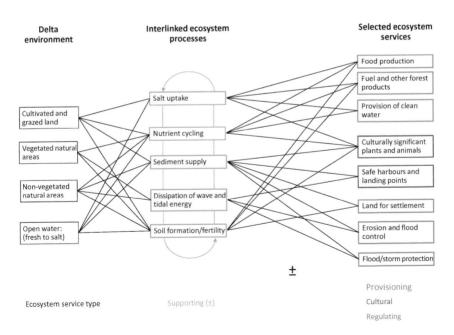

Fig. 1.2 Principal ecosystem processes and services in delta environments

The diversity and mobility of such services have significant implications for the distribution of well-being and poverty and how best to intervene to prevent or alleviate poverty and other social goals.

1.5 Social Drivers, Constraints and Dynamics in the Realisation of Well-Being from Ecosystem Services

Ecosystem service science is critical for the effective management of delta environments. Any decline in ecosystem services is likely to adversely impact the well-being of poor populations; this raises fundamental issues of development priorities as discussed in Chap. 2. A range of social processes attenuate or reinforce the benefits from ecosystem services. Secure and broad rights to ecosystem services, for example, have been shown to buffer against seasonal fluctuations and income shocks. Social relations, through a moral economy of hierarchical structures, reciprocity and compliance with informal rules, allow those without formal property rights to ecosystem services to gain access. However, in coastal Bangladesh informal access has not been upheld when it comes to commercial development of land; many people have lost informal access to water and land through enclosures for the development of brackish shrimp aquaculture. Credit is an essential means of obtaining the capital required to access ecosystem benefits, but often comes with exploitative conditions. Migration is used as a livelihood risk-spreading strategy to access alternative labour markets during seasonal fluctuations, as a short-term coping mechanism against shocks and to overcome chronic livelihood insecurity, but is again subject to exploitative practices of labour. Thus any attempt to consider improving the benefits associated with ecosystem services needs to consider the rights associated with the services and the system that support the engagement of the poor with those services (e.g. credit) and the degree to which they can be protected.

The processes that mediate between aggregate well-being and ecosystem are outlined in Fig. 1.3. It represents the interactions between elements of well-being, resource productivity and the social and economic structures that constrain livelihoods, highlighting that all these elements

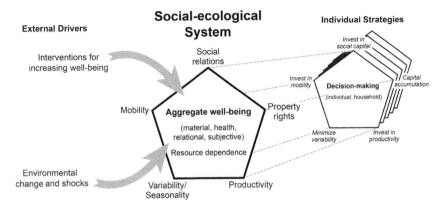

Fig. 1.3 Social-ecological systems and decision-making over ecosystem services

are dynamic and interrelated. The pentagon in Fig. 1.3 represents the processes that link ecosystem services to well-being through five components: the productivity of ecosystems, property rights, access and social relations, seasonality and climate variability and mobility.

The issues of property rights to ecosystem services and the ability of groups to maintain them, their appropriate assignment and their distribution are fundamental aspects of the role of ecosystem services in the provision of well-being because secure and broad access to ecosystem services provides security and allows consumption to be predictable. Property rights have implications for the sustainability of ecosystem service use: the way that decisions are made on resources has a large impact on who is able to benefit from them, how much and for how long. The relative merits of different property rights, ranging from private, to state-controlled, to open access resources have been studied extensively, showing that sustainability requires equitable and transparent systems of control and compliance (Dietz et al. 2003). Adhikari et al. (2004), for example, demonstrate how wealth, landholding and social status determine unequal patterns of access to commonly held forest resources. The way that natural resources are managed influences the ability of people to access the benefits they provide, as well as affecting the sustainability of that resource. These issues of access capabilities and an ability to function in society are central to defining poverty (Leach et al. 1999).

Social relations structure economic relations and create what Scott (1977) refers to as a moral economy of hierarchical structures, reciprocity and compliance with informal rules. Thus potential benefits from ecosystem services accrue to local power holders. Institutional arrangements such as debts and loans and sharecropping, gained through and creating exploitative social relations, are the most common way of smoothing these fluctuations. However, other patron-client relations include reducing rents during poor harvests, paying for education and other social payments. These arrangements have been shown to further the interests of patrons as they ensure the continuation of the community as a whole and ensure support that maintains privileged positions in society (Wood 2003). For those with limited rights, social mechanisms such as patron, patron-client and other reciprocity provide access to ecosystem services. However, many of these social mechanisms are a major constraint on capital accumulation and hence constitute poverty traps.

However, the Millennium Ecosystem Assessment (2005) systematically demonstrated that the *productivity* of ecosystem services is central to continued well-being and that degradation of such services has direct knock-on effects to sustaining material, relational and cultural dimensions. The productivity of all agricultural systems, for example, are directly determined by the interaction of ecosystem functions, and the process of agriculture is, in effect, the management of those functions, in optimising outputs for food, fibre and sustainable landscapes (Zhang et al. 2007). Thus while social processes affect the ability of the poor to access benefits of ecosystem services, levels of well-being are also contingent on the initial productivity of the system. The provision of different ecosystem services is directly connected by the underlying ecological functions.

Seasonality and inter-annual variability in ecosystem services have a significant impact on the transitory or chronic nature of poverty. There is a large body of evidence that shows that the phenomenon of seasonal poverty is widespread in agrarian economies, driven by weather and crop failure and by knock-on effects on seasonal demand for labour. The phenomenon of seasonal poverty is masked by annual average measures of consumption (Dercon and Krishnan 2000) and implies large numbers of households are vulnerable to seasonal poverty and drops in consumption across agricultural-based economies (McKay and Lawson 2003). In Ethiopia, for example,

Dercon and Krishnan (2000) showed high variability in poverty from surveying 1,400 households over two years, both reflected in income and consumption, through direct effects of seasons and weather and in indirect effect in prices and labour demand.

The relationship between migration and ecosystem services is complex. Migration can act as a route out of poverty, but has a diversity of impacts and feedbacks on ecosystem services (Black et al. 2011). Migration can cause degradation of ecosystem services through migration to forest and other frontiers, particularly when migrants are unfamiliar with new agro-ecosystems and risks (Winkels 2008). Migration can also cause the degradation of ecosystem services through the investment of remittances in capital-intensive enterprises such as shrimp farming or cash crops (Adger et al. 2002; Naylor et al. 2002). At the same time, shocks to ecosystem services directly affect migration, with evidence from Bangladesh, for example, showing that crop failures induce additional temporary migration, but there are 'significant barriers to migration for vulnerable households' (Gray and Mueller 2012 p. 6000). Mobility is central to much ecosystem service use, not least because people migrate to pursue ecosystem services, for example, by accessing alternative labour markets in different regions where crops are harvested at different times. Ecosystem services associated with fisheries are themselves mobile, and so individuals may choose to migrate to follow that resource (Kramer et al. 2002). Mobility can also reduce the ability of institutions to manage ecosystem services as common pool resources, as people from outside the local area may extract ecosystem services without respecting, or being aware of, local arrangements for their sustainable management.

These five factors variously act as social and environmental constraints on well-being at the system level. They are the arenas by which individuals and households seek to overcome constraints to their well-being, by investing in social relations through social capital, productivity including human capital through education, mobility or reducing variability through smoothing consumption and production processes (right panel of Fig. 1.3).

Hence there are a set of complex relationships between ecosystem services and livelihood and development processes. In Fig. 1.3, well-being is highlighted as having material, relational and subjective dimensions relating to agency and quality of life (Gough et al. 2007). Thus poverty, as a lack of well-being, has those multiple dimensions, including outcomes such as ill health, perceived insecurity and social marginalisation and exclusion, as well as direct lack of material well-being and deprivation. Specific research has focused on determining where and when the contribution of ecosystem services is important for those with low levels of well-being, particularly in resource-dependent societies and populations in poverty. This is the point from which the research activities and their outputs, described in this book, makes a distinct contribution.

1.6   Contributions of this Book

The book builds on current knowledge of ecosystem services, well-being and environments in an integrated and policy-relevant analysis of the processes and potential futures of societies living in deltas. It does so by focusing on the GBM delta. This is the world's most populous delta, with a critical role in the lives of a significant proportion of the population of Bangladesh and West Bengal (India). In particular, it focuses on a study site in coastal Bangladesh as explained in more detail in Chap. 4. While the analysis is particular to that study site and the wider delta, the generic lessons for the future of deltas and coastal environments are considered. The integrative analysis is structured into seven parts and comprises 29 chapters. Part 1 comprises Chaps. 1, 2, 3 and 4 and summarises the major conclusions of the work, including the fundamental relationship between ecosystem services and human well-being, and the links to policy processes and development. Chapter 4 also provides an overview of the methods and approach of the study, including consideration of the research questions. Part 2 comprises Chaps. 5, 6, 7 and 8 and introduces the study area and some of its key characteristics. Part 3 comprises Chaps. 9, 10, 11 and 12 and develops a set of biophysical and social-economic scenarios that facilitates policy-relevant analysis of the future of the study area. Part 4 comprises Chaps. 13, 14, 15, 16, 17, 18, 19, 20, 21, 22 and 23 and analyses a wide range of elements of the delta social-ecological systems using observations and models from the scale of the catchments and Bay of Bengal down to micro-level processes. Part 5 comprises Chaps. 24, 25, 26 and 27 and considers the implications of these changes for specific ecosystem services such as the Sundarbans mangroves and capture fisheries. Part 6 considers the integration of the preceding material into a policy-relevant integrated framework of analysis culminating in an integrated systems model for the delta. This allows an analysis of the biophysical and social future of the study area to inform policy. Part 6 also considers the science policy interface of the emerging knowledge and how it is used by stakeholders and wider society.

The core objective of the research is to integrate knowledge and to generate specific ideas for managing resources and implementing policy and practice in deltas. Figure 1.4 shows how these elements contribute to the overall vision. The chapters build on process and empirical insights into biophysical systems and ecosystem services, along with social and economic systems around health, economic activity and demographic change. They do so in order to generate policy insights, with chapters on elements of current governance and potential futures, directly engaging with the perceptions of key stakeholders.

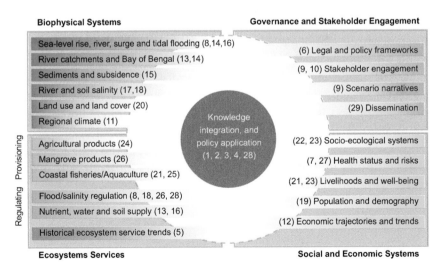

Biophysical Systems

Sea-level rise, river, surge and tidal flooding (8,14,16)
River catchments and Bay of Bengal (13,14)
Sediments and subsidence (15)
River and soil salinity (17,18)
Land use and land cover (20)
Regional climate (11)

Agricultural products (24)
Mangrove products (26)
Coastal fisheries/Aquaculture (21, 25)

Flood/salinity regulation (8, 18, 26, 28)
Nutrient, water and soil supply (13, 16)

Historical ecosystem service trends (5)

Regulating   Provisioning

Ecosystems Services

Knowledge integration, and policy application (1, 2, 3, 4, 28)

Governance and Stakeholder Engagement

(6) Legal and policy frameworks
(9, 10) Stakeholder engagement
(9) Scenario narratives
(29) Dissemination

(22, 23) Socio-ecological systems
(7, 27) Health status and risks
(21, 23) Livelihoods and well-being
(19) Population and demography
(12) Economic trajectories and trends

Social and Economic Systems

Fig. 1.4 Guide to the individual and integrated research discussed in this book

1.7   Conclusion

This book builds on insights from a multidisciplinary perspective on eco-system services and their importance to human well-being and applies these ideas in deltas in an integrated manner to inform decision-making for poverty alleviation. There is, of course, a wide variety of approaches and unresolved questions and relationships between elements of well-being and the underlying ecosystem processes (Norgaard 2010; Pascual et al. 2017; Suich et al. 2015). But at their core, the key scientific issues relate to how ecosystems bring multiple benefits to society, both in mate-rial terms and through other pathways. The benefits from ecosystem ser-vices include those associated with direct economic use, with protecting health and mitigation of hazards.

The book explores the issues outlined in this chapter in detail for delta environments in order to give context to the broad assessment of the sustainability of a range of possible future trajectories within deltas, focusing on both the biophysical processes of their productivity and the prospects for securing ecosystem services for poverty alleviation objec-tives. The following chapters also, for the first time, explore in a system-atic manner how social processes such as migration, access and property rights, and social relations interact with ecosystem services to result in the distribution of well-being in deltas. The book highlights the leverage points for action on these mechanisms that have been uncovered through integrated modelling and an increased understanding of delta social-ecological systems. Integration across diverse knowledge domains and model simulation is a key and novel aspect of the research, and as such the findings in this book, linking biophysical changes to human well-being within a coupled model framework. This allows the exploration of possible futures in a participatory and policy-relevant manner that can engage with national stakeholders.

References

Adger, W.N., P.M. Kelly, A. Winkels, L.Q. Huy, and C. Locke. 2002. Migration, remittances, livelihood trajectories, and social resilience. *Ambio* 31 (4): 358–366. https://doi.org/10.1639/0044-7447(2002)031[0358:mrl tas]2.0.co;2.

Adhikari, B., S. Di Falco, and J.C. Lovett. 2004. Household characteristics and forest dependency: Evidence from common property forest management in Nepal. *Ecological Economics* 48 (2): 245–257. doi: https://doi.org/10.1016/j.ecolecon.2003.08.008.

Agardy, T., J. Alder, P. Dayton, S. Curran, A. Kitchingman, M. Wilson, A. Catenazzi, J. Restrepo, C. Birkeland, S. Blaber, S. Saifullah, G. Branch, D. Boersma, S. Nixon, P. Dugan, N. Davidson, and C. Vorosmarty. 2005. Chapter 19: Coastal ecosystems. In *Ecosystems and human well-being: Current states and trends*, ed. R. Hassan, R. Scholes, and N. Ash, vol. 1, 513–549. Washington, DC: Island Press.

Akter, S., and B. Mallick. 2013. The poverty–vulnerability–resilience nexus: Evidence from Bangladesh. *Ecological Economics* 96: 114–124. doi: https://doi.org/10.1016/j.ecolecon.2013.10.008.

Alkire, S. 2007. Choosing dimensions: The capability approach and multidimensional poverty. In *The many dimensions of poverty*, ed. N. Kakwani and J. Silber, 89–119. London: Palgrave.

Amoako Johnson, F., C.W. Hutton, D. Hornby, A.N. Lázár, and A. Mukhopadhyay. 2016. Is shrimp farming a successful adaptation to salinity intrusion? A geospatial associative analysis of poverty in the populous Ganges–Brahmaputra–Meghna Delta of Bangladesh. *Sustainability Science* 11 (3): 423–439. https://doi.org/10.1007/s11625-016-0356-6.

Arkema, K.K., G.M. Verutes, S.A. Wood, C. Clarke-Samuels, S. Rosado, M. Canto, A. Rosenthal, M. Ruckelshaus, G. Guannel, J. Toft, J. Faries, J.M. Silver, R. Griffin, and A.D. Guerry. 2015. Embedding ecosystem services in coastal planning leads to better outcomes for people and nature. *Proceedings of the National Academy of Sciences of the United States of America* 112 (24): 7390–7395. https://doi.org/10.1073/pnas.1406483112.

Balmford, A., A. Bruner, P. Cooper, R. Costanza, S. Farber, R.E. Green, M. Jenkins, P. Jefferiss, V. Jessamy, J. Madden, K. Munro, N. Myers, S. Naeem, J. Paavola, M. Rayment, S. Rosendo, J. Roughgarden, K. Trumper, and R.K. Turner. 2002. Ecology – Economic reasons for conserving wild nature. *Science* 297 (5583): 950–953. https://doi.org/10.1126/science.1073947.

Banerjee, A., E. Duflo, N. Goldberg, D. Karlan, R. Osei, W. Pariente, J. Shapiro, B. Thuysbaert, and C. Udry. 2015. A multifaceted program causes lasting progress for the very poor: Evidence from six countries. *Science* 348 (6236). https://doi.org/10.1126/science.1260799.

Barbier, E.B. 2015. Climate change impacts on rural poverty in low-elevation coastal zones. *Estuarine Coastal and Shelf Science* 165: A1–A13. https://doi.org/10.1016/j.ecss.2015.05.035.

Barrett, C.B., and M.A. Constas. 2014. Toward a theory of resilience for international development applications. *Proceedings of the National Academy of Sciences of the United States of America* 111 (40): 14625–14630. https://doi.org/10.1073/pnas.1320880111.

Bennett, E.M., G.D. Peterson, and L.J. Gordon. 2009. Understanding relationships among multiple ecosystem services. *Ecology Letters* 12 (12): 1394–1404. https://doi.org/10.1111/j.1461-0248.2009.01387.x.

Bianchi, T.S. 2016. *Deltas and humans: A long relationship now threatened by global change.* Oxford: Oxford University Press.

Black, R., W.N. Adger, N.W. Arnell, S. Dercon, A. Geddes, and D. Thomas. 2011. The effect of environmental change on human migration. *Global Environmental Change* 21 (Supplement 1): S3–S11. https://doi.org/10.1016/j.gloenvcha.2011.10.001.

Bourguignon, F., and S.R. Chakravarty. 2003. The measurement of multidimensional poverty. *The Journal of Economic Inequality* 1 (1): 25–49. https://doi.org/10.1023/A:1023913831342.

Costanza, R., M. Kemp, and W. Boynton. 1995. Scale and biodiversity in estuarine ecosystems. In *Biodiversity loss: Economic and ecological issues,* ed. C. Perrings, Karl-Göran Mäler, C. Folke, and Bengt-Owe Jansson, 84–125. Cambridge: Cambridge University Press.

Costanza, R., R. d'Arge, R. de Groot, S. Faber, M. Grasso, B. Hannon, K. Limburg, S. Naeem, R.V. O'Neill, J. Paruelo, and R.G. Raskin. 1997. The value of the world's ecosystem services and natural capital. *Nature* 387: 253–260.

Costanza, R., R. de Groot, P. Sutton, S. van der Ploeg, S.J. Anderson, I. Kubiszewski, S. Farber, and R.K. Turner. 2014. Changes in the global value of ecosystem services. *Global Environmental Change* 26: 152–158. https://doi.org/10.1016/j.gloenvcha.2014.04.002.

Daily, G.C., T. Söderqvist, S. Aniyar, K. Arrow, P. Dasgupta, P.R. Ehrlich, C. Folke, A. Jansson, B.-O. Jansson, N. Kautsky, S. Levin, J. Lubchenco, K.-G. Mäler, D. Simpson, D. Starrett, D. Tilman, and B. Walker. 2000. The value of nature and the nature of value. *Science* 289 (5478): 395.

Dasgupta, S., M.I. Sobhan, and D. Wheeler. 2016. Impact of climate change and aquatic salinization on mangrove species and poor communities in the Bangladesh Sundarbans. *Policy Research Working Paper 7736*. World Bank. http://documents.worldbank.org/curated/en/452761467210045879/pdf/WPS7736.pdf. Accessed 09 Jan 2017.

Daw, T., K. Brown, S. Rosendo, and R. Pomeroy. 2011. Applying the ecosystem services concept to poverty alleviation: The need to disaggregate human well-being. *Environmental Conservation* 38 (4): 370–379. https://doi.org/10.1017/s0376892911000506.

Day, J.W., J. Agboola, Z. Chen, C. D'Elia, D.L. Forbes, L. Giosan, P. Kemp, C. Kuenzer, R.R. Lane, R. Ramachandran, J. Syvitski, and A. Yañez-Arancibia. 2016. Approaches to defining deltaic sustainability in the 21st century. *Estuarine, Coastal and Shelf Science* 183: 275–291. doi: https://doi.org/10.1016/j.ecss.2016.06.018. Part B.

Dercon, S., and P. Krishnan. 2000. Vulnerability, seasonality and poverty in Ethiopia. *Journal of Development Studies* 36 (6): 25–53. https://doi.org/10.1080/00220380008422653.

Diener, E.D., R.A. Emmons, R.J. Larsen, and S. Griffin. 1985. The satisfaction with life scale. *Journal of Personality Assessment* 49: 71–75.

Dietz, T., E. Ostrom, and P.C. Stern. 2003. The struggle to govern the commons. *Science* 302 (5652): 1907.

Ericson, J.P., C.J. Vorosmarty, S.L. Dingman, L.G. Ward, and M. Meybeck. 2006. Effective sea-level rise and deltas: Causes of change and human dimension implications. *Global and Planetary Change* 50 (1–2): 63–82. https://doi.org/10.1016/j.gloplacha.2005.07.004.

Fisher, B., R.K. Turner, and P. Morling. 2009. Defining and classifying ecosystem services for decision making. *Ecological Economics* 68 (3): 643–653. https://doi.org/10.1016/j.ecolecon.2008.09.014.

Fisher, J.A., G. Patenaude, K. Giri, K. Lewis, P. Meir, P. Pinho, M.D.A. Rounsevell, and M. Williams. 2014. Understanding the relationships between ecosystem services and poverty alleviation: A conceptual framework. *Ecosystem Services* 7: 34–45. https://doi.org/10.1016/j.ecoser.2013.08.002.

Gough, I., J.A. McGregor, and L. Camfield. 2007. Theorising wellbeing in international development. In *Wellbeing in developing countries: From theory to research*, ed. I. Gough and J.A. McGregor. Cambridge: Cambridge University Press.

Gray, C.L., and V. Mueller. 2012. Natural disasters and population mobility in Bangladesh. *Proceedings of the National Academy of Sciences of the United States of America* 109 (16): 6000–6005. https://doi.org/10.1073/pnas.1115944109.

Hallegatte, S., A. Vogt-Schilb, M. Bangalore, and J. Rozenberg. 2017. *Unbreakable: Building the resilience of the poor in the face of natural disasters, climate change and development series.* Washington, DC: World Bank.

Hossain, M.S., F. Eigenbrod, F. Amoako Johnson, and J.A. Dearing. 2017. Unravelling the interrelationships between ecosystem services and human wellbeing in the Bangladesh delta. *International Journal of Sustainable Development & World Ecology* 24 (2): 120–134. https://doi.org/10.1080/135 04509.2016.1182087.

Kramer, R.A., S.M.H. Simanjuntak, and C. Liese. 2002. Migration and fishing in Indonesian coastal villages. *Ambio* 31 (4): 367–372. https://doi. org/10.1639/0044-7447(2002)031[0367:mafiic]2.0.co;2.

Larson, J.S. 1993. The measurement of social well-being. *Social Indicators Research* 28 (3): 285–296. https://doi.org/10.1007/BF01079022.

Leach, M., R. Mearns, and I. Scoones. 1999. Environmental entitlements: Dynamics and institutions in community-based natural resource management. *World Development* 27 (2): 225–247. https://doi.org/10.1016/s0305-750x(98)00141-7.

McKay, A., and D. Lawson. 2003. Assessing the extent and nature of chronic poverty in low income countries: Issues and evidence. *World Development* 31 (3): 425–439. https://doi.org/10.1016/s0305-750x(02)00221-8.

MEA. 2005. Ecosystems and human well-being: Synthesis. *Millennium Ecosystem Assessment (MEA).* Island Press http://www.millenniumassessment. org/documents/document.356.aspx.pdf. Accessed 01 Aug 2016.

Naidoo, R., A. Balmford, R. Costanza, B. Fisher, R.E. Green, B. Lehner, T.R. Malcolm, and T.H. Ricketts. 2008. Global mapping of ecosystem services and conservation priorities. *Proceedings of the National Academy of Sciences* 105 (28): 9495–9500. https://doi.org/10.1073/pnas.0707823105.

Naylor, R.L., K.M. Bonine, K.C. Ewel, and E. Waguk. 2002. Migration, markets, and mangrove resource use on Kosrae, Federated States of Micronesia. *Ambio: A Journal of the Human Environment* 31 (4): 340–350. https://doi. org/10.1579/0044-7447-31.4.340.

Neumayer, E. 2003. *Weak versus strong sustainability: Exploring the limits of two opposing paradigms.* Cheltenham: Edward Elgar.

Nicholls, R.J., C.W. Hutton, A.N. Lázár, A. Allan, W.N. Adger, H. Adams, J. Wolf, M. Rahman, and M. Salehin. 2016. Integrated assessment of social and environmental sustainability dynamics in the Ganges-Brahmaputra-Meghna delta, Bangladesh. *Estuarine, Coastal and Shelf Science* 183: 370–381. https://doi.org/10.1016/j.ecss.2016.08.017. Part B.

Norgaard, R.B. 2010. Ecosystem services: From eye-opening metaphor to complexity blinder. *Ecological Economics* 69 (6): 1219–1227. https://doi.org/10.1016/j.ecolecon.2009.11.009.

Pascual, U., J. Phelps, E. Garmendia, K. Brown, E. Corbera, A. Martin, E. Gomez-Baggethun, and R. Muradian. 2014. Social equity matters in payments for ecosystem services. *Bioscience* 64 (11): 1027–1036. https://doi.org/10.1093/biosci/biu146.

Pascual, U., I. Palomo, W.M. Adams, K.M.A. Chan, T.M. Daw, E. Garmendia, E. Gomez-Baggethun, R.S. de Groot, G.M. Mace, B. Martin-Lopez, and J. Phelps. 2017. Off-stage ecosystem service burdens: A blind spot for global sustainability. *Environmental Research Letters* 12 (7): 075001. https://doi.org/10.1088/1748-9326/aa7392.

Polidoro, B.A., K.E. Carpenter, L. Collins, N.C. Duke, A.M. Ellison, J.C. Ellison, E.J. Farnsworth, E.S. Fernando, K. Kathiresan, N.E. Koedam, S.R. Livingstone, T. Miyagi, G.E. Moore, N.N. Vien, J.E. Ong, J.H. Primavera, S.G. Salmo, J.C. Sanciangco, S. Sukardjo, Y.M. Wang, and J.W.H. Yong. 2010. The loss of species: Mangrove extinction risk and geographic areas of global concern. *PLoS One* 5 (4). https://doi.org/10.1371/journal.pone.0010095.

Raudsepp-Hearne, C., G.D. Peterson, M. Tengö, E.M. Bennett, T. Holland, K. Benessaiah, G.K. MacDonald, and L. Pfeifer. 2010. Untangling the environmentalist's paradox: Why is human well-being increasing as ecosystem services degrade? *Bioscience* 60 (8): 576–589. https://doi.org/10.1525/bio.2010.60.8.4.

Rockström, J., W. Steffen, K. Noone, Å. Persson, F.S. Chapin, E.F. Lambin, T.M. Lenton, M. Scheffer, C. Folke, H.J. Schellnhuber, B. Nykvist, C.A. de Wit, T. Hughes, S. van der Leeuw, H. Rodhe, S. Sorlin, P.K. Snyder, R. Costanza, U. Svedin, M. Falkenmark, L. Karlberg, R.W. Corell, V.J. Fabry, J. Hansen, B. Walker, D. Liverman, K. Richardson, P. Crutzen, and J.A. Foley. 2009. A safe operating space for humanity. *Nature* 461 (7263): 472–475. https://doi.org/10.1038/461472a.

Scott, J.C. 1977. *The moral economy of the peasant: Rebellion and subsistence in Southeast Asia*. New Haven: Yale University Press.

Suich, H., C. Howe, and G. Mace. 2015. Ecosystem services and poverty alleviation: A review of the empirical links. *Ecosystem Services* 12: 137–147. https://doi.org/10.1016/j.ecoser.2015.02.005.

Syvitski, J.P.M. 2008. Deltas at risk. *Sustainability Science* 3 (1): 23–32. https://doi.org/10.1007/s11625-008-0043-3.

Syvitski, J.P.M., A.J. Kettner, I. Overeem, E.W.H. Hutton, M.T. Hannon, G.R. Brakenridge, J. Day, C. Vorosmarty, Y. Saito, L. Giosan, and R.J. Nicholls. 2009. Sinking deltas due to human activities. *Nature Geoscience* 2 (10): 681–686. https://doi.org/10.1038/ngeo629.

Waycott, M., C.M. Duarte, T.J.B. Carruthers, R.J. Orth, W.C. Dennison, S. Olyarnik, A. Calladine, J.W. Fourqurean, K.L. Heck, A.R. Hughes, G.A. Kendrick, W.J. Kenworthy, F.T. Short, and S.L. Williams. 2009. Accelerating loss of seagrasses across the globe threatens coastal ecosystems. *Proceedings of the National Academy of Sciences of the United States of America* 106 (30): 12377–12381. https://doi.org/10.1073/pnas.0905620106.

Wilson, C.A., and S.L. Goodbred, Jr. 2015. Construction and maintenance of the Ganges-Brahmaputra Meghna Delta: Linking process, morphology, and stratigraphy. *Annual Review of Marine Science* 7: 67–88.

Winkels, A. 2008. Rural in-migration and global trade – Managing the risks of coffee farming in the central highlands of Vietnam. *Mountain Research and Development* 28 (1): 32–40. https://doi.org/10.1659/mrd.0841.

Wood, G. 2003. Staying secure, staying poor: The "Faustian bargain". *World Development* 31 (3): 455–471. https://doi.org/10.1016/s0305-750x(02)00213-9.

Woodroffe, C.N., R.J. Nicholls, Y. Saito, Z. Chen, and S.L. Goodbred. 2006. Landscape variability and the response of Asian megadeltas to environmental change. In *Global change and integrated coastal management: The Asia-Pacific region*, ed. N. Harvey, 277–314. Berlin: Springer.

Zhang, W., T.H. Ricketts, C. Kremen, K. Carney, and S.M. Swinton. 2007. Ecosystem services and dis-services to agriculture. *Ecological Economics* 64 (2): 253–260. https://doi.org/10.1016/j.ecolecon.2007.02.024.

2

Ecosystem Services Linked to Livelihoods and Well-Being in the Ganges-Brahmaputra-Meghna Delta

Helen Adams, W. Neil Adger, and Robert J. Nicholls

2.1 Introduction

This chapter addresses one of the main aims of the research that lies at the core of intellectual effort to discern how ecosystem services relate to poverty and its alleviation, to provide an assessment of whether and how development efforts for poverty reduction can be achieved alongside maintaining and building the integrity of the environment and ecosystem services.

H. Adams (✉)
Department of Geography, King's College London, London, UK

W. Neil Adger
Geography, College of Life and Environmental Sciences,
University of Exeter, Exeter, UK

R. J. Nicholls
Faculty of Engineering and the Environment and Tyndall Centre for Climate
Change Research, University of Southampton, Southampton, UK

This dilemma is common at all scales and in all ecoregions of the world and has been at the heart of sustainable development challenges and discourses for decades, as described briefly in Chap. 1. The challenges and trade-offs between development and maintaining a healthy environment have been recognised and analysed from all major theoretical perspectives. This includes issues of environmental entitlements (Leach et al. 1999), dilemma of the commons (Ostrom 1990), capability approaches and the development of changing livelihoods based on capital (Scoones 1998).

Policies focused on agricultural reform and the green revolution have played a major part in alleviating poverty, raising living standards and increasing food security across the developing world throughout later twentieth century (Hartmann and Boyce 1983; Hayami and Kikuchi 2000) but have simultaneously putting pressure on the underlying ecosystem resource. As such, there is a rich body of theory on the relationship between poor, natural resource-dependent people and their environment that comes to startlingly different conclusions regarding the causes of persistent poverty. These theoretical approaches include entitlement theory (Leach et al. 1999), political ecology (Robbins 2011), resilience theory (Gunderson and Holling 2002), social vulnerability (Adger 1999) and governmentality (e.g. Agrawal 2005).

The research findings reported in this book consider multiple perspectives. They draw first on new insights into the role of the environment as a set of ecosystem services and on new knowledge on environmental and ecological processes within marine, coastal and aquatic environments (described in Chap. 1). The analysis then contextualises that emerging knowledge within development trajectories and interventions for Bangladesh and South Asia. In doing so, it is possible to demonstrate how and whether these ecosystem service approaches provide new directions and insights into the struggle to promote well-being and sustainability. Third, the analysis draws on a wide range of perspectives on what constitutes well-being and poverty to expand the definition of poverty alleviation beyond maximising income. Finally, it places systems thinking at the core of the research, integrating knowledge from multiple disciplines across time and space to evaluate the implications of future interventions for ecosystem services, associated livelihoods and human well-being.

The analysis in this book therefore focuses on the interactions between ecosystem services and social dynamics, both in the present and potentially in the future. As a result, five key results emerge and are discussed in turn: (i) social mechanisms vary with bundles of ecosystem services to create defined social-ecological systems, (ii) subjective and material well-being indicators vary with social-ecological system, (iii) the nature of the interaction between subjective and material well-being and ecosystem services varies over time, (iv) trade-offs exist between different social-ecological systems and parallel flows of labour and (v) ecosystem services for well-being must be contextualised within changing rural economies.

2.2 Key Findings from Systems Perspectives on Well-Being and Ecosystem Services

The integrated multi-method approach demonstrates that the relationship between ecosystem services and well-being in coastal Bangladesh is highly contingent and differentiated a result of its distinct social-ecological systems (see Chap. 22). Hence, the focus shifts to understanding how ecosystem services can reduce particular types of poverty for specific groups of people over different timescales.

Rural livelihoods are diverse over space and time, and populations rely on more than one provisioning ecosystem service for their income. Ecosystem services commonly occur in bundles (sets of services that repeatedly appear together) and as such certain groups of services are more accessible than others. The adoption of a social-ecological system approach allows diversification of ecosystem service use to be considered, including being dependent on a subsidiary service in a particular zone (e.g. fishing in an agricultural area) (Sect. 2.2.1). Thus, the probability of being poor varies in space with the available bundles of ecosystem services and proximity to certain geographical features such as the coast or major rivers, or access to roads and cities.

The ability of ecosystem services to create well-being is dynamic and path dependent. Current productivity is a result of policy decisions made regarding infrastructure (e.g. coastal embankments) and the prioritisation of ecosystem services (e.g. monoculture rice agriculture versus open access fisheries) which has implications for future benefits. For example, high levels of shrimp monoculture productivity are unsustainable where supporting services have already been eroded by high salinity, but are resilient against reversal due to sea-level rise and the near impossibility of large scale desalination of soil (Sect. 2.2.3).

This example also highlights trade-offs between different bundles of ecosystem services across time and space, affecting the provision of benefits to the poorest in rural settings. There has been a steady concentration of ecosystem services into agriculture and aquaculture that tend to benefit those with access to land, to the detriment of open access provisioning services (e.g. fishing) and supporting services such as water quality—that are crucial for the poor. Thus, while ecosystem services are alleviating poverty through the export of shrimp, for example, this approach is neither sustainable nor pro-poor (Sect. 2.2.4).

However, it is crucial to note that existing inequalities within villages that keep the poor trapped in poverty are unlikely to be redressed by ecosystem service-based interventions, especially in a monetised rural economy that is becoming progressively less dependent on local ecosystem services (see Chaps. 12 and 28). For example, currently, a third of the population studied in this research have no access to ecosystem services at all for income, and even fewer have access to land to cultivate (Sect. 2.2.5).

Finally, the opportunities and losses occurring in the region should be analysed in the context of the market economy. While traditional ecosystem-based and social mechanisms of survival and subsistence have been undermined by market-based approaches, many of the opportunities that could emerge—namely, more sophisticated off-farm activities—have not materialised. Thus, while ecosystem services are increasingly monetised, the subsistence activities they undermine have not been replaced. Migration to alternative labour markets and debt tend to fill any gaps in income.

The means by which ecosystem services generate well-being in Bangladesh is therefore in transition, moving from subsistence-based approaches that provide safety nets, but without the potential for poverty alleviation, to market-based approaches where economic benefits are greater but tend to accrue to fewer people living in these rural areas and those who already have resources. Concurrently, rural livelihoods have become less and less dependent on local ecosystem services, with off-farm work and migration to urban areas or alternative labour markets contributing a growing share to household incomes.

2.2.1   New Analysis of Ecosystems as Critical to Poverty and Development

The approach adopted in this research to understanding poverty-environment linkages is novel in four key ways. First, it takes an integrated, systems approach that considers interactions, feedbacks and trade-offs, which is missing in most analyses (Dempsey and Robertson 2012). Second, the research considers many different epistemic approaches including the consideration of poverty-environment linkages from multiple methodological and theoretical standpoints (Nicholls et al. 2016). Some of these are integrated within the modelling framework, while others provide richness and understanding to the findings. Third, the analysis is future oriented. It is not sufficient to understand present ecological determinants of well-being, without understanding the capacity for these systems or services to continue to generate well-being into the future under various political, social-economic and environmental scenarios. Finally, the outcome of the analysis is not an answer to a single question, but rather a process which provides key insights concerning associative and causal linkages that have the potential to untangle and answer a wide range of questions on poverty and the environment (Chap. 28).

The research also describes a range of plausible future trajectories derived in a participatory manner. Hence, while the individual components of the analysis are interesting, the integration of these components is ground-breaking—for example, the integration of social differentiation in rural settlements with biophysical outputs to model poverty through time based on changes in the natural environment.

2.2.2   Social Mechanisms Co-vary with the Bundles of Ecosystem Services

This research dynamically analyses the two most important ecosystem services in terms of livelihoods in the delta: agriculture (including aquaculture where appropriate Chap. 24) and fisheries (focusing on offshore capture fisheries Chap. 25) under future environmental change and management scenarios. The area and species distribution of the Sundarbans mangrove forest are also modelled with a preliminary assessment of ecosystem services including protection against storm surges (Chap. 26). These three key provisioning services were operationalised using seven social-ecological systems, defined as freshwater and brackish aquaculture, irrigated and non-irrigated agriculture, riverine and char environments, the coastal zone and the Sundarbans dependent zone (see Chap. 22, Adams et al. 2013, 2016).

The social-ecological system classification recognises that although in certain regions a specific type of service may dominate livelihoods, households usually have more than one type of income source and that these sources may change through the year depending on the character of the ecosystem (Raudsepp-Hearne et al. 2010). Households select different ecosystem services from within the bundle at different times of the year. Social-ecological systems are thus the result of human activities to mediate the negative impacts of environmental variability and to manage bundles of ecosystem services (Martín-López et al. 2012). Social systems dictate the rules of access to resources and influence the winners and losers of trade-offs between different benefits (Walker et al. 2004), ultimately affecting the relationship between ecosystem service dependence and poverty outcomes.

The relationship between ecosystem services and poverty changes because social mechanisms and other factors co-vary with bundles of ecosystem services. For example, the presence of opportunities for supplementing incomes with open access resources (e.g. fisheries, forest products), land ownership, opportunities for sharecropping and leasing land, agricultural labour, access to off-farm income opportunities, the level of exposure to extreme events, the impacts of cyclones and storm surges, the negative impacts on agriculture from aquaculture and the presence of landlords on whom the poor can rely for assistance through patron-client relationships all vary between social-ecological systems (see Adams and Adger 2016).

2.2.3 Spatial Variation in Ecosystem Services within Delta Environments

Assets, income, nutrition- and blood pressure-related health indicators and subjective well-being vary with location. Waterlogging, high salinity and access are significantly associated with poverty in the study area with different spatial patterns apparent for these three variables (see Chap. 21 and Amoako Johnson et al. 2016). Soil salinity is significantly associated with poverty around the Sundarbans, waterlogging in the centre of the study area, while the lack of access dominates in the east of the study area. For example, the factors associated with asset poverty vary across the study area. Considering all social-ecological systems, the probability of being materially and subjectively poor decreases as household dependence on ecosystem services for income increases. However, the irrigated agricultural zone showed the opposite relationship, with increasing dependence on ecosystem services being associated with a higher probability of being materially poor.

Similar spatial and social-ecological system-based differences are found in the health indicators (Chap. 27) and in how individuals perceive their own well-being. Levels of malnourishment are higher than the national average but vary across the study area. For example, food consumption varies across the study area and by social-ecological system. Irrigated agriculture areas show the lowest protein intake and one of the lowest calorie intakes. In comparison, households living in the char and rain-fed agricultural social-ecological systems also have low calorie intake levels, but the protein consumption (from fish) is much higher and child undernutrition lower. This indicates that fish consumption appears crucial to health in some social-ecological system.

2.2.4 Temporal Variations in Well-Being from Ecosystems

Ecosystem services vary by season and across years with implications for chronic and seasonal poverty. When examining past trends, three factors suggests that maintaining current productivity of agriculture and aquaculture will be challenging. First, historic analysis shows that recent increases in these provisioning services have been accompanied by concomitant decreases in underlying supporting services (see Chap. 5 and Hossain et al. 2016). Second, infrastructural interventions to facilitate increases in productivity of provisioning services (e.g. coastal embankments and polders) have caused a rigidity trap reducing flexibility in adaptation to future climate change (Adams et al. 2013). Third, seasonal changes (wet/dry seasons) in household livelihoods reflect changing work opportunities, leading to different long-term poverty trajectories (Lázár et al. 2016).

Since the 1950s, production of rice, shrimp and fisheries has increased consistently with gross domestic product (GDP) and per capita income (Hossain et al. 2016). However, this has been accompanied by a decrease in the quality of supporting services such as water quality and availability, natural hazard and erosion protection and maintenance of biodiversity, as well as availability of forest products. Thus, although provisioning ecosystem services of rice and agriculture have supported national level growth, it has been at the expense of the systems that support them (including potentially irrigation-induced salinisation of soil) and therefore may not be sustainable into the future (see Chap. 24).

During the 1990s, many provisioning and supporting ecosystem services declined (Hossain et al. 2016) linked to the modification of the natural functioning of rivers and their interaction with the floodplain. This includes tidal sediment deposition outside of the polders and drainage congestion within the polders (Islam 2006), while the interiors of polders have lost substantial elevation (e.g. Hoque and Alam 1997). It was these polders that initially enabled an increase in productivity by protecting the floodplain from inundation and, in turn, allowed the development of multi-cropping and aquaculture.

The longer-term future implications of such past and irreversible changes to the natural environment are problematic. For example, while productivity increases were enabled by this infrastructure, continued increases will be challenged under a future changing climate (Adams et al. 2014). Some of these problems may be ameliorated through upgrading the embankments but a more fundamentally sustainable long-term management technique such as controlled sedimentation within polders to build elevation, termed 'tidal river management' in Bangladesh (Amir et al. 2013; Auerbach et al. 2015), may be beneficial on a large scale. Looking into the future it is unclear whether an increase in GDP will eventually lead to the environmental investment necessary (i.e. following the environmental Kuznets curve, Hossain et al. 2016) to halt the further degradation of supporting services for agriculture.

Agricultural models of the delta (Chaps. 24 and 28) show that dry season productivity is currently constrained by salinity. Crop productivity may be maintainable to 2050 due to the positive impacts of projected increases in rainfall, temperature (within the range of rice) and CO_2 fertilisation. The constraining factors are fertility and heat stress if the monsoon season rains remain sufficient to remove salinity that has accumulated during the dry season. This, however, will be impacted in the longer term by sea-level rise, subsidence, dry season decreases in upstream flow and human water management (Chaps. 13, 16 and 17). Again tidal river management may be applied. A second crop could provide an additional income although income from wet season rice cultivation is constrained by low market prices.

Fisheries are second only to agriculture as a source of income in Bangladesh and form the main source of protein (Chap. 27). Offshore capture fisheries models (Chap. 25) project small decreases in overall fisheries productivity with climate change. It remains to be seen whether such decreases can be offset with sustainable management practices. However, the two most important fish species (Bombay duck and especially Hilsa) are susceptible to a potential collapse due to unsustainable fishing practices. This would intensify livelihood stress for subsistence fishers and emphasise the importance of sustainable exploitation of these resources.

A final way that temporal dimensions can provide answers as to why and when ecosystem services may be able to alleviate poverty emerges from analysis of long-term poverty trajectories, driven by day-to-day coping strategies and seasonal livelihood diversification. These trajectories have been characterised for a range of livelihood diversification strategies in a quantitative model (Lázár et al. 2016) based on survey data collected as part of this research (see Chap. 23 and Adams et al. 2016), household income and expenditure data and census data. Modelling livelihood trajectories for different archetypal households, with different seasonal livelihood strategies and multiple coping strategies during periods of low income, shows the transient nature of poverty and the ways in which farm and off-farm employment combine to create more or less stable well-being pathways. The analysis (see Chap. 28) shows that, while land ownership is crucial to avoid poverty, the poverty outcomes of small landowners are highly variable. Differences in micro-level choices therefore accumulate to create different outcomes for households with similar livelihood and poverty characteristics. The analysis also shows that most households have incomes that do not come from ecosystem services. Many households have two or three income sources, and almost all households show seasonal changes in their income type. Strategies vary, with different variations in income between seasons and different diversification strategies to maintain income. Modelling poverty trajectories in this way allows these seasonal drivers of long-term poverty dynamics to be integrated with other biophysical models to understand drivers of poverty at different scales and how poverty may change in the future under different interventions. Simulation results reveal the poverty alleviation role of off-farm income types and the importance of the quality of that off-farm employment, since households relying on small-scale, cottage industries are most likely to be poor and stay poor. These results support other studies that indicate land ownership is a necessary stepping stone out of poverty as it provides households with the capital to access high end off-farm income opportunities.

This research has therefore confirmed the need to consider ecosystem services for poverty alleviation in the wider context of agrarian reform. Many of the barriers to creating pro-poor ecosystem services-based livelihoods emerge from processes put into place during the 'Green Revolution'—for example, polders (Adams et al. 2014), and, more recently, the Blue Revolution of the aquaculture industry (Amoako Johnson et al. 2016). The judgement is whether any environmental degradation is justified for food security, national wealth objectives and the fair distribution of benefits.

2.2.5   Ecosystem Service Trade-offs between Social-Ecological Systems and Labour Mobility

Whether or not ecosystem services are a force for good in the diverse and dynamic delta environment of coastal Bangladesh relates to the nature of the trade-offs between different social-ecological systems. Trade-offs are an important part of the ecosystem service framework; for each service prioritised, another service will be diminished (Rodriguez et al. 2006). The same is true in the delta. Analysing past trends in the delta shows trade-offs between provisioning ecosystem services (that have been increasing) and the systems that support them (that have shown a consistent decline) (Hossain et al. 2016).

In the study area, trade-offs tended to work in a way that further concentrates rights to ecosystem services to those that already have them. For example, agriculture and aquaculture practices contribute to the degradation of the open access resources on which the landless depend, leaving them even further marginalised. This can be conceptualised by looking at the nature of the property rights system. Social-ecological systems where private property rights dominate, such as aquaculture, are most destructive to other systems. Ecosystem services from open access systems more readily co-exist. People react to changes in ecosystem services and there is a livelihood mobility dimension to any trade-offs between social-ecological systems. People move to alternative systems, or change jobs, to counteract the seasonality and irregularity of income from ecosystem services in one social-ecological system and when systems are degraded over the long term or labour is no longer required (e.g. as agricultural land is converted to agriculture employment declines significantly). Thus, migration of people between systems and livelihoods counters to the availability of ecosystem services.

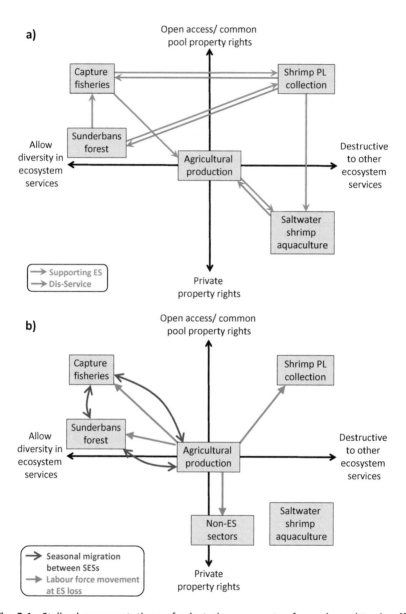

Fig. 2.1 Stylised representations of selected movements of people and trade-offs between ecosystem services, between social-ecological systems in the coastal zone of Bangladesh, geographically and between seasons. The y-axis represents the degree to which the services of that social-ecological system are privately owned. The x-axis represents whether the productivity of the social-ecological system is dependent on the degradation of another. Arrows show flows of labour, materials and process linkages. The diagrams illustrate the inter-dependent nature of social-ecological systems in the delta and the potential trade-offs in productivity and rural employability

Figure 2.1 illustrates some of the ways in which ecosystem services and benefits are transferred between different social-ecological systems across space and time, and the concurrent migration flows following livelihood opportunities. Ecosystem benefits, and thus people, move from one social-ecological system to another because of land use change and the transformation of one system to another, and seasonally. For example, where embankments protect the flood plain from inundation, those dependent on capture fisheries for livelihoods (e.g. traditional fishermen, boatmen for transportation) rely more heavily on the other social-ecological systems, move into off-farm opportunities or leave the area in search of economic opportunities. Therefore, labour is constantly moving between these different systems based on the season or the availability of resources.

Ecosystem services and benefits, or the capacity of a system to be productive, are also 'moving' between each system, the productivity of each social-ecological system being affected by the productivity of the others. The movement of ecosystem services between the systems is exemplified in the supply of wild shrimp larvae for pond aquaculture, sourced in the Sundarbans forest. While subsistence fishing exists without any detriment to the Sundarbans, this fry collection is a destructive process, not only to the Sunderbans where fish productivity is reduced because of bycatch but also in offshore fisheries, since the Sundarban forest supports nurseries for offshore fisheries by providing a supply of shrimp fry (Islam and Islam 2011). Thus the increase in productivity of shrimp farms has been at the expense of the Sundarbans biodiversity and productivity and the offshore fish catch.

2.3   Ecosystem Services in Changing Delta Agricultural Economies

The ability of rural populations in deltas to use ecosystem services for poverty alleviation is a function of the productivity of the ecosystem services, access and entitlements to those services, prior infrastructural or policy interventions and the dynamics of the diverse social-ecological systems. Specific ecosystem services are important for poverty alleviation because they have higher market value or are more abundant. However, a route out of poverty depends on the combination of pre-existing access levels and entitlements and the mix of different bundles of ecosystem services available over time and space.

The decades of rural development aimed at increasing agricultural productivity across deltas in South Asia have left certain populations behind. There are poor rural populations who have been unable to access the economic benefits that have accompanied integration into the market economy (e.g. Rigg 2006) and thus rely on increasingly degraded open access resources (e.g. fisheries) or precarious forms of off-farm employment (e.g. small-scale manufacturing or cottage industries). Households living in rural areas without the safety net provided by ecosystem services are doubly vulnerable: exposed to volatile markets and globalisation processes, but without the safety net of basic subsistence. Ecosystem services for poverty alleviation therefore must be analysed in the wider context of agrarian change. Worldwide, agriculture and resource-based local economies are diversifying (Bebbington 2000; Rigg 2006). As a result, solutions to poverty alleviation within rural societies typical of the Bangladesh delta are unlikely to lie solely within the realms of ecosystem services.

For the Ganges-Brahmaputra-Meghna delta, perhaps by contrast and as shown throughout this research, reliance on ecosystem services remains important both for poverty prevention and potentially for poverty alleviation. The lack of development of a range of sophisticated off-farm opportunities (World Bank 2016) may contribute to this as the poorest in the study area are those dependent on small-scale manufacturing ('cottage' industries) (Chap. 23). Furthermore, levels of migration, the means by which rural lives have been able to continue across much of south Asia, are low within the population surveyed (around 15 per cent of households had a migrant currently or in the past four months). Agriculture continues to be the lynchpin of the rural economy in Bangladesh (World Bank 2016), something that the findings of this research support. Thus, perhaps those in poverty are experiencing the worst of both worlds—the loss of traditional, subsistence-based ecosystem service use, without the benefits of market integration.

2.4   Conclusion

Access to ecosystem services continues to be the lynchpin of well-being in deltaic rural areas in Bangladesh, yet access to these services is diminishing for the poor. Current patterns of winners and losers from development processes are persistent, and ecosystem services are unlikely to lift the rural poor out of poverty without a complete restructuring of social and economic relations in rural areas. If this is the case, future ecosystem service research should focus on pro-poor, environmentally sustainable and rural development opportunities.

Yet, dynamic changes in the relationship between ecosystem services and poverty are apparent, and thus past or present relationships may not be a good guide to understanding the future (see Chap. 28). The availability of ecosystem services may change radically due to external forces, potentially remote from the delta's ecosystem services, such as (i) upstream flows from India and beyond (Chap. 13), (ii) global sea-level rise (Chap. 14), shifts in global and regional markets (Chap. 12) and (iii) changing demography (Chap. 19). Furthermore, Bangladesh is moving towards a more urbanised future, where a diminishing proportion of the population of delta areas will be directly reliant on ecosystem services for their livelihoods (Banks et al. 2011). Thus, future problems may revolve around enabling and improving access of the poor, urban populations to ecosystem services, either within the city or virtually through, for example, food networks to ensure minimal levels of well-being.

Fundamentally, the ability of ecosystem services to meet poverty alleviation objectives must be placed within broader questions of sustainability of rural areas, alternative economic systems, population growth and the impacts of extreme environmental change. Systems-based and modelling approaches, as exemplified in this research, are well-suited to explore such dynamic, multi-scalar and potentially non-linear relationships.

References

Adams, H., and W.N. Adger. 2016. Mechanisms and dynamics of wellbeing-ecosystem service links in the southwest coastal zone of Bangladesh. UK Data Service Reshare. https://doi.org/10.5255/UKDA-SN-852356.

Adams, H., W.N. Adger, H. Huq, R. Rahman, and M. Salehin. 2013. Wellbeing-ecosystem service links: Mechanisms and dynamics in the southwest coastal zone of Bangladesh. ESPA Deltas working paper 2. Southampton: University of Southampton. www.espadelta.net/resources_/working_papers/. Accessed 12 Dec 2016.

———. 2014. Transformations in land use in the southwest coastal zone of Bangladesh: Resilience and reversibility under environmental change. In *Proceedings of transformation in a changing climate*, June 19–21. Oslo, Norway.

Adams, H., W.N. Adger, S. Ahmad, A. Ahmed, D. Begum, A.N. Lázár, Z. Matthews, M.M. Rahman, and P.K. Streatfield. 2016. Spatial and temporal dynamics of multidimensional well-being, livelihoods and ecosystem services in coastal Bangladesh. *Scientific Data* 3: 160094. https://doi.org/10.1038/sdata.2016.94.

Adger, W.N. 1999. Social vulnerability to climate change and extremes in coastal Vietnam. *World Development* 27 (2): 249–269.

Agrawal, A. 2005. Environmentality: Technologies of government and the making of subjects. In *New ecologies for the twenty-first century*, ed. A. Escobar and D. Rocheleau. Durham: Duke University Press.

Amir, M.S.I.I., M.S.A. Khan, M.M.K. Khan, M.G. Rasul, and F. Akram. 2013. Tidal river sediment management-A case study in southwestern Bangladesh. *International Journal of Environmental, Chemical, Ecological, Geological and Geophysical Engineering* 7 (3): 176–185.

Amoako Johnson, F., C.W. Hutton, D. Hornby, A.N. Lázár, and A. Mukhopadhyay. 2016. Is shrimp farming a successful adaptation to salinity intrusion? A geospatial associative analysis of poverty in the populous Ganges–Brahmaputra–Meghna Delta of Bangladesh. *Sustainability Science* 11 (3): 423–439. https://doi.org/10.1007/s11625-016-0356-6.

Auerbach, L.W., S.L. Goodbred, D.R. Mondal, C.A. Wilson, K.R. Ahmed, K. Roy, M.S. Steckler, C. Small, J.M. Gilligan, and B.A. Ackerly. 2015. Flood risk of natural and embanked landscapes on the Ganges-Brahmaputra tidal delta plain. *Nature Climate Change* 5 (2): 153–157. https://doi.org/10.1038/nclimate2472.

Banks, N., M. Roy, and D. Hulme. 2011. Neglecting the urban poor in Bangladesh: Research, policy and action in the context of climate change. *Environment and Urbanization* 23 (2): 487–502. https://doi.org/10.1177/0956247811417794.

Bebbington, A. 2000. Reencountering development: Livelihood transitions and place transformations in the Andes. *Annals of the Association of American Geographers* 90 (3): 495–520. https://doi.org/10.1111/0004-5608.00206.

Dempsey, J., and M.M. Robertson. 2012. Ecosystem services: Tensions, impurities, and points of engagement within neoliberalism. *Progress in Human Geography* 36 (6): 758–779. https://doi.org/10.1177/0309132512437076.

Gunderson, L.H., and C.S. Holling, eds. 2002. *Panarchy: Understanding transformations in human and natural systems*. Washington, DC: Island Press.

Hartmann, B., and J.K. Boyce. 1983. *A quiet violence: View from a Bangladesh village*. London: Zed Books.

Hayami, Y., and M. Kikuchi. 2000. *A rice village saga: Three decades of green revolution in the Philippines*. London/Lanham: Macmillan Press/Barnes & Noble.

Hoque, M., and M. Alam. 1997. Subsidence in the lower deltaic areas of Bangladesh. *Marine Geodesy* 20 (1): 105–120. https://doi.org/10.1080/01490419709388098.

Hossain, M.S., J.A. Dearing, M.M. Rahman, and M. Salehin. 2016. Recent changes in ecosystem services and human well-being in the Bangladesh coastal zone. *Regional Environmental Change* 16 (2): 429–443. https://doi.org/10.1007/s10113-014-0748-z.

Islam, M.R. 2006. Managing diverse land uses in coastal Bangladesh: Institutional approaches. In *Environment and livelihoods in tropical coastal zones: Managing agriculture-fishery-aquaculture conflicts*, ed. C.T. Hoanh, T.P. Tuong, J.W. Gowing, and B. Hardy. Wallingford: CAB International.

Islam, K.M., and M.N. Islam. 2011. Economics of extraction of products from Sundarbans reserve forest. *Bangladesh Journal of Agricultural Economics* 34 (1/2): 29–53.

Lázár, A.N., H. Adams, W.N. Adger, and R.J. Nicholls. 2016. *Characterising and modelling households in coastal Bangladesh for model development*. Project working paper. ESPA Deltas, University of Southampton. www.espadelta.net.

Leach, M., R. Mearns, and I. Scoones. 1999. Environmental entitlements: Dynamics and institutions in community-based natural resource management. *World Development* 27 (2): 225–247. https://doi.org/10.1016/s0305-750x(98)00141-7.

Martín-López, B., I. Iniesta-Arandia, M. García-Llorente, I. Palomo, I. Casado-Arzuaga, D. García Del Amo, E. Gómez-Baggethun, E. Oteros-Rozas,

I. Palacios-Agundez, B. Willaarts, J. González, F. Santos-Martín, M. Onaindia, C. López-Santiago, and C. Montes. 2012. Uncovering ecosystem service bundles through social preferences. *PLoS One* 7 (6): e38970. https://doi.org/10.1371/journal.pone.0038970.

Nicholls, R.J., C.W. Hutton, A.N. Lázár, A. Allan, W.N. Adger, H. Adams, J. Wolf, M. Rahman, and M. Salehin. 2016. Integrated assessment of social and environmental sustainability dynamics in the Ganges-Brahmaputra-Meghna delta, Bangladesh. *Estuarine, Coastal and Shelf Science* 183: 370–381. https://doi.org/10.1016/j.ecss.2016.08.017. Part B.

Ostrom, E. 1990. *Governing the commons: The evolution of institutions for collective action*. Cambridge: Cambridge University Press.

Raudsepp-Hearne, C., G.D. Peterson, and E.M. Bennett. 2010. Ecosystem service bundles for analyzing tradeoffs in diverse landscapes. *Proceedings of the National Academy of Sciences of the United States of America* 107 (11): 5242–5247. https://doi.org/10.1073/pnas.0907284107.

Rigg, J. 2006. Land, farming, livelihoods, and poverty: Rethinking the links in the Rural South. *World Development* 34 (1): 180–202. https://doi.org/10.1016/j.worlddev.2005.07.015.

Robbins, P. 2011. *Political ecology: A critical introduction*. 2nd ed. Chichester: Wiley.

Rodriguez, J.P., T.D. Beard, E.M. Bennett, G.S. Cumming, S.J. Cork, J. Agard, A.P. Dobson, and G.D. Peterson. 2006. Trade-offs across space, time, and ecosystem services. *Ecology and Society* 11 (1): 28. www.ecologyandsociety.org/vol11/iss1/art28/.

Scoones, I. 1998. Sustainable rural livelihoods: A framework for analysis IDS Working paper 72. Brighton: Institute of Development Studies (IDS). http://www.ids.ac.uk/files/dmfile/Wp72.pdf. Accessed 11 July 2017.

Walker, B., C.S. Holling, S.R. Carpenter, and A. Kinzig. 2004. Resilience, adaptability and transformability in socio-ecological systems. *Ecology and Society* 9 (2): 5. www.ecologyandsociety.org/vol9/iss2/art5/.

World Bank. 2016. Dynamics of rural growth in Bangladesh : Sustaining poverty reduction. Washington, DC: World Bank. https://openknowledge.worldbank.org/handle/10986/24369. Accessed 11 July 2017.

3

An Integrated Approach Providing Scientific and Policy-Relevant Insights for South-West Bangladesh

Robert J. Nicholls, Craig W. Hutton, Attila N. Lázár, W. Neil Adger, Andrew Allan, Paul G. Whitehead, Judith Wolf, Md. Munsur Rahman, Mashfiqus Salehin, Susan E. Hanson, and Andres Payo

3.1 Introduction

As explained in Chap. 1, deltas are vulnerable to sea-level rise and climate change, reflecting their low elevation which is largely controlled by the present sea level (Ericson et al. 2006). Bangladesh, in particular, has been identified as an impact hotspot for sea-level rise in multiple studies over several decades (e.g. Milliman et al. 1989; Warrick et al. 1993; Huq et al. 1995; World Bank 2010). However, a range of other drivers are important in all

R. J. Nicholls (✉) • A. N. Lázár • S. E. Hanson
Faculty of Engineering and the Environment and Tyndall Centre for Climate Change Research, University of Southampton, Southampton, UK

C. W. Hutton
Geodata Institute, Geography and Environment, University of Southampton, Southampton, UK

W. Neil Adger
Geography, College of Life and Environmental Sciences, University of Exeter, Exeter, UK

© The Author(s) 2018
R. J. Nicholls et al. (eds.), *Ecosystem Services for Well-Being in Deltas*,
https://doi.org/10.1007/978-3-319-71093-8_3

deltas including Bangladesh, such as regional factors (especially catchment management, socio-economic development and governance quality) and numerous delta plain factors (e.g. sediment starvation, subsidence, land use change, ill-planned interventions) (see Woodroffe et al. 2006; Syvitski et al. 2009; Tessler et al. 2015; Day et al. 2016). Collectively, this threatens an increase in a range of hazards such as flooding, inundation, salinisation and erosion, in turn impacting delta residents' vulnerability, livelihoods and food security. To fully analyse these risks an integrated or systems analysis of the delta is required (Nicholls et al. 2016; Chap. 4).

Such an analysis of south-west coastal Bangladesh (henceforth coastal Bangladesh) situated within the Ganges-Brahmaputra-Meghna (GBM) delta (Chap. 4, Figs. 4.1 and 4.2) is described in this book. The overall aim of this research is twofold. Firstly, it is to understand coastal Bangladesh through the lens of ecosystem services and associated livelihoods. Secondly, it is to make this information available in a form that is suitable for relevant decision makers. This requires linking science to policy at the landscape scale. Chapter 2 analysed how ecosystem services relate to poverty and its alleviation in deltas, and its development and environmental management implications. The aim of this chapter is to

A. Allan
School of Law, University of Dundee, Dundee, UK

P. G. Whitehead
School of Geography and the Environment, University of Oxford, Oxford, UK

J. Wolf
National Oceanographic Centre, Liverpool, UK

Md. Munsur Rahman • M. Salehin
Institute of Water and Flood Management, Bangladesh University of Engineering and Technology, Dhaka, Bangladesh

A. Payo
British Geological Survey, Keyworth, Nottingham, UK

review the scientific insights concerning the delta system and consider their relevance and utility to national policy. The issue of transferability of this understanding to other deltas is also briefly considered.

Integration and policy relevance was at the heart of the research design, necessitating ongoing discussions among the research team from the commencement of this research. Making it a core theme contributed to the success of integration. Consistency is also important and coordination across the project ensures that all the components follow a common conceptual model and set of narratives about future conditions. Chapter 4 outlines the integrated methodology which included scenario development, stakeholder analysis and engagement (Chaps. 9, 10, 11 and 12), socio-economic analysis and bespoke household surveys (Chaps. 19, 20, 21, 22 and 23) and biophysical modelling of the range of processes (Chaps. 13, 14, 15, 16, 17 and 18) and supporting ecosystem services (Chaps. 24, 25 and 26).

Building on this integrated framework, a dedicated integrated assessment model was built to (i) harmonise across the different scales and models, (ii) provide practical run times and (iii) address feedbacks by tightly coupling the individual assessment models. This led to the development of the Delta Dynamic Integrated Emulator Model (ΔDIEM) (Fig. 3.1 and Chap. 28). ΔDIEM is designed to assess the potential social-ecological trajectories of coastal Bangladesh, including the role of different development and adaptation choices. Hence policy choices are an explicit input to the analysis. While ΔDIEM is a sophisticated software tool, its development has also facilitated major discussion and learning between project participants from diverse disciplinary backgrounds. ΔDIEM and the wider project research encapsulate learning, coordination and cooperation amongst the diverse project partners. There is also ongoing engagement with national level stakeholders engaged in strategic planning. The premise of this engagement is that, in this case, influencing national policy maximises impact of the research, and this required us to understand this national policy context. Hence the research theory of change revolves around stakeholder engagement with national stakeholders to promote understanding, usefulness, trust and ownership concerning the approach and results. As explained in Chap. 9, these social dimensions of integration are expressed as an 'iterative learning process' (see also Chap. 4, Fig. 4.5) where

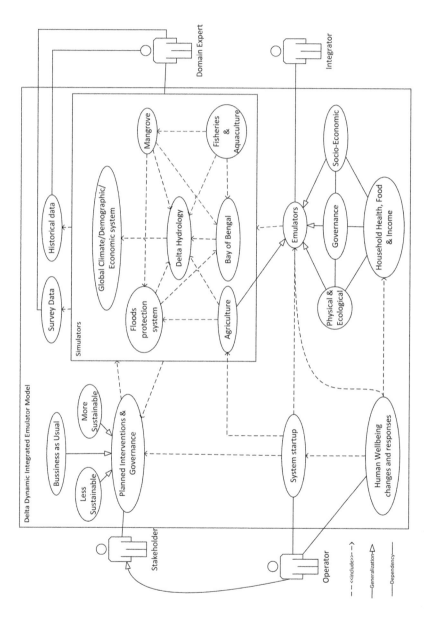

Fig. 3.1 Conceptual view of the integrated model, including the roles of stakeholders, domain experts, specialist integrators within the research team and operators

the model design and development is coupled to repeated stakeholder engagement and co-production of scenarios and results.

The dynamic nature of the policy environment within Bangladesh also needs to be recognised and addressed as this influences the approach and outcome of the research described (see Chap. 6). Bangladesh has an aspiration to be a middle-income country by 2030 and follows a five-year planning cycle coordinated by the Planning Commission, Government of Bangladesh with the last plan covering the period 2016 to 2020, inclusive (GED 2015). A new Perspective Plan coincided with the research engagement. That Plan laid out a national development pathway to 2021 to coincide with the 50th anniversary of the founding of Bangladesh (GED 2012). Subsequently, the policy environment evolved rapidly with the development of the Bangladesh Delta Plan 2100 (BDP2100) as a living and integrated national planning document (BanDuDeltAS 2014). It is inspired by international experience, especially the Delta Commissie (2008) and the subsequent creation of the Delta Commission and associated plans in the Netherlands (Van Alphen 2016). Globally, the Sustainable Development Goals (SDGs) have also been developed (UN 2015) within the lifetime of the project and are being widely applied in Bangladesh (e.g. GED 2017). With the Prime Minister of Bangladesh becoming a member of the United Nations High-Level Panel on Water (for SDG 6), Bangladesh has bolstered its SDG-related activities, including developing concept papers and action plans. The stakeholder engagement strategy was kept flexible and the emergence of the BDP2100 in particular created a stronger environment of stakeholder engagement facilitated by the Planning Commission, enhancing the development and application of this research.

This chapter is structured as follows. Section 3.2 considers key insights that emerged from the research described in this book and their policy implications. It considers the present situation, potential trends and ecosystem services, including the integrated assessment, policy implications, and briefly the SDGs. Section 3.3 discusses the implications for the future of coastal Bangladesh and sets the research into a broader context. Section 3.4 provides conclusions.

3.2   Key Insights Concerning Coastal Bangladesh

3.2.1   Present Status

Bangladesh is perceived internationally to be significantly constrained by challenges linked to climate change and sea-level rise (Roy et al. 2016). Certainly temperatures and sea levels are slowly rising and the delta is subsiding, which are adverse trends (Hijioka et al. 2014). Further, there is the longstanding threat of cyclones, although preparedness has improved dramatically since the 1991 cyclone, greatly reducing loss of life (Lumbruso et al. 2017). But the trajectory of development and well-being in Bangladesh is the outcome of demographic change, the urbanisation trends, the rise of manufacturing and regional geopolitics, coupled to environmental constraints or challenges, as recognised at the core of the analysis in this book.

The national population of Bangladesh increased fourfold between 1950 and 2013, from 38 to 157 million and is projected to exceed 200 million by 2050 (UN 2013). There is strong ongoing urbanisation exemplified by growth of Dhaka from about 300,000 people in 1950 to a world city of over 18 million people in Greater Dhaka today. Most data sources show significant increases in per capita GDP: for example, according to the World Bank from US$410 in 2000 to US$1,360 in 2016. The direct sources of economic growth are changes to the structure of the economy: growth in (i) manufactured exports especially clothing, (ii) overseas remittances and (iii) non-farm employment (Chap. 12). The historical trends and trade-offs between provisioning and regulating ecosystem services in Bangladesh (Chap. 5) have already been emphasised in Chaps. 1 and 2. Production of agriculture and fisheries has boomed, but at the cost of environmental quality and regulation. This demonstrates the strong temporal dynamics of ecosystem services and important trade-offs that need to be considered in environmental management. Linked to this is one significant land use change: the significant expansion of saline shrimp farming replacing rice fields north of the Sundarbans (Chap. 20). The growth of shrimp-based aquaculture appears to be associated with both high dry season salinity and the proximity of a substantial settlement with good access and processing facilities (in this case Khulna). These changes might be viewed as a useful adaptation to increasing salinity intrusion in the region, with shrimp providing a new export commodity. However, brackish water aquaculture also has environmental impacts and consequences for land tenure, livelihood displacements and income loss, food insecurity and health, rural unemployment, social unrest, conflicts and forced migration (Chap. 20).

Even though the national population is rising in Bangladesh, the coastal population is declining, reflecting net migration to urban areas. Chapter 19 estimates that the present study area population of 14 million could decline to 11 to 13.5 million by 2050. This reflects multiple drivers of migration and sea-level rise, and climate change do not appear to be direct drivers to date (Roy et al. 2016), although anecdotally this is often reported as such in the media.

3.2.2   Potential Trends for Coastal Bangladesh

A range of potential trends are apparent for coastal Bangladesh over the next 50 to 100 years, with climate change being a major driver of these changes (Table 3.1). However, other human-induced trends are occurring, and these might be more important than climate change in some cases. For example, climate change will intensify the monsoon (Chap. 11) with greater run-off during the wet season and less freshwater flow during the dry season (Whitehead et al. 2015a, b). The development of major catchment infrastructure in neighbouring upstream countries (e.g. inter-linking river project in India (National Water Development Agency 2017)) may lead to further reduction in dry season flows. The potential magnitude of these reductions is much larger than those due to climate change alone, raising challenges of intergovernmental coordination between states on the GBM basin (Yasuda et al. 2017).

In addition to water supply, this research suggests that due to climate change alone the supply of fluvial sediment being delivered to the GBM delta is likely to increase by around 50 per cent by 2100 (Chap. 15). However, catchment changes are likely to have the opposite effect, and a net reduction in the catchment sediment supply to the delta appears

Table 3.1 Key potential trends for coastal Bangladesh to 2100

	Key trend
Climate	The monsoon is expected to intensify with higher peak run-off down the Ganges, Brahmaputra and Meghna. Similarly dry season flows are expected to decline due to climate change
	Climate-induced sea-level rise will be enhanced by subsidence, giving greater amounts of relative sea-level rise
	Fluvial flooding (during the monsoon) and surge flooding (due to cyclones) are expected to intensify due to climate change and relative sea-level rise unless flood management is upgraded and improved
Salinisation	Increased salinisation of the coastal waters is expected due to relative sea-level rise
Sediments/ freshwater supply	The large sediment supply to the delta will increase significantly due to climate change
	However, upstream catchment dams and diversions have a greater potential to reduce freshwater flow and sediment delivery, with particular concern for freshwater supply during the dry season
Ecosystem services	Changes in the primary productivity in the Bay of Bengal appear to be small
	Agriculture will become increasingly stressed in the region, especially during the dry season. Salinisation and higher temperatures play an important role
	While primary productivity and fisheries can be maintained with small reductions under sustainable management, fisheries are in danger of collapse under less sustainable management regimes
	The Sundarbans are more resilient to sea-level rise than suggested in earlier assessments. Even under the highest relative sea-level rise scenario (about 1.7 m), it is estimated that more than 60% of the area will survive to 2100. However, the constituent mangrove species in the Sundarbans will change towards more salt-tolerant species and become less diverse

more likely. This is a critical factor as sediment has built the delta over millennia and reduction in sediment supply is seen to threaten many deltas globally (Ericson et al. 2006; Syvitski et al. 2009), for example, the Mekong delta (Chapman et al. 2016). More effort to understand the future catchment sediment budget and delta level morphodynamics in the GBM delta could pay dividends, especially in terms of managing sediments and sedimentation in beneficial ways at the delta scale (Angamuthu 2017).

Flooding is a defining issue in coastal Bangladesh and nationally (Chaps. 8 and 16). There are several sources of flooding, including fluvial, tidal and cyclone-induced floods, as well as compound floods from multiple sources at the same time. Given the importance of compound floods, basic research on understanding compound flooding in Bangladesh may well bring dividends in adaptation planning (cf. Wahl et al. 2015). In addition to these flood sources, the extensive polder system and their embankments across the study area must be considered. The potential magnitude of land inundation due to all forms of flooding is aggravated over the twenty-first century by relative sea-level rise and in some cases larger upstream river flows. For the scenarios considered, the changes in cyclone-induced flooding are greater than those for fluvial flooding. The flood analysis demonstrates that the effects of sea-level rise are not simple inundation as assumed in earlier analyses (e.g. Milliman et al. 1989; Huq et al. 1995). This reflects the existence of extensive polders and embankment systems as well as the areas outside the polders, such as the Sundarbans, which can accrete sediment with sea-level rise (see Chap. 26).

Salinity is a defining issue in coastal Bangladesh as high salinities of the rivers and estuaries penetrate inland degrading the water resources and agriculture (Chaps. 17 and 18). There are spatial and temporal gradients with the saltiest river waters in the western estuarine section and the Sundarbans forest and a strong seasonal signal in the freshwater distribution controlled by variable river discharge from the monsoon to the dry season. Under all future projections of freshwater input and oceanic changes, salinity is predicted to increase in the river channels. This increase is more pronounced in the central and western estuarine section with important implications for agriculture, shrimp farming and local wellbeing. For soil salinity, inter-seasonal climate variability appears to determine if this increases or declines relative to the present (Payo et al. 2017).

3.2.3   Trends in Ecosystem Services

Agriculture is the single most important provider of ecosystem services to the people of coastal Bangladesh. Income from agriculture is currently constrained by the limited availability of good quality irrigation water in the dry season, access to markets and the cost of fertilisers. It is reviewed further using the ΔDIEM results in Chap. 28. However, future monsoon variability and high temperatures during the dry season can hinder crop productivity more than salinity up to 2050 (Chap. 24).

Climate change is projected to reduce fish production in the Bangladesh Exclusive Economic Zone by up to ten per cent (Chaps. 14 and 25). These impacts are larger for the two major species (Hilsa shad and Bombay duck). Overfishing combined with climate change could reduce Hilsa catches by up to 90 per cent by 2050. In contrast, good management can maintain sustainable catches over the same period.

The Sundarbans mangrove forest is critical in providing ecosystem services to the poor (Chap. 26). The Sundarbans also provide significant buffering from floods to landward regions during cyclones augmenting traditional flood defences. Mangroves decline in area and show species change, but remain an important land cover providing important benefits under all the scenarios considered to 2100. This is a more optimistic outcome than previously widely reported (Hijioka et al. 2014).

3.3   Integrated Assessment Using the Delta Dynamic Integrated Emulator Model

ΔDIEM has simulated the ecosystem service and livelihood futures of coastal Bangladesh to 2050 for a range of scenarios (Chap. 28). This shows a more complex set of behaviours than the uncoupled analysis. The region develops under all the scenarios considered, but the rate of development depends on the future population dynamics and the economic situation. Present agriculture-based livelihoods are constrained by land availability, high dry season salinity and agro-economics. Monsoon rains supply adequate water to grow one main season rice crop, but farmers' incomes are constrained by the market price for rice compared with the direct agriculture costs (i.e. this allows little or no profit). Lack of fresh water is a constraint for irrigated crops in the drier months. The use of low irrigation water quality (i.e. too high salt content) contributes to soil salinisation reducing farm production potential. Future dry season crop production is also likely to be constrained by drought (i.e. lack of water and high air temperature). Fortunately, the development of drought- and salt-tolerant varieties of rice and other crops is progressing and is essential under the future scenarios considered here.

For the average household, socio-economics is decoupled from climate and environmental change, although this is strongly related to future population dynamics and the poorest remain dependent on environmental services. This suggests that non-ecosystem services-based income sources can maintain well-being in coastal Bangladesh even if the traditional fish- and farm-based livelihoods become less profitable. Policies setting advantageous market conditions for farmers and fishers, creating new off-farm job opportunities and establishing new infrastructure to connect people to natural resources and markets would have a significant impact on the future welfare of the study area. However, even though the economy and human well-being can grow under appropriate governance and social policies, hazards and environmental change will still have a detrimental effect on food security in Bangladesh. For example, soil salinity, together with household socio-economic characteristics, is one of the main explanatory variables of food security for coastal households (i.e. increasing salinity leads to reduced food security) regardless of wealth levels (Szabo et al. 2015). As already noted, soil salinisation will continue to increase after 2050, and these changes will likely accelerate. Furthermore, cyclones are widely expected to intensify in the future (Church et al. 2013) (Chap. 14) which is highly detrimental when coupled with relative sea-level rise. Hence in addition to development to 2050, strategic adaptation measures to deal with climate change are essential in the longer term.

3.4 Policy Implications

Table 3.2 summarises some of the key policy implications of this research. There are several immediate priorities related to the current status of ecosystem services. There are also a number of longer-term measures where research and demonstration activities are required now, so that full implementation is available in a few decades. Working with sedimentation and

Table 3.2 Key policy implications of changing ecosystem services and related issues for coastal Bangladesh

	Policy implication
Short term	Current trends in ecosystem services are not sustainable, and an immediate policy response is required, especially for regulating ecosystem services
	Fisheries management requires urgent attention
	Training of farmers and fishers for good management practices are essential
	Implementation of policy across multiple sectors and scales will be challenging in the light of limited legal and coordinated institutional capacity
Longer term (although research and demonstration activities are required now)	The potential increases in fluvial and coastal flooding over the twenty-first century require a strategic adaptation response
	The possible role of sediment-based adaptation should be considered—that is controlled sedimentation to build land elevation mimicking the natural processes that build deltas. This provides a strategy which can maintain coastal Bangladesh into an uncertain future providing sufficient sediment is available
	Continued development of climate-resistant crop varieties (salt, temperature, etc.) will be essential to sustain agriculture in coastal Bangladesh
	The resilience of the Sundarbans to sea-level rise suggests more exploration of nature-based solutions, such as green mangrove belts would be prudent

nature-based solutions is innovative and offers great potential. The scale of implementation should also be noted as this will be needed at the delta scale. These ideas are fully consistent with the Bangladesh Delta Plan 2100. These conclusions should also be seen in the context of the existing legal and institutional situation that faces challenges in coordinating policy implementation across ecosystem services and livelihood enhancement in coastal Bangladesh (Chap. 6).

3.5   Relevance to the Sustainable Development Goals

In the period of this research, the SDGs have been developed and adopted globally as development targets for 2030 (UN 2015). They comprise a set of 17 'Global Goals' with 169 targets between them and are encouraging global activity on how to achieve and how to measure these achievements (Szabo et al. 2016), including in Bangladesh. The SDGs have essential and ambitious targets that are important today. This is reflected in the short timescale aiming to achieve most of these goals by 2030. However, 2030 is too close to experience significant climate and environmental changes, and therefore, the long-term needs for adaptation are not explicitly embedded in the SDGs.

While not designed explicitly to support the SDGs, the approaches described in this book have great potential in this regard. In particular they can provide interdisciplinary insights on several SDGs and consider both short-term and long-term trends and their context. This was discussed and supported in a two-day meeting presenting the results in this book, held at the Royal Society, London in November 2016. These approaches already provide a wide range of indicators across the societal and environmental domains, which may be more suitable than many simple measured indicators used previously. Trends in river flows, sediment fluxes and coastal salinity are three relevant trends discussed in this book that relate to various SDGs, and especially SDG 6 (Clean water and sanitation). Furthermore, the methods are flexible and can be developed and targeted as required.

3.6   Discussion

The integrated approach that has been adopted in this research has yielded both scientific and policy-relevant insights. It shows that there is a need for adaptation to climate change in coastal Bangladesh, as widely recognised, but this adaptation needs to be set into the wider context of the developing Bangladesh society. Given the large flood plain areas, adaptation and planning for increased flooding is essential. More consideration needs to be given to understanding and predicting embankment breaching as this will become more likely over time if no planning is undertaken. Upgrading the embankments, as already proposed in the Bangladesh Delta Plan 2100, is one approach, combined with other actions such as improved warning systems (Dasgupta et al. 2014). More fundamentally there is a need to develop more sustainable long-term delta-scale management techniques that utilise sediments and sedimentation in new ways. This is large-scale controlled (or engineered) sedimentation within polders as well as working with morphodynamic processes to build land elevation and sustain natural ecological functions such as fisheries spawning (Amir et al. 2013; Auerbach et al. 2015). 'Tidal river management' has only been employed locally to date, and to be successful would require both fundamental scientific investigation of morphodynamic processes and the development of innovative large-scale engineering application. In addition to the technical dimensions of these policies, the social dimensions will also need to be considered such as education and training, infrastructure development and targeted financial support to the poorest of the poor. Ignoring these social dimensions is likely to impede or stop their application.

In addition to the insights obtained, the method itself is worthy of consideration. The hybrid integrated framework and method has allowed a move away from an ad hoc external expert or purely indicator-based approach. It provides an opportunity not only to explore the interactions between domains of knowledge as diverse as oceanographic modelling and perception-based assessments of well-being but also to incorporate the views of stakeholders throughout in ways which maintain the relevance and credibility of the findings, despite their complexity. The assumptions are explicit and have been debated and challenged as part of this co-production process, and changed as knowledge has grown and the detailed questions being posed have evolved with this understanding. Hence, it provides an explicit framework to analyse the problem and provide spatially explicit output. In particular, it forces the users to identify, consider and explore the limits to knowledge of the relevant system. Such reflection is critical in problems where gaps and limits in current knowledge need to be acknowledged and addressed. It should be noted that ΔDIEM and its associated approaches continue to be developed and applied in partnership with the Bangladesh Delta Plan 2100, including assessing specific proposals.

This approach depends upon systems analysis and simulation modelling. Given the difficulty of predicting change in all of the systems considered here, such modelling could be regarded as being almost naïve. The limits to what is represented are recognised in the models, but the aim was to represent all the relevant processes and their interactions; many of the insights emerged within the integrated modelling. This facilitated development of conceptual ideas, promoted detailed discussion across discipline boundaries and resulted in usable algorithms and software. As increased understanding is obtained, algorithms can be modified or replaced. Even though this particular funded research project is complete, the process is not at an end. As experience is gained, the exploration of the complexities, interdependencies and uncertainties of coastal Bangladesh, and more widely, will continue. This includes considering a wide range of possible strategies for development linking to the Bangladesh Delta Plan, for example.

Building these types of co-produced analytical tools represents a significant amount of effort and resource, but the new insights, capacity building, scientific and policy applications and understanding generated justify this approach fully. Further, the ambitious policy processes such as the Bangladesh Delta Plan 2100 demand the development of these types of tools. In the Netherlands, the Delta Commissie (2008) led to the development of the national Delta Model (Prinsen et al. 2015; Van Alphen 2016), while in the Mississippi delta, restoration activities have led to the Coastal Master Plan Model (Coastal Protection and Restoration Authority 2013). While less common in developing country settings, these tools are still required to support developing management and development plans. Hence, these methods could be applied more widely across other deltas, as many issues are common. The methods could be extended to consider other issues such as cities and urbanisation, which are critical issues in most populated deltas. In principle the methods described are not delta-specific and could be applied in other coastal and even non-coastal contexts where strong social-ecological coupling exists. As such, the approach is flexible to user needs.

3.7   Conclusions

This chapter has considered the insights provided by a new integrated framework for analysing the future of coastal Bangladesh. The approach is modular allowing incremental improvement. It is also based on a participatory approach, and this was a key element in the successful engagement with national stakeholders and capturing their thinking within the analysis. The results show three key high-level interacting challenges for the future trajectory of coastal Bangladesh over the coming 30 years: continued potential decline in ecosystem services, persistent poverty and the interacting challenges of climate change with adaptive responses and measures.

First, the modelling results and the informed perspective of key stakeholders suggest that the future is more sensitive to policy interventions and responses to them than climate change; favourable development scenarios could transform Bangladesh towards sustainable development paths in a 30-year timescale. However, severe climate events such as cyclones are likely to represent a potential brake on such development. Second, ecosystem services diminish as a proportion of the economy with time, continuing historic trends. Third, significant poverty persists in some locations under all scenarios with specific populations remaining vulnerable, left behind and in need to direct assistance and policy targeting. These include populations without ownership to land that are directly dependent on informal labour markets and those dependent on the informal economy more generally (see also Chap. 2).

Beyond 2050, the effect of climate change may be much larger, and concerns about higher temperatures, changing hydrology and run-off, sea-level rise and more intense cyclones are important. Strategic adaptation to these challenges is essential, and preparation is required now. In this regard, innovative measures such as working with sedimentation processes and the creation of mangrove buffer zones deserve development and trial implementation at the large scale.

The Government of Bangladesh is promoting the application of these approaches and is evaluating several of proposed interventions with the Bangladesh Delta Plan 2100. This type of practical application is the best way to learn about application and is continuing the iterative learning loop on which the project methodology is based.

References

Amir, M.S.I.I., M.S.A. Khan, M.M.K. Khan, M.G. Rasul, and F. Akram. 2013. Tidal river sediment management-A case study in southwestern Bangladesh. *International Journal of Environmental, Chemical, Ecological, Geological and Geophysical Engineering* 7 (3): 176–185.

Angamuthu, B. 2017. The morphodynamic characteristics of a mega tidal delta over decadal to centennial timescales: A model-based analysis. Unpublished PhD thesis, University of Southampton.

Auerbach, L.W., S.L. Goodbred, D.R. Mondal, C.A. Wilson, K.R. Ahmed, K. Roy, M.S. Steckler, C. Small, J.M. Gilligan, and B.A. Ackerly. 2015. Flood risk of natural and embanked landscapes on the Ganges-Brahmaputra tidal delta plain. *Nature Climate Change* 5 (2): 153–157. https://doi.org/10.1038/nclimate2472.

BanDuDeltAS. 2014. Inception report. Bangladesh Delta plan 2100 formulation project. Dhaka: General Economics Division (GED), Planning Commission, Government of the People's Republic of Bangladesh.

Chapman, A.D., S.E. Darby, H.M. Hồng, E.L. Tompkins, and T.P.D. Van. 2016. Adaptation and development trade-offs: Fluvial sediment deposition and the sustainability of rice-cropping in An Giang Province, Mekong Delta. *Climatic Change* 137 (3): 593–608. https://doi.org/10.1007/s10584-016-1684-3.

Church, J.A., P.U. Clark, A. Cazenave, J.M. Gregory, S. Jevrejeva, A. Levermann, M.A. Merrifield, G.A. Milne, R.S. Nerem, P.D. Nunn, A.J. Payne, W.T. Pfeffer, D. Stammer, and A.S. Unnikrishnan. 2013. Sea level change. In *Climate change 2013: The physical science basis. Contribution of working group I to the fifth assessment report of the intergovernmental panel on climate change*, ed. T.F. Stocker, D. Qin, G.-K. Plattner, M. Tignor, S.K. Allen, J. Boschung, A. Nauels, Y. Xia, V. Bex, and P.M. Midgley. Cambridge/New York: Cambridge University Press.

Coastal Protection and Restoration Authority. 2013. 2017 Coastal master plan. Model improvement plan. Version II (Revised March 2014). Baton Rouge: The Water Institute of the Gulf. http://coastal.la.gov/wp-content/uploads/2013/09/MIP-Overview-3-12-14.pdf. Accessed 28 Aug 2017.

Dasgupta, S., M. Huq, Z.H. Khan, M.M.Z. Ahmed, N. Mukherjee, M.F. Khan, and K. Pandey. 2014. Cyclones in a changing climate: The case of Bangladesh. *Climate and Development* 6 (2): 96–110. https://doi.org/10.1080/17565529.2013.868335.

Day, J.W., J. Agboola, Z. Chen, C. D'Elia, D.L. Forbes, L. Giosan, P. Kemp, C. Kuenzer, R.R. Lane, R. Ramachandran, J. Syvitski, and A. Yañez-Arancibia. 2016. Approaches to defining deltaic sustainability in the 21st century. *Estuarine, Coastal and Shelf Science* 183: 275–291. https://doi.org/10.1016/j.ecss.2016.06.018. Part B.

Delta Commissie. 2008. Working together with water. A living land builds for its future. Findings of the Delta Commissie. The Netherlands: Delta Commissie. http://www.deltacommissie.com/doc/deltareport_full.pdf. Accessed 28 Aug 2017.

Ericson, J.P., C.J. Vorosmarty, S.L. Dingman, L.G. Ward, and M. Meybeck. 2006. Effective sea-level rise and deltas: Causes of change and human dimension implications. *Global and Planetary Change* 50 (1–2): 63–82. https://doi.org/10.1016/j.gloplacha.2005.07.004.

GED. 2012. Perspective plan of Bangladesh 2010–2021: Making vision 2021 a reality. General Economics Division (GED), Planning Commission. Dhaka: General economics division (GED), Planning Commission, Government of the People's Republic of Bangladesh. http://www.plancomm.gov.bd/perspective-plan/. Accessed 20 July 2016.

———. 2015. Seventh five year plan FY2016 – FY2020: Accelerating growth, empowering citizens. Final draft. Dhaka: General Economics Division (GED), Planning Commission, Government of the People's Republic of Bangladesh. http://plancomm.gov.bd/wp-content/uploads/2015/11/7FYP_after-NEC_11_11_2015.pdf. Accessed 6 Jan 2017.

———. 2017. Data gap analysis of sustainable development goals (SDGs): Bangladesh perspective. Dhaka: General Economics Division (GED), Planning Commission, Government of the People's Republic of Bangladesh. http://www.plancomm.gov.bd/perspective-plan/. Accessed 19 Aug 2017.

Hijioka, Y., E. Lin, J.J. Pereira, R.T. Corlett, X. Cui, G.E. Insarov, R.D. Lasco, E. Lindgren, and A. Surjan. 2014. Asia. In *Climate change 2014: Impacts, adaptation, and vulnerability. Part B: Regional aspects. Contribution of working group II to the fifth assessment report of the intergovernmental panel on climate change*, ed. V.R. Barros, C.B. Field, D.J. Dokken, M.D. Mastrandrea, K.J. Mach, T.E. Bilir, M. Chatterjee, K.L. Ebi, Y.O. Estrada, R.C. Genova, B. Girma, E.S. Kissel, A.N. Levy, S. MacCracken, M.D. Mastrandrea, and L.L. White, 1327–1370. Cambridge, UK: Cambridge University Press.

Huq, S., S.I. Ali, and A.A. Rahman. 1995. Sea-level rise and Bangladesh: A preliminary analysis. *Journal of Coastal Research* SI 14: 44–53.

Lumbruso, D.M., N.R. Suckall, R.J. Nicholls, and K.D. White. 2017. Enhancing resilience to coastal flooding from severe storms in the USA:

International lessons. *Natural Hazards and Earth Systems Sciences*. https://doi.org/10.5194/nhess-17-1-2017.

Milliman, J.D., J.M. Broadus, and G. Frank. 1989. Environmental and economic implications of rising sea level and subsiding deltas: The Nile and Bengal examples. *Ambio* 18 (6): 340–345.

National Water Development Agency. 2017. National perspectives for water resources development National Water Development Agency, Ministry of Water Resources, Government of India. http://www.nwda.gov.in/index2.asp?slid=108&sublinkid=14&langid=1. Accessed 28 Aug 2017.

Nicholls, R.J., C.W. Hutton, A.N. Lázár, A. Allan, W.N. Adger, H. Adams, J. Wolf, M. Rahman, and M. Salehin. 2016. Integrated assessment of social and environmental sustainability dynamics in the Ganges-Brahmaputra-Meghna delta, Bangladesh. *Estuarine, Coastal and Shelf Science* 183: 370–381. https://doi.org/10.1016/j.ecss.2016.08.017. Part B.

Payo, A., A.N. Lázár, D. Clarke, R.J. Nicholls, L. Bricheno, S. Mashfiqus, and A. Haque. 2017. Modeling daily soil salinity dynamics in response to agricultural and environmental changes in coastal Bangladesh. *Earth's Future*. https://doi.org/10.1002/2016EF000530.

Prinsen, G., F. Sperna Weiland, and E. Ruijgh. 2015. The Delta model for fresh water policy analysis in the Netherlands. *Water Resources Management* 29 (2): 645–661. https://doi.org/10.1007/s11269-014-0880-z.

Roy, M., J. Hanlon, and D. Hulme. 2016. *Bangladesh confronts climate change: Keeping our heads above water*. London: Anthem Press.

Syvitski, J.P.M., A.J. Kettner, I. Overeem, E.W.H. Hutton, M.T. Hannon, G.R. Brakenridge, J. Day, C. Vorosmarty, Y. Saito, L. Giosan, and R.J. Nicholls. 2009. Sinking deltas due to human activities. *Nature Geoscience* 2 (10): 681–686. https://doi.org/10.1038/ngeo629.

Szabo, S., D. Begum, S. Ahmad, Z. Matthews, and P.K. Streatfield. 2015. Scenarios of population change in the coastal Ganges Brahmaputra Delta (2011–2051). *Asia Pacific Population Journal* 30 (2): 51–72.

Szabo, S., R.J. Nicholls, B. Neumann, F.G. Renaud, Z. Matthews, Z. Sebesvari, A. AghaKouchak, R. Bales, C.W. Ruktanonchai, J. Kloos, E. Foufoula-Georgiou, P. Wester, M. New, J. Rhyner, and C. Hutton. 2016. Making SDGs work for climate change hotspots. *Environment: Science and Policy for Sustainable Development* 58 (6): 24–33. https://doi.org/10.1080/00139157.2016.1209016.

Tessler, Z.D., C.J. Vörösmarty, M. Grossberg, I. Gladkova, H. Aizenman, J.P.M. Syvitski, and E. Foufoula-Georgiou. 2015. Profiling risk and sustainability in coastal deltas of the world. *Science* 349 (6248): 638. https://doi.org/10.1126/science.aab3574. Licence.

UN. 2013. World population prospects, the 2012 revision. Department of Economic and Social Affairs, Population Division. https://esa.un.org/unpd/wpp/Publications/. Accessed 30 May 2014.

———. 2015. Transforming our world: The 2030 agenda for sustainable development. Resolution adopted by the general assembly on 25 September 2015. New York: United Nations.

Van Alphen, J. 2016. The delta programme and updated flood risk management policies in the Netherlands. *Journal of Flood Risk Management* 9 (4): 310–319. https://doi.org/10.1111/jfr3.12183.

Wahl, T., S. Jain, J. Bender, S.D. Meyers, and M.E. Luther. 2015. Increasing risk of compound flooding from storm surge and rainfall for major US cities. *Nature Climate Change* 5 (12): 1093–1097. https://doi.org/10.1038/nclimate2736.

Warrick, R.A., E.M. Barrow, and T.M.L. Wigley, eds. 1993. *Climate and sea level change: Observations, projections, implications.* Cambridge: Cambridge University Press.

Whitehead, P.G., E. Barbour, M.N. Futter, S. Sarkar, H. Rodda, J. Caesar, D. Butterfield, L. Jin, R. Sinha, R. Nicholls, and M. Salehin. 2015a. Impacts of climate change and socio-economic scenarios on flow and water quality of the Ganges, Brahmaputra and Meghna (GBM) river systems: Low flow and flood statistics. *Environmental Science-Processes and Impacts* 17 (6): 1057–1069. https://doi.org/10.1039/c4em00619d.

Whitehead, P.G., S. Sarkar, L. Jin, M.N. Futter, J. Caesar, E. Barbour, D. Butterfield, R. Sinha, R. Nicholls, C. Hutton, and H.D. Leckie. 2015b. Dynamic modeling of the Ganga river system: Impacts of future climate and socio-economic change on flows and nitrogen fluxes in India and Bangladesh. *Environmental Science-Processes and Impacts* 17 (6): 1082–1097. https://doi.org/10.1039/c4em00616j.

Woodroffe, C.N., R.J. Nicholls, Y. Saito, Z. Chen, and S.L. Goodbred. 2006. Landscape variability and the response of Asian megadeltas to environmental change. In *Global change and integrated coastal management: The Asia-Pacific region*, ed. N. Harvey, 277–314. Dordrecht: Springer.

World Bank. 2010. Economics of adaptation to climate change : Bangladesh. Volume 1 main report. Washington, DC: World Bank Group. https://openknowledge.worldbank.org/handle/10986/12837. Accessed 09 Jan 2017.

Yasuda, Y., D. Aich, D. Hill, P. Huntjens, and A. Swain. 2017. *Transboundary water cooperation over the Brahmaputra River. Legal political economy analysis of current and future potential cooperation.* The Hague: The Hague Institute for Global Justice.

4

Integrative Analysis for the Ganges-Brahmaputra-Meghna Delta, Bangladesh

Robert J. Nicholls, Craig W. Hutton, W. Neil Adger,
Susan E. Hanson, Md. Munsur Rahman,
and Mashfiqus Salehin

4.1 The Ganges-Brahmaputra-Meghna Delta and Study Area

The Ganges-Brahmaputra-Meghna (GBM) delta is one of the world's most dynamic and significant deltas. Geologically, the delta covers most of Bangladesh and parts of West Bengal (India) and has a total population exceeding 100 million people (Ericson et al. 2006; Woodroffe et al. 2006). The Ganges and Brahmaputra rivers rise in the Himalayas (collectively with catchments in five countries: China, Nepal, India, Bhutan, Bangladesh) and ultimately deposit their freshwater and sediment in the

R. J. Nicholls (✉) • S. E. Hanson
Faculty of Engineering and the Environment and Tyndall Centre for Climate Change Research, University of Southampton, Southampton, UK

C. W. Hutton
Geodata Institute, Geography and Environment, University of Southampton, Southampton, UK

© The Author(s) 2018
R. J. Nicholls et al. (eds.), *Ecosystem Services for Well-Being in Deltas*,
https://doi.org/10.1007/978-3-319-71093-8_4

GBM delta and the Bay of Bengal (Fig. 4.1). The Meghna River has a smaller Bangladesh and Indian catchment. The delta has historically supported high population densities through provision ecosystem services—highly productive farming and fishing systems.

The aggregate population of Bangladesh increased four-fold between 1950 and 2013, from 38 to 157 million, and is projected to exceed 200 million by 2050 with continued significant urbanisation (UN 2013). Land use and population distribution in the delta have also been changing rapidly over the past few decades, in particular a growing urban population in major cities such as Dhaka and Chittagong (and Kolkata in neighbouring India) and land use changes such as expansion of export-led saline shrimp aquaculture. Gross domestic product (GDP) per capita has risen significantly from US$410 in 2000 to US$1,360 by 2016, driven primarily by industrial-led growth in manufacturing, including garments and remittance income from international migration and from economic growth across service and non-farm sectors.

The study area considered for this research, shown in Fig. 4.2, represents a politically, socially and environmentally coherent region, within the defined Coastal Zone of Bangladesh. It is the seaward part of the delta within Bangladesh, south of Khulna and west of the Meghna River to the Indian border. There are numerous islands near the Meghna River, isolating many communities, although transport links are being improved. It also includes the Bangladeshi portion of the Sundarbans, the largest mangrove forest in the world. Excluding the cities of Khulna and Barisal, the study area is largely rural with extensive agriculture, aquaculture and capture fisheries.

W. Neil Adger
Geography, College of Life and Environmental Sciences,
University of Exeter, Exeter, UK

Md. Munsur Rahman • M. Salehin
Institute of Water and Flood Management, Bangladesh University of
Engineering and Technology, Dhaka, Bangladesh

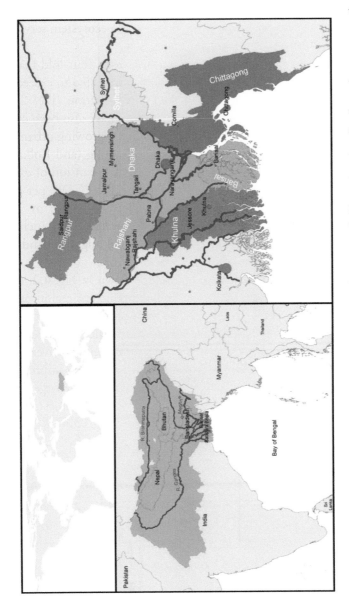

Fig. 4.1 Catchment area of the Ganges, Brahmaputra and Meghna rivers (left), and major distributaries, seven national Divisions (of Bangladesh) and urban centres (red dots) (right)

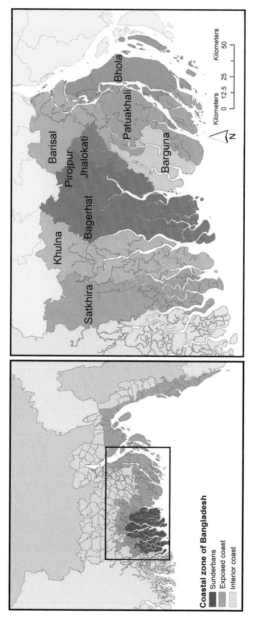

Fig. 4.2 The study area is located within the Coastal Zone of Bangladesh (left) and is comprised of nine districts (right)

Comprising nine districts within two Divisions (part of Khulna and all of Barisal Divisions), the population of the study area is approximately 14 million based on the 2011 census, representing approximately ten per cent of the national population (BBS 2012). Demographic projections suggest that by 2050 the area will have an ageing stable or declining population, somewhere between 11 and 14 million people with migration and mobility critical drivers of demographic outcomes. Migration trends in the delta mirror well-established global trends, with a general increase in urban populations in coastal areas due to migration to coastal cities (de Sherbinin et al. 2011; Seto 2011). Although the study area is essentially rural, it does gain population from this trend. The rural to urban transition sits alongside episodes of involuntary out-migration as both direct and indirect consequences of environmental risks such as salinisation and cyclonic storm surges (Black et al. 2011; Gray and Mueller 2012; Martin et al. 2014). The incidence of poverty is similar or higher than the national average (31 per cent), being 32 and 39 per cent, respectively, in the Khulna and Barisal Divisions (BBS 2011). Savings or access to finance is limited for most of Bangladesh's population (Mujeri 2015), making households vulnerable to economic shocks.

The study area covers one of the world's largest lowlands with an elevation up to three metres—one metre above normal high tides—and it is subject to tidal exchange along numerous north-south channels. It is the area within Bangladesh most exposed to sea-level rise (Milliman et al. 1989; Huq et al. 1995; World Bank 2010), and settlements are subject to tidal flooding, riverine flooding, arsenic in local water supplies, seasonal and longer-term salinisation of water supplies and soils and water logging. Cyclones and associated storm surges have historically caused substantial damage. For example, Cyclone 'Sidr' in 2007 caused several thousand fatalities and significant economic consequences.

Dynamic processes such as flows of water, floods, elevation and the supply of sediments shape the delta. In addition to global sea-level rise, there is a broad regional subsidence of two to three millimetres a year, with localised hotspots exhibiting higher subsidence rates (Brown and Nicholls 2015). Some areas of land are eroded while accretion increases others: there has been a net overall *gain* of land in Bangladesh over the past few decades reflecting a continued current sediment supply, despite upstream damming over the past century (Brammer 2014; Wilson and Goodbred 2015). River floods mainly occur during the wet season (monsoon), when a large volume of water is received from the upstream catchments. This typically results in 20–60 per cent inundation of Bangladesh annually (Salehin et al. 2007), although in the study area such flooding is limited by an extensive system of coastal embankments and polders. Flood risk is projected to increase over the incoming decades, given increased overall discharge projections for the Ganges and other major Himalayan rivers. For the Ganges and Brahmaputra, Lutz et al. (2014), for example, show increased run-off till mid-twenty-first century due to a combination of increased precipitation and glacier melt in the high upper catchments.

Cyclones and tropical storms regularly make landfall in Bangladesh, typically more than once per year, leading to high winds, extreme sea levels and saltwater flooding (Alam and Dominey-Howes 2015; Mutahara et al. 2016). This can damage crops and properties with consequences for life and livelihoods, creating social vulnerability for the study area population. The economic and health consequences of these events are highly significant, and virtually all analyses show that the consequences for economies and societies will increase with projected climate change, with impacts falling disproportionally on poor and vulnerable populations and acting as a brake on development (Hallegatte et al. 2017).

Yet interventions and governance processes can make dramatic differences to vulnerability, poverty and the trajectories of development in the delta. Improved disaster risk management, for example, through the growth of flood warnings and cyclone shelters has greatly reduced the death toll during extreme floods and cyclones compared to the large mortality in 1970 and 1991 (Shaw et al. 2013; Lumbruso et al. 2017). The aspirations of the Bangladesh government are towards economic growth and poverty alleviation, as directly articulated in their planning processes:

> a Bangladesh which, by 2021, will be a middle income country where poverty will be drastically reduced where, our citizens will be able to meet every basic need and where development will be on fast track, with ever-increasing rates of inclusive growth. (Government Vision 2021, (CPD 2007))

The Bangladesh government has also initiated the Bangladesh Delta Plan 2100 (BanDuDeltAS 2014) as part of the strategy for long-term economic sustainability. The Plan is an integrated process that takes a long-term view of the development of Bangladesh and recognises the interconnectedness of environmental sustainability and social progress. The Plan is part inspired by similar integrated planning processes such as for the challenges of sustaining large proportions of low-lying land in the Netherlands (Delta Commissie 2008; Van Alphen 2016).

4.2 Research Challenges in the Ganges-Brahmaputra-Meghna Delta

Ecosystem services in the delta are strongly interconnected, and there are significant trade-offs between different ecosystem services such as the trade-off and even conflict between freshwater agriculture, especially rise paddy, and brackish shrimp aquaculture. Coastal Bangladesh has a system of polders started in the 1960s, where the land is surrounded by coastal embankments with flapgate drains to regulate water levels and enhance agriculture. However, over time, polderisation both prevents sedimentation and promotes subsidence due to drainage (Auerbach et al. 2015), and there has been sedimentation of the riverbed. This degrades soil quality unless artificial fertilisers are used, encourages waterlogging and increases potential flood depths if the embankments fail. Controlled flooding and sedimentation termed 'tidal river management' is an experimental response which might find widespread application (Amir et al. 2013; Nowreen et al. 2014). The balance between sea water and freshwater is a critical issue in the study area (see Chaps. 24 and 28; Clarke et al. 2015; Lázár et al. 2015). This varies seasonally and saltwater pushes far inland during the low river flow period between the annual monsoon rains, and cyclones can also cause saltwater flooding by generating extreme sea levels. If the land becomes too saline, traditional agriculture is degraded. If this persists there are limited options: moving to salt-tolerant crops (which are being continuously developed but add cost) or converting to shrimp aquaculture which is usually for export and providing much lower levels of employment (Ali 2006; Islam et al. 2015; Amoako Johnson et al. 2016).

Upstream water use and diversion to irrigation and other uses also enhances salinisation, as exemplified by the effects of the Farakka Barrage on dry season flows down the Ganges into Bangladesh. The Sundarbans are an important buffer against cyclones, but are threatened by sea-level rise and other stresses (e.g. pollution). They provide a range of ecosystem goods which are available to the poorest, as well as tourism based around the Bengal tiger, an iconic but endangered species.

Explaining social outcomes of ecosystem service use within the GBM delta requires consideration of (i) the magnitude and mobility of ecosystem services and associated populations, (ii) seasonality and other short-term temporal dynamics of ecosystems, (iii) social structures such as the debt economy, (iv) capital accumulation and reciprocity in economic relations and (v) the distribution issues associated with ownership and access to land and resources such as fisheries. These mechanisms are persistent and engrained in social-ecological systems and their governance. They have been used to explain the continued presence of poverty, social exclusion and patterns of uneven development in many contexts (Hartmann and Boyce 1983; Bebbington 1999; Ribot and Peluso 2003). The social mechanisms are manifest in measurable outcomes: notably the material well-being and incomes of populations, their nutritional status and health outcomes and, in so-called subjective well-being, how people perceive their present and futures (Camfield et al. 2009).

This scientific body of evidence further shows that the well-being and health status of populations coming from ecosystem services do not depend on individual elements of ecosystems, but rather on bundles of ecosystems that collectively produce desirable and socially useful outcomes. The people, ecosystems, services and mechanisms used to access these services together combine to create distinct social-ecological systems, unique to each bundle of services. Hence, social-ecological systems form the basis for much of the analysis. The characteristics of co-production of ecosystem services at the landscape scale lead to significant trade-offs between types of ecosystem services (Raudsepp-Hearne et al. 2010). In the GBM delta, such trade-offs are apparent, with Hossain et al. (2016) demonstrating how land use intensification over the past 50 years has significantly increased provisioning ecosystem services per capita, but with a concurrent decline in natural habitats and regulating services.

The dynamics of deltaic social-ecological systems are such that trends are not easily identifiable if simple deterministic relationships are assumed. For the GBM case, for example, populations in poverty persist despite the presence of diverse, highly productive ecological systems. This raises the question of whether access to ecosystem services really can alleviate poverty as opposed to simply maintain a baseline level of welfare (Adams et al. 2013). Similarly, land conversion to shrimp and prawn aquaculture produces high-value commercial products, yet has not transformed the economic fortunes of the localities in which it is practised. Rather aquaculture is co-located with areas of persistent poverty, with the health and economic well-being of associated populations being negatively affected by salinisation (Amoako Johnson et al. 2016).

4.3   Aims and Objectives of This Research

The research described in this book builds on the diverse facets of the current challenges for low-lying deltas such as the GBM, new knowledge on ecosystem services and on human well-being in Bangladesh with an explicit aim of generating findings relevant for long-term planning and policy processes in the delta.

The overall aim is to provide substantive and rigorous results as well as usable and accessible methods and tools to evaluate ecosystem services and livelihoods in coastal Bangladesh, emphasising where and when policy processes and interventions can make a difference. The three main supporting objectives are (i) to engage ecosystem service science for policy analysis and application to coastal and delta system; (ii) to integrate social, physical and ecosystem sciences for the first comprehensive analysis of delta systems and the challenges of the future and (iii) to advance new integrated models linking ecosystem services to all aspects of human well-being and poverty.

The research effort aims to embed participatory and co-production processes from inception through to generating policy lessons. Lessons from diverse stakeholder processes show that key parameters for successful engagement (as indicated by long-term continued use of scientific evidence) include recognising the position and constraints on existing governance institutions, the limitations of scientific evidence for policy, and the critical role of autonomous action and engagement by the governance institutions themselves (Sterling et al. 2017). The research reported here builds on those principles, striving to make the science as useful and salient as well as practical. Hence, stakeholders help to shape and identify problem definition, scenario development and results assessment and interpretation. This gives a strong sense of ownership of the research in Bangladesh.

Integration is also embedded from the start, and the construction of an integrated model is a key aim and outcome of the project. This model aims to couple a range of biophysical processes together with household livelihood information derived from the analysis of census and detailed household survey results. The coupled nature of the integrated model is unusual if not unique as the influence of biophysical changes on livelihoods and human well-being are calculated directly.

As this research developed so the Bangladesh Delta Plan 2100 (BanDuDeltAS 2014) emerged as a key policy process which informed this research. It defined issues such as timescales of analysis, a main focus to 2050 but broader interest in possible biophysical trends to 2100. It also defined possible development interventions, such as the benefits of the proposed Ganges Barrage which is designed to increase dry season flows into the study area.

4.4   The Research Approach

To answer the research questions, key insights from the science of ecosystem services and system analysis in general form a foundation. In deltas, ecosystem services include the processes that bring freshwater, sediments, productive and biologically diverse wetlands and fisheries, and productive land for agriculture. Based on the categories identified in the Millennium Ecosystem Assessment (MEA 2005), a host of (i) supporting, (ii) provisioning, (iii) regulating and (iv) cultural ecosystem services can be identified (Table 4.1 and see Chap. 1, Fig. 1.2). Here the major focus is on the provisioning ecosystem services, although other types of service are considered where relevant, such as the buffering of storms provided by mangroves (a regulating service).

Table 4.1 The ecosystem services considered in this book (with chapter numbers), based on the Millennium Ecosystem Assessment (2005)

Ecosystem service	Related issues addressed in this book (and relevant chapter numbers)
Supporting services	
Nutrient cycling	Agriculture, sediment/morphology (15, 18, 20, 24)
Soil formation	Agriculture, sediment/morphology (15, 18, 20, 24)
Primary production	Agriculture, sediment/morphology, fisheries (14, 15, 18, 20, 24, 25)
Provisioning services	
Food	Fisheries, agriculture (20, 24, 25)
Fuel (wood)	Mangroves (26)
Freshwater	Catchments, salinity (13, 17, 18)
Regulating services	
Water regulation	Legal and policy frameworks, catchments (6, 13)
Erosion control	Sediment/morphology (15)
Water purification	Catchments (13, 17, 18)
Disease control, human	Household survey (22, 23)
Disease control, pest	Agriculture (24)
Natural hazard regulation	Catchments, floods, sediment/morphology, salinity, subsidence, mangroves (8, 13, 15, 16, 17, 18, 21, 26)
Cultural/aesthetic services	
Recreation/ecotourism	Mangroves (26)
Cultural diversity	Household survey (22, 23)

Analysing the future of ecosystem services and human livelihoods in coastal Bangladesh is a complex multi-disciplinary problem. The core issues when integrating the social, physical and ecological dynamics of deltas is the identification and measurement of the mechanisms by which the system components interact to produce human well-being and, importantly, whether these relationships are stable and hence predictable over time. The conceptual framework of the research is, in essence, the diverse social-ecological systems within the GBM delta, and an explanation of how social phenomena and environmental drivers combine to constrain well-being health and pathways of development. A framework that focuses on the mechanisms that link ecosystem services with social outcomes is therefore developed. These mechanisms are core to all the research tasks, including the design of the integrated model, the Delta Dynamic Integrated Emulator Model (ΔDIEM) (Chap. 28).

The analysis is inevitably multi-scale, reflecting the multiple processes shaping the delta. An integrated framework is adopted to be able to represent individual processes, their interactions and how they affect populations and livelihoods. Each topic is considered independently, and this is drawn together using the common framework and scenarios. Four distinct scales are considered: (i) global concerning issues such as climate change; (ii) regional, including the river basin and Bay of Bengal; (iii) national (Bangladesh) and (iv) sub-national, focussing on the study area and the political units that make it up, down to Unions (Fig. 4.3).

Figure 4.4 summarises the overall approach adopted. Governance analysis and stakeholder engagement are continuous throughout and are essential due to the research's participatory nature. This incorporates issue identification, scenario development and discussion of results, including consideration of possible responses and development measures to explore in the integrated assessment (see Chap. 10). This involved an innovative scenario development process combining climate emissions and social-economic change which aimed to generate a dialogue across institutions and sectors to creating a shared future vision and addressing challenges in a holistic, shared and integrated way (Chap. 9).

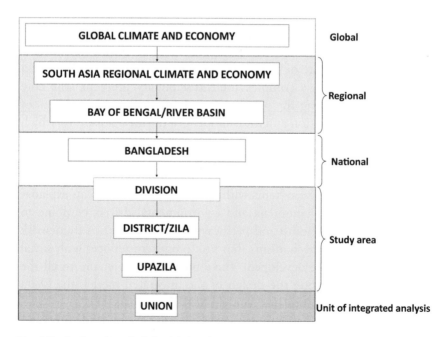

Fig. 4.3 Scales of analysis in this book

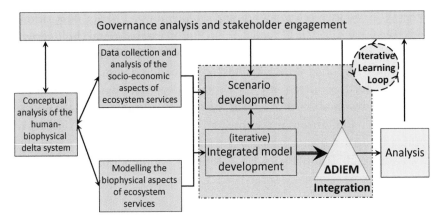

Fig. 4.4 The overall approach and flow of information in the analysis

A preliminary qualitative analysis was essential to test and develop the detailed questions and approaches that were required. In particular, the relationship between ecosystem services and livelihoods was poorly understood at the beginning of the analysis. This process identified a set of seven distinct social-ecological systems (SESs) within the study area, reflecting both the nature of the ecosystem services and human access and their exploitation: (i) rainfed agriculture, (ii) irrigated agriculture, (iii) freshwater prawn aquaculture, (iv) saltwater shrimp aquaculture, (v) eroding islands (charlands), (vi) Sundarbans mangrove forest and (vii) offshore fisheries (Adams and Adger 2016; see also Chap. 22).

This was followed by more detailed social-economic analysis and data collection, including an innovative household survey which collects empirical seasonal data on ecosystem services and livelihoods (Adams et al. 2016; see also Chap. 23). In parallel there is a major effort to analyse and simulate a range of biophysical and social-economic processes with consistent assumptions. This included catchments (Chap. 13), the Bay of Bengal (Chap. 14), sedimentary and morphodynamic processes (Chap. 15), mangrove processes (Chap. 26), flooding and various hydrological processes (Chap. 16), including salinisation (Chaps. 17 and 18) and census-based mapping (Chap. 21). Using their own expertise but co-working where possible, research into specific ecosystem services is also undertaken: health (Chap. 27), fisheries (Chap. 25), agriculture (Chap. 24) and mangroves (Chap. 26). To support these activities and encourage integration, a range of exogenous and endogenous scenarios were developed (see Chaps. 9, 10, 11, and 12), including extensive stakeholder participation in the endogenous scenario development (Chaps. 9 and 10). Preliminary scenarios were available for the early phases of analysis, and developed further to feed into the overall integrated assessment. Hence, integration was built in throughout the project culminating in the development and application of the integrated model (ΔDIEM). ΔDIEM is applied using a range of scenarios to explore possible trajectories of the study area to 2050 and discuss possible policy and development interventions (Chap. 3).

4.5   Linking to the Policy Process

The research was explicitly linked to national scale policy on the basis that policy influence at this scale could have widespread benefits. This includes the innovative iterative learning process where stakeholders

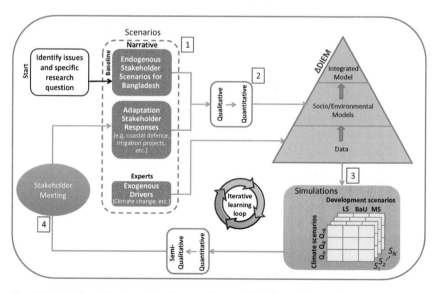

Fig. 4.5 An iterative learning loop using ΔDIEM for policy analysis, comprising (1) scenario development, including adaptation responses, (2) qualitative to quantitative translation to ΔDIEM inputs, (3) simulations using ΔDIEM and (4) stakeholder review of the simulations, which can lead to a new cycle of analysis (Reprinted with permission from Nicholls et al. 2016)

(from government to civil society) are involved in all stages of the research ensuring stakeholder trust, maintain interest in and are willing to participate (see Fig. 4.5). This also helps to provide a process for decision makers to engage and adaptively explore outcomes of the implementation of individual policies or rafts of policies into the future. The iterative loop can be repeated many times, allowing co-learning concerning problems and possible solutions, including trade-offs. This process is considered in more detail in Chaps. 9 and 28. In addition, to help increase understanding and inform the wider population beyond the policy community, research outcomes are disseminated using multiple methods, including a bespoke 'Pot Song' (Chap. 29).

References

Adams, H., and W.N. Adger. 2016. Mechanisms and dynamics of wellbeing-ecosystem service links in the southwest coastal zone of Bangladesh. UK Data Service Reshare. https://doi.org/10.5255/UKDA-SN-852356.

Adams, H., W.N. Adger, H. Huq, M. Rahman, and M. Salehin. 2013. Transformations in land use in the southwest coastal zone of Bangladesh: Resilience and reversibility under environmental change. In *Proceedings of transformation in a changing climate international conference*, University of Oslo, Oslo, Finland.

Adams, H., W.N. Adger, S. Ahmad, A. Ahmed, D. Begum, A.N. Lázár, Z. Matthews, M.M. Rahman, and P.K. Streatfield. 2016. Spatial and temporal dynamics of multidimensional well-being, livelihoods and ecosystem services in coastal Bangladesh. *Scientific Data* 3: 160094. https://doi.org/10.1038/sdata.2016.94. Licence Data Descriptor.

Alam, E., and D. Dominey-Howes. 2015. A new catalogue of tropical cyclones of the northern Bay of Bengal and the distribution and effects of selected landfalling events in Bangladesh. *International Journal of Climatology* 35 (6): 801–835. https://doi.org/10.1002/joc.4035.

Ali, A.M.S. 2006. Rice to shrimp: Land use/land cover changes and soil degradation in Southwestern Bangladesh. *Land Use Policy* 23 (4): 421–435. https://doi.org/10.1016/j.landusepol.2005.02.001.

Amir, M.S.I.I., M.S.A. Khan, M.M.K. Khan, M.G. Rasul, and F. Akram. 2013. Tidal river sediment management-A case study in southwestern Bangladesh. *International Journal of Environmental, Chemical, Ecological, Geological and Geophysical Engineering* 7 (3): 176–185.

Amoako Johnson, F., C.W. Hutton, D. Hornby, A.N. Lázár, and A. Mukhopadhyay. 2016. Is shrimp farming a successful adaptation to salinity intrusion? A geospatial associative analysis of poverty in the populous Ganges–Brahmaputra–Meghna Delta of Bangladesh. *Sustainability Science* 11 (3): 423–439. https://doi.org/10.1007/s11625-016-0356-6.

Auerbach, L.W., S.L. Goodbred, D.R. Mondal, C.A. Wilson, K.R. Ahmed, K. Roy, M.S. Steckler, C. Small, J.M. Gilligan, and B.A. Ackerly. 2015. Flood risk of natural and embanked landscapes on the Ganges-Brahmaputra tidal delta plain. *Nature Climate Change* 5 (2): 153–157. https://doi.org/10.1038/nclimate2472.

BanDuDeltAS. 2014. Inception report. Bangladesh Delta Plan 2100 Formulation Project. Dhaka: General Economics Division (GED), Planning Commission, Government of the People's Republic of Bangladesh.

BBS. 2011. *Report of the household income and expenditure survey 2010*. Dhaka: Bangladesh Bureau of Statistics (BBS).

———. 2012. Bangladesh population and housing census 2011 – Socio-economic and demographic report. Report national series, Volume 4. Dhaka: Bangladesh Bureau of Statistics (BBS) and Statistics and Informatics Division (SID), Ministry of Planning, Government of the People's Republic of Bangladesh. http://www.bbs.gov.bd/PageSearchContent.aspx?key=census%20 2011. Accessed 7 July 2016.

Bebbington, A. 1999. Capitals and capabilities: A framework for analyzing peasant viability, rural livelihoods and poverty. *World Development* 27 (12): 2021–2044. https://doi.org/10.1016/S0305-750X(99)00104-7.

Black, R., W.N. Adger, N.W. Arnell, S. Dercon, A. Geddes, and D. Thomas. 2011. The effect of environmental change on human migration. *Global Environmental Change* 21 (Supplement 1): S3–S11. https://doi.org/10.1016/j.gloenvcha.2011.10.001.

Brammer, H. 2014. Bangladesh's dynamic coastal regions and sea-level rise. *Climate Risk Management* 1: 51–62. https://doi.org/10.1016/j.crm.2013.10.001.

Brown, S., and R.J. Nicholls. 2015. Subsidence and human influences in mega deltas: The case of the Ganges-Brahmaputra-Meghna. *Science of the Total Environment* 527: 362–374. https://doi.org/10.1016/j.scitotenv.2015.04.124.

Camfield, L., K. Choudhury, and J. Devine. 2009. Well-being, happiness and why relationships matter: Evidence from Bangladesh. *Journal of Happiness Studies* 10 (1): 71–91. https://doi.org/10.1007/s10902-007-9062-5.

Clarke, D., S. Williams, M. Jahiruddin, K. Parks, and M. Salehin. 2015. Projections of on-farm salinity in coastal Bangladesh. *Environmental Science-Processes and Impacts* 17 (6): 1127–1136. https://doi.org/10.1039/c4em00682h.

CPD. 2007. Bangladesh Vision 2021. Prepared under the initiative of the Nagorik Committee 2006. Dhaka: Centre for Policy Dialogue (CPD).

de Sherbinin, A., M. Castro, F. Gemenne, M.M. Cernea, S. Adamo, P.M. Fearnside, G. Krieger, S. Lahmani, A. Oliver-Smith, A. Pankhurst, T. Scudder, B. Singer, Y. Tan, G. Wannier, P. Boncour, C. Ehrhart, G. Hugo, B. Pandey, and G. Shi. 2011. Preparing for resettlement associated with climate change. *Science* 334 (6055): 456–457. https://doi.org/10.1126/science.1208821.

Delta Commissie. 2008. Working together with water: A living land builds for its future. Findings of the Deltacommissie. The Hague: Deltacommissie. http://www.deltacommissie.com/doc/deltareport_full.pdf. Accessed 16 Mar 2017.

Ericson, J.P., C.J. Vorosmarty, S.L. Dingman, L.G. Ward, and M. Meybeck. 2006. Effective sea-level rise and deltas: Causes of change and human dimension implications. *Global and Planetary Change* 50 (1–2): 63–82. https://doi.org/10.1016/j.gloplacha.2005.07.004.

Gray, C.L., and V. Mueller. 2012. Natural disasters and population mobility in Bangladesh. *Proceedings of the National Academy of Sciences of the United States of America* 109 (16): 6000–6005. https://doi.org/10.1073/pnas.1115944109.

Hallegatte, S., A. Vogt-Schilb, M. Bangalore, and J. Rozenberg. 2017. *Unbreakable: Building the resilience of the poor in the face of natural disasters, climate change and development series*. Washington, DC: World Bank.

Hartmann, B., and J.K. Boyce. 1983. *A quiet violence: View from a Bangladesh village*. London: Zed Books.

Hossain, M.S., J.A. Dearing, M.M. Rahman, and M. Salehin. 2016. Recent changes in ecosystem services and human well-being in the Bangladesh coastal zone. *Regional Environmental Change* 16 (2): 429–443. https://doi.org/10.1007/s10113-014-0748-z.

Huq, S., S.I. Ali, and A.A. Rahman. 1995. Sea-level rise and Bangladesh: A preliminary analysis. *Journal of Coastal Research* SI 14: 44–53.

Islam, G.M.T., A. Islam, A.A. Shopan, M.M. Rahman, A.N. Lázár, and A. Mukhopadhyay. 2015. Implications of agricultural land use change to ecosystem services in the Ganges delta. *Journal of Environmental Management* 161: 443–452. https://doi.org/10.1016/j.jenvman.2014.11.018.

Lázár, A.N., D. Clarke, H. Adams, A.R. Akanda, S. Szabo, R.J. Nicholls, Z. Matthews, D. Begum, A.F.M. Saleh, M.A. Abedin, A. Payo, P.K. Streatfield, C. Hutton, M.S. Mondal, and A.Z.M. Moslehuddin. 2015. Agricultural livelihoods in coastal Bangladesh under climate and environmental change – A model framework. *Environmental Science-Processes and Impacts* 17 (6): 1018–1031. https://doi.org/10.1039/c4em00600c.

Lumbruso, D.M., N.R. Suckall, R.J. Nicholls, and K.D. White. 2017. Enhancing resilience to coastal flooding from severe storms in the USA: International lessons. *Natural Hazards and Earth Systems Sciences.* https://doi.org/10.5194/nhess-17-1-2017.

Lutz, A.F., W.W. Immerzeel, A.B. Shrestha, and M.F.P. Bierkens. 2014. Consistent increase in high Asia's runoff due to increasing glacier melt and precipitation. *Nature Climate Change* 4 (7): 587–592. https://doi.org/10.1038/nclimate2237. Licence Letter.

Martin, M., M. Billah, T. Siddiqui, C. Abrar, R. Black, and D. Kniveton. 2014. Climate-related migration in rural Bangladesh: A behavioural model. *Population and Environment* 36 (1): 85–110. https://doi.org/10.1007/s11111-014-0207-2.

MEA. 2005. Ecosystems and human well-being: Synthesis. Millennium Ecosystem Assessment (MEA). Washington, DC: Island Press. http://www.millenniumassessment.org/documents/document.356.aspx.pdf. Accessed 01 Aug 2016.

Milliman, J.D., J.M. Broadus, and F. Gable. 1989. Environmental and economic implications of rising sea level and subsiding deltas: The Nile and Bengal examples. *Ambio* 18 (6): 340–345.

Mujeri, M.K. 2015. Improving access of the poor to financial services. Background paper for the seventh Five Year Plan. Dhaka: General Economics Division, Bangladesh Planning Commission, Government of the People's Republic of Bangladesh. http://www.plancomm.gov.bd/wp-content/uploads/2015/02/1_Improving-Access-of-the-Poor-to-Financial-Services.pdf. Accessed 04 Aug 2016.

Mutahara, M., A. Haque, M.S.A. Khan, J.F. Warner, and P. Wester. 2016. Development of a sustainable livelihood security model for storm-surge hazard in the coastal areas of Bangladesh. *Stochastic Environmental Research and Risk Assessment* 30 (5): 1301–1315. https://doi.org/10.1007/s00477-016-1232-8.

Nicholls, R.J., C.W. Hutton, A.N. Lázár, A. Allan, W.N. Adger, H. Adams, J. Wolf, M. Rahman, and M. Salehin. 2016. Integrated assessment of social and environmental sustainability dynamics in the Ganges-Brahmaputra-Meghna delta, Bangladesh. *Estuarine, Coastal and Shelf Science* 183: 370–381. https://doi.org/10.1016/j.ecss.2016.08.017. Part B.

Nowreen, S., M.R. Jalal, and M.S.A. Khan. 2014. Historical analysis of rationalizing South West coastal polders of Bangladesh. *Water Policy* 16 (2): 264–279. https://doi.org/10.2166/wp.2013.172.

Raudsepp-Hearne, C., G.D. Peterson, and E.M. Bennett. 2010. Ecosystem service bundles for analyzing tradeoffs in diverse landscapes. *Proceedings of the National Academy of Sciences of the United States of America* 107 (11): 5242–5247. https://doi.org/10.1073/pnas.0907284107.

Ribot, J.C., and N.L. Peluso. 2003. A theory of access. *Rural Sociology* 68: 153–181.

Salehin, M., A. Haque, M.R. Rahman, M.S.A. Khan, and S.K. Bala. 2007. Hydrological aspects of 2004 floods in Bangladesh. *Journal of Hydrology and Meteorology* 4 (1): 33–44.

Seto, K.C. 2011. Exploring the dynamics of migration to mega-delta cities in Asia and Africa: Contemporary drivers and future scenarios. *Global Environmental Change* 21: S94–S107. https://doi.org/10.1016/j.gloenvcha.2011.08.005.

Shaw, R., F. Mallick, and A. Islam, eds. 2013. *Disaster risk reduction approaches in Bangladesh, disaster risk reduction*. Tokyo: Springer.

Sterling, E.J., E. Betley, A. Sigouin, A. Gomez, A. Toomey, G. Cullman, C. Malone, A. Pekor, F. Arengo, M. Blair, C. Filardi, K. Landrigan, and A.L. Porzecanski. 2017. Assessing the evidence for stakeholder engagement in biodiversity conservation. *Biological Conservation* 209: 159–171. https://doi.org/10.1016/j.biocon.2017.02.008.

UN. 2013. World population prospects, the 2012 revision. Department of Economic and Social Affairs, Population Division. https://esa.un.org/unpd/wpp/Publications/. Accessed 30 May 2014.

Van Alphen, J. 2016. The delta programme and updated flood risk management policies in the Netherlands. *Journal of Flood Risk Management* 9 (4): 310–319. https://doi.org/10.1111/jfr3.12183.

Wilson, C.A., and S.L. Goodbred, Jr. 2015. Construction and maintenance of the Ganges-Brahmaputra Meghna Delta: Linking process, morphology, and stratigraphy. *Annual Review of Marine Science* 7: 67–88. https://doi.org/10.1146/annurev-marine-010213-135032.

Woodroffe, C.N., R.J. Nicholls, Y. Saito, Z. Chen, and S.L. Goodbred. 2006. Landscape variability and the response of Asian megadeltas to environmental change. In *Global change and integrated coastal management: The Asia-Pacific region*, ed. N. Harvey, 277–314. Dordrecht: Springer.

World Bank. 2010. Economics of adaptation to climate change: Bangladesh. Volume 1 main report. Washington, DC: World Bank Group. https://openknowledge.worldbank.org/handle/10986/12837. Accessed 09 Jan 2017.

Part 2

Present Status of the Ganges-Brahmaputra-Meghna Delta

5

Recent Trends in Ecosystem Services in Coastal Bangladesh

John A. Dearing and Md. Sarwar Hossain

5.1   Introduction

The Bangladesh coastal zone is a highly dynamic system. Recorded statistics list 174 natural disasters during the period 1974–2007 (Rahman et al. 2010), and an estimated one million deaths as a consequence of cyclones in the period 1960–1990 (Ericksen et al. 1996). Floods in 1998 caused losses of buildings and infrastructure worth two billion United States dollars (USD) (Chowdhury 2001). The area of land lost through riverbank erosion through the period 1996–2000 led to financial losses totalling 540 million USD (Salim et al. 2007). The world's largest mangrove forest, the

---

The original version of this chapter was revised. The second author's name was incorrect. The chapter has been updated with the correct name of the author. An erratum to this chapter can be found at https://doi.org/10.1007/978-3-319-71093-8_30

J. A. Dearing (✉)
Geography and Environment, Faculty of Social, Human and Mathematical Sciences, University of Southampton, Southampton, UK

Md. Sarwar Hossain
Institute of Geography, University of Bern, Bern, Switzerland

© The Author(s) 2018
R. J. Nicholls et al. (eds.), *Ecosystem Services for Well-Being in Deltas*,
https://doi.org/10.1007/978-3-319-71093-8_5

Sundarbans, provides a livelihood for around three million people (Iftekhar and Islam 2004; Iftekhar and Saenger 2008) and protects more than ten million people from cyclonic storms. But the forest is vulnerable to cyclone damage, as in 2007 when around 36 per cent of the mangrove area was severely damaged leading to losses of livelihood (CEGIS 2007). Food security has also been severely compromised by climate extremes. Around one million tons of food grain were lost to drought in 1997 and 50 per cent of all grain in 1982 was damaged by flood (Islam et al. 2011). These catastrophic impacts serve to underline the vulnerability of the zone to extreme events but give only a partial picture of the longer-term interactions between society and the environment that may affect the zone's resilience. For example, there are major concerns about the individual and combined effects of shrimp farms, irrigation and flood dykes on regional salinity levels that may in turn constrain freshwater availability and crop productivity.

Hence, long-term perspectives are a prerequisite for understanding social-ecological system dynamics, developing and testing simulation models, and defining safe operating spaces for the region as it moves towards meeting the Sustainable Development Goals (SDGs—UN 2015) by 2030. This chapter reviews historical trends in ecosystem services and changes in human development over the past few decades that underlie these statistics (Hossain et al. 2016a, b). The chapter summarises social-ecological dynamics including preliminary analyses of trade-offs between economic growth, well-being and ecosystem services.

5.2   Location and Data Sources

The south-west coastal zone as defined in this study comprises the modern districts of Satkhira, Khulna, Bagerhat, Pirojpur, Barguna, Jhalokati, Patuakhali, Barisal and Bhola within the divisions of Barisal and Khulna (see Chap. 4, Figs. 4.1 and 4.2). The zone covers an area of ~25,500 km^2 with a total population of around 14 million and population densities ranging from 400 to 800 people/km^2 (BBS 2012). The Sundarbans mangrove forest (a UNESCO World Heritage Site) extends across 6,000 km^2 of the southern parts of this area.

Indicators of ecosystem services across the zone (Table 5.1 and Fig. 5.1) are classified as provisioning services or regulating and habitat services. Different

Table 5.1 Data sources for ecosystem services and specific indicators. Where possible, aggregate district-level data collected since 1985 from the nine districts are aggregated into three sets that are equivalent to the three larger 'greater districts' or 'regions' of Barisal, Khulna and Patuakhali, which existed before 1985. T = traditional varieties. HYV = high-yielding varieties (see Hossain et al. 2016a for data sources)

Ecosystem services	Indicators	Temporal scale	Data sources
Provisioning services			
Food production	Rice (T-and HYV-Aus, T-and HYV-*Aman*, T-and HYV-*Boro*) Vegetables Potato Sugarcane, jute, onion, spices (garlic, ginger, turmeric and coriander) Fish Shrimp Honey	Total rice 1948–2010 Rice varieties 1969–2010	Bangladesh Bureau of Statistics
Forest products	Timber types (Glopata, Goran, Gewa) Beeswax	1974–2010	Department of Forest, Khulna, Bangladesh; Zmarlicki 1994; Chaffey et al. 1985
Regulating services			
Water quality	Surface water salinity Soil salinity	1964–2006	Uddin and Haque 2010; Islam 2008
Local climate	Temperature and precipitation	1949–2007	Bangladesh Meteorological Department
Water availability	River discharge Groundwater level	1934–2010	Bangladesh Water Development Board
Natural hazard protection	Crop damage (due to cyclones, flooding, water logging and excessive rainfall)	1970–2010	Bangladesh Bureau of Statistics
Erosion protection	Fluvial erosion and accretion	1963–2009	Mapped data; Rahman et al. 2011
Habitat services			
Maintenance of biodiversity	Mangrove density Mangrove volume Mangrove area Mangrove floristic composition Tiger numbers Deer numbers	1960–1997	MoEF 2002; FAO 1999; Khan 2005; MoEF 2010; FAO 2007; Dey 2007; Chowdhury 2001; Helalsiddiqui 1998; Hendrichs 1975; Gittins 1980
Cultural services			
Recreational services	Number of tourist visitors	2000–2009	IUCN 1997; Department of Forest, Khulna

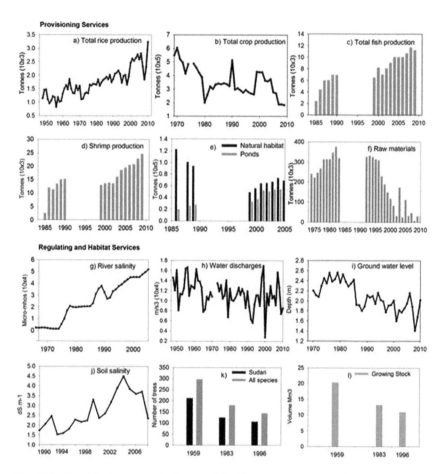

Fig. 5.1 Provisioning, regulating and habitat ecosystem services over recent decades: (a) total rice production, (b) total crop production, (c) total fish production, (d) shrimp production, (e) aquacultural production from natural wetlands and artificial ponds, (f) production of natural forest materials, (g) monitored salinity in the Poshur river at Mongla (Khulna), (h) mean annual river discharge in the Lower Meghna-Ganges river at Hardinge Bridge, (i) monitored depth to groundwater at Dacope (Khulna), (j) average monitored soil salinity at seven sites in Khulna and Patuakhali, (k) monitored tree numbers at an experimental plot in the Sundarbans, (l) estimated volume of growing mangrove in the Sundarbans (see Hossain et al. 2016a for data sources)

food and forest produce, such as rice, fish and timber, are selected to show the trajectories of provisioning services. Water salinity, river discharge, groundwater levels, crop damage and erosion and accretion rates are used as indicators of regulating services. Tree density and growing stock data for main two mangrove species (*Heritiera fomes* and *Excoecaria agallocha* known locally as *Sundri* and *Gewa*, respectively) and all species of the mangrove are selected for biodiversity indicators along with the numbers of two globally endangered mammal species (Royal Bengal tigers and deer) in the mangrove forest. External environmental drivers (Fig. 5.3) include mean annual temperature and rainfall, sea level and data for cyclone frequency and magnitude. Data are drawn from annual official data available for the nine districts and the Sundarbans National Park, and partly from *ad hoc* time-series of monitored point data or sequences of mapped data from hydrological, climatological and agricultural organisations and other scientific studies.

The list of human well-being indicators (Table 5.2 and Fig. 5.2) are selected according to the Millennium Development Goals for Bangladesh (MDG 2015). Data for income, sanitation, electricity, safe drinking water, crop production and production cost at the household level are drawn from Household Income and Expenditure Survey (HIES) datasets collected by the Bangladesh Bureau of Statistics and World Bank in 1995/96, 2000, 2005 and 2010. Data for child and infant mortality, maternal health, education and access to media come from the Demographic and Health Surveys collected in 1993, 1996, 2000, 2004, 2007 and 2010.[1] Although the Demographic and Health Survey data are at divisional level, two divisions (Khulna and Barisal) have been used for analysing the indicators as the study area covers all districts of Barisal division and around 65 per cent of the total area of Khulna division. Continuous series data are normalised (z-scores) to allow comparisons of curve shapes. Full details of these and other data analyses are available in Hossain et al. (2016a, b).

5.3   Ecosystem Services and Drivers

5.3.1   Provisioning Services

Food production is mainly derived from agricultural and aquacultural goods with a small amount directly from natural goods (e.g. wild fish, honey). Since 1970, the cultivated crop area has fluctuated between

Table 5.2 Human well-being indicators based on MEA (2005) and OECD (2013) classifications (see Hossain et al. 2016a for details of data sources)

| Dimension | | | Temporal | |
MA	OECD	Indicators	scale	Data source
Health	Quality of life	Child mortality (probability of dying before the first birth day) Infant mortality (probability of dying between the first and fifth birth day) Proportion of births attended by skilled health personnel (%)	1993–2011	Demographic health surveys
Material	Material condition	Sector-wise household income Production cost	1995–2010	Authors' own calculations from household income and expenditure survey Income from shrimp farms and production cost; (Islam 2007)
		Gross domestic product (GDP)	1974–2005	Bangladesh Bureau of Statistics
		Poverty—percentage of people living below the upper and lower poverty thresholds	1983–2010	Bangladesh Bureau of Statistics (Wodon 1997)
Security	Quality of life (personal security)	Access to electricity, sanitation, drinking water source	1995–2010	Authors' own calculations from Household Income and Expenditure Survey data
Freedom of choice and action	Quality of life	Education (% of people completed primary education)— man and women Access to mass media (television and newspaper)	1993–2011	Demographic health surveys

Drivers

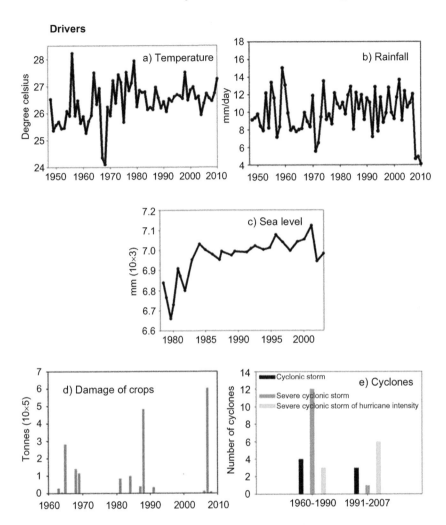

Fig. 5.2 Selected indicators of environmental drivers: **(a)** mean annual tempera-ture, **(b)** mean annual rainfall, **(c)** averaged annual sea level curve (Hiron Point and Khepupara; Brammer 2014), **(d)** damage to crops, **(e)** cyclone frequency and intensity (see Hossain et al. 2016a for data sources)

10,000 and 12,000 km² and in 2009 represents around 47 per cent of the total land area. Despite this relatively stable crop area, regional rice pro-duction (Fig. 5.1a) rose five-fold between 1969 and 2010 as traditional rice varieties were replaced by high-yielding varieties. Similar records for sugarcane and jute production show declining trends since the 1970s, whilst production figures for both summer and winter vegetables and spices have risen since the 1990s. About 100,000 tonnes (t) of fish and shrimp are produced each year, around 30 per cent from rivers and estuaries, five per cent from the natural wetlands, with the majority (65 per cent) from artificial ponds and fish farms. However, since the 1990s, data show that annual shrimp production has increased from 5,000 t to more than 25,000 t, and fish production from ponds has risen from 22,000 to 60,000 t (Fig. 5.1c, d). Production from rivers and estuaries has declined dramatically, reflecting the increased production from artificial ponds and shrimp farms that continues to the present (Fig. 5.1e). Current shrimp and fish production figures are significantly higher in Khulna (>20,000 t) than Barisal (<50 t) and Patuakhali (<900 t). Trends in timber and other natural materials from the Sundarbans reached a maximum in the 1980s before declining steeply after 1997 (Fig. 5.1f). Honey collection shows relatively low values in the 1970s and 1980s, but has increased by 1.7 t per year over the period 1974–2010.

5.3.2   Regulating and Habitat Services

Current concerns about regulating and habitat services mainly revolve around water quality, especially salinity and nutrient levels, water availability in terms of availability as river discharge, and water drainage in terms of high groundwater levels. Additionally, some riverside and coastal areas are losing land to erosion processes, and there are concerns that biodiversity, especially across the Sundarbans, is declining. Although only a few long-term records of salinity are available, making extrapolation or averaging of data across the region difficult, some datasets point to general trends. For example, annual mean river water salinity in the Poshur river at Mongla in Khulna (Fig. 5.1g) increased steadily from approximately <5 deciSiemens per metre (dS/m) in the early 1970s to approximately >50 dS/m in 2005. Similar salinity levels were found in rivers in the Rampal and Paikgacha areas until 1995 (Hossain et al. 2016a), but these were followed by a steep decline in the late 1990s. Soil salinity measured during dry (Dec.–May) and wet seasons (June–Nov.) in Khulna and Patuakhali since the 1990s shows contrasting records (Hossain et al. 2016b). In Khulna, three sites show salinity values increasing from <5 dS/m to reach peak values >10 dS/m in the 2000s, with maximum annual figures exceeding 20 dS/m. Four sites in the Patuakhali region show lower absolute values (<5 dS/m), even though it is closer to the coast than Khulna, with only slightly rising trends towards the present. Averaging all the records suggests rising salinity values from the 1990s until 2004 with recent values falling back (Fig. 5.1j). There are no data for actual fertiliser applications, but modelled nutrient loading (total fertiliser and manure input) for three locations in the study area (using the Global NEWS model[2]) suggests more than a doubling in the application of fertiliser and manure between 1970 and 2000: 18 t/km^2/year in 1970 to 39 t/km^2/year in 2000 (Hossain et al. 2016b). Construction of the Farakka Barrage on the Ganges in India in 1975 has caused mean annual discharge measured downstream at Hardinge Bridge to the north of the region (Fig. 5.1) to gradually decline (Fig. 5.1h). In particular, dry season river discharge has effectively halved from 2,000–4,000 m^3/s before 1975 to <1,000–2,000 m^3/s afterwards: an overall rate change only reversed temporarily by extreme flood events in 2000–2001 and 2007. Groundwater levels show markedly different trends. At one site in Khulna (Fig. 5.1i), and another in Barisal, the groundwater levels have been rising (depth below ground decreasing) since the 1980s and now come within 1.5 m of the surface in the post-monsoon season. But at two other sites, groundwater levels are fairly constant at one and gradually deepening at another (Hossain et al. 2016b). In terms of land losses, geomorphological mapping along the major rivers since the 1970s shows that erosion has been greater than accretion in all decades except the 1980s. Mean erosion rates have decreased from 23 km^2/year in the 1970s to <10 km^2/year in the 2000s, while mean accretion rates have also declined from 9 to 3 km^2/year over the same period. The most recent estimate of net land loss in the study area adjacent to the rivers is around 6 km^2/year. However, there is also evidence that accretion rates around the Meghna estuary show a land gain of approximately 3 km^2/year since 2001. The total area of mangroves has increased over the period 1959–2000, while mangrove tree density and numbers (Fig. 5.1k) appear to have declined substantially leading to a halving in the estimated growing stock

(Fig. 5.1l). Numbers of deer in the mangrove forest show fluctuating figures (between 50,000 and 95,000) since 1975, but numbers of tigers seem to have peaked at 450 in 1982 and 2004 with a decline to 200 in 2007. Data for the total number of tourists visiting the Sundarbans between 1996 and 2004 show around 5,000 visitors/year from 1996 to 2004 growing to 25,000 per year in the period 2010–2011.

5.3.3   Environmental Drivers

Local climate has changed significantly over past decades. Mean annual temperature data from three stations since 1948 show a significantly increasing trajectory in all four seasons (Fig. 5.2a). Averaged mean daily rainfall figures show no major trend and are generally in the range 8–12 mm/day (Fig. 5.2b) with extreme lows (<5 mm/day) in Khulna in 1971 and 2008–2010 and extreme highs (>45 mm/day) in Barisal in 1955 and 1960, and in Patuakhali in 1983. Monsoon and post-monsoon seasonal rainfall trends generally increased from the mid-1960s to the 1990s followed by sharply declining trends since 2007. There is some evidence for a shift in rainfall from the monsoon to the post-monsoon period after the 1990s, but all seasons show a decline after 2007 (Hossain et al. 2016b). The mean sea level from two coastal sites shows a gradual rise since the 1980s (Fig. 5.2c) threatening the ecosystem services at the coast particularly in concert with cyclones. The doubling in the frequency of severe cyclonic storms over the period 1991–2007 compared to 1960–1990 (Fig. 5.2e) is particularly noteworthy. Crop damage from natural disasters (Fig. 5.2d), including flooding, cyclones and waterlogging, since 1963 shows recorded damage in 35 per cent of the years with three clusters around 1963–1969, 1981–1991 and 2006–2009. The last two of these disasters are linked to the impacts of the extreme floods in 1987–1988 and super-cyclone Sidr in 2007.

5.4   Social Trends and Well-Being

The population of the coastal zone of Bangladesh was approximately 14 million in the 2011 census with 80 per cent designated as rural population. The total population decreased marginally between 1974 and 1991 to around eight million before rising to the current level. Total gross domestic product (GDP) across the zone has also increased from 74 million USD to around 1025 million USD in the period 1978–2005 with two periods of sharply rising trends in the 1980s and late 1990s (Fig. 5.3a). Barisal, Khulna and Patuakhali all show similar trends, but Patuakhali currently has a significantly lower GDP (less than 400 million USD) than the others (>1,000 million USD). Per capita income has also risen over the past decades with figures for 2005 showing highest income in Khulna (559 USD/person) followed by Patuakhali (393 USD/person) and Barisal (358 USD/person). Despite crop production accounting for the largest share of total GDP (Fig. 5.3b), fishery and non-ecosystem-based livelihoods (e.g. manufacturing) have seen the largest rises in median income (76 and 8 per cent, respectively), while median income from solely agricultural livelihoods has decreased by 18 per cent. Rising levels of per capita income have resulted in a 17 per cent reduction in the numbers of people classified as living below the upper poverty line since 1995 (Fig. 5.3c). Levels of poverty for Khulna and Barisal in 2010 were 33 and 40 per cent, respectively, of the total population, reductions from 60 and 47 per cent in 1983.

Infant and child mortality have declined 50 and 75 per cent, respectively, in less than two decades (Fig. 5.3d), reflecting the improvement in medical care at births (Fig. 5.3e), access to improved sanitation facilities (Fig. 5.3f) and electricity. Universal access to improved drinking water sources has, however, remained largely unchanged since 1995 (Fig. 5.3g); about five per cent of tube-wells that have been tested across the region show the presence of arsenic, and rising levels of salinity in river and groundwater (see Chaps. 17 and 18) may also be affecting the availability of drinking water (Hossain et al. 2016a). Using the data sources identified in Table 5.2, there has been a significant increase in both male and female education (Fig. 5.3h), although societal participation (as indicated by the percentage of women who read newspapers weekly) has declined two per cent between 1993 and 2010.

Where there are observed improvements in human well-being, they are often associated with the poverty alleviation efforts of government and NGOs that have provided access to sanitation, safe drinking water, better health facilities and free primary education (Chowdhury et al. 2013, 2011). The Bangladesh government received more than 8,500 million USD from 1981 to 2012 in foreign aid to develop health, education

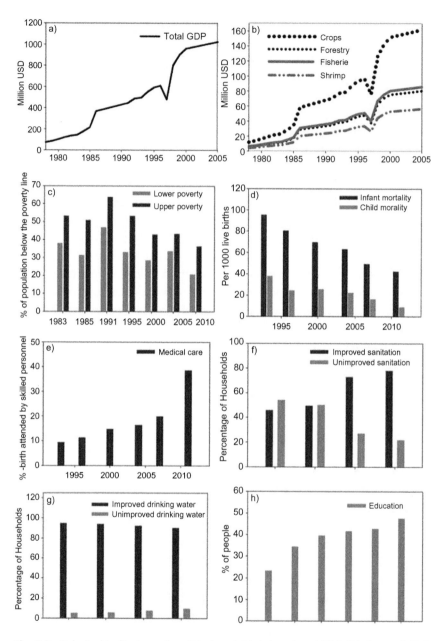

Fig. 5.3 Selected indicators of well-being: **(a)** total regional GDP, **(b)** share of GDP across land production sectors, **(c)** lower and higher poverty level, **(d)** infant and child mortality, **(e)** trained medical care at births, **(f)** improved and unimproved sanitation, **(g)** improved and unimproved drinking water, **(h)** uptake of education (see Hossain et al. 2016a for data sources)

and family planning sectors and now allocates around 20 per cent of the annual national budget each year to these sectors. Other government investment in commercial sectors such as agriculture, industry, energy and rural development has also supported the eradication of poverty (UNDP 2014).

5.4.1   Links Between Ecosystem Services and Well-Being

The importance of natural ecosystem products (e.g. timber, honey, wild caught fish) to well-being is probably low and declining yet is an important safety net for the poorest households, as described in Chap. 2. Forest Department records of the total revenue derived from the harvested product (e.g. timber, honey, etc.) of the Sundarbans show a steep decline since the 1990s which, although estimated to represent a small fraction of the value added income from agriculture and aquaculture, represents a significant safety net for poorer populations. Wild fish catches in rivers have also declined substantially in contrast to the rising trend of fish cultivation in ponds (FAO 2014). Over the whole period, provisioning services have increased due to rising agricultural and fisheries productivity. Poverty has however reduced, with many hypothesised explanations emphasising changing access to migration, literacy and health outcomes. The land use shift to value added crops and shrimp farming may have also played a role in poverty reduction. But, as observed previously, in recent times there are contrasts between rising provisioning services and livelihood incomes in different sectors.

For agriculture, the decrease of nearly one-third in the average cultivated area at the household level coupled to a two-fold increase in crop yield points strongly to agricultural intensification. Farmers have adopted high-yielding crops which demand more fertiliser, pesticides and other inputs compared to the traditional local rice varieties (Hossain et al. 2016a, 2013; Ali 1999) resulting in a seven-fold increase in production costs. This increase in production costs and the lack of fair pricing, it is argued, are the main reasons for the falling median incomes of agricultural households (Hossain et al. 2013; IFPRI 2013). However, the situation is complicated. The salinity increase due to shrimp farming also reduces crop production and creates unemployment for landless farmers because of the low labour demand in shrimp farming compared to crop production (Swapan and Gavin 2011). As a result, farmers are more likely to belong to the poor income groups. This suggests strongly that regional poverty alleviation has not been driven by natural ecosystem products or agricultural intensification but rather incomes generated from aquaculture, especially shrimp farming, non-ecosystem service sectors and the national investment in infrastructure, education and health care.

To what extent these multi-decadal changes have been driven by external environmental changes is difficult to quantify or determine with certainty. Within the region there are multiple interacting drivers that include global climate change, the take up of new agricultural methods, national infrastructural developments and local-national policy. Some interactions are strong, direct and linear, while others are weaker, nonlinear and may include feedback loops. For example, correlation analysis suggests that climate has directly helped to promote the uptake of high-yielding rice varieties as these are adaptable to rising temperatures and the declining rainfall trend in the pre-monsoon season; traditional varieties are now less viable. In contrast, the conversion of rice fields into productive brackish shrimp farming appears to have been favoured by the rising salinity levels of both soils and water sources. Salinisation is related to the relative rise in sea level and eastwards migration of the main river courses, but has been accelerated by polderisation (Swapan and Gavin 2011; Islam 2006; Mirza and Ericksen 1996) and flood control projects from the 1960s onwards particularly the Farakka Barrage (Mirza 1997, 1998). In turn, shrimp farming tends to increase local soil salinity levels thus generating a positive feedback loop between land use change and environmental degradation (Rahman et al. 2013).

5.4.2 Links Between Provisioning and Regulating Services

Even though poverty levels among farmers remain high, indicators for average well-being and provisioning services paint a fairly positive picture. In contrast, there is ample evidence to argue that non-food ecosystem services such as water availability, water quality, biodiversity and land stability have declined, at least in comparison with conditions in the 1960s.

These opposing trends (Fig. 5.4) constitute a trade-off between wealth creation and poverty alleviation and unsustainable land use. It is also clear that this relationship (known as an environmental Kuznets curve, see Dinda 2004) shows no turning point towards lower levels of environmental degradation while poverty continues to lessen as seen in many middle- and high-income countries (Hossain et al. 2016a). Such unsustainable conditions coupled with the clear evidence for later monsoons and higher temperatures should signal a need for proactive adaptive strategies. As Raudsepp-Hearne et al. (2010) have argued, the growing losses of regulating (and supporting) services may eventually be expected to feedback negatively on essential provisioning services, which in coastal Bangladesh would be expected to drive declines in rice, shrimp and fish production. This would impact first on the poorest sector, the rural

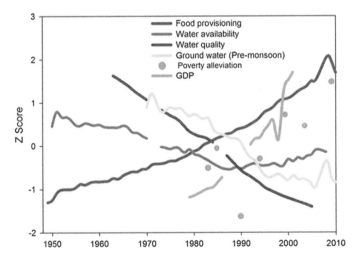

Fig. 5.4 A schematic summary of provisioning services, regulating services, wealth and poverty alleviation 1950–2010. The contrasting curves of normalised data represent a 'trade-off' between rising levels of food production, average wealth and well-being across the region against a background of deteriorating aspects of water (see also Hossain et al. 2016a)

farmers, still dependent on both wage income and subsistence products, before gradually affecting other sectors, the larger landowners and associated processing industries.

5.5 Regional Resilience

In dynamical terms, the most striking feature about the data is the dominance of trending rather than stationary curves. The rise in GDP is the most dramatic as it occurs in less than half the timescale of the others. In longer time-series, the shifts towards deteriorating trends in climate drivers and regulating services occurred in the mid-1970s (water availability), mid-1980s (water quality), early 1980s and 2007 (rainfall), mid-1970s temperature and early 1990s (groundwater level). Shifts towards improving trends of provisioning services occurred in the 1980s. The presence of trends over decadal timescales indicates that many variables have an underlying 'slow' component possibly driven by positive feedback loops. But some variables also show fast or high-frequency components over annual and shorter timescales, particularly water quality, rainfall and water availability. System resilience depends upon the interactions between fast and slow variables. The observation that many of the slowly changing non-food ecosystem services that regulate and support key ecosystem processes are deteriorating may indicate weakening resilience within the dependent agriculture and aquaculture sectors. This is especially important because historical analysis shows that flooding and severe cyclones, particularly in 1987–1988, 1991 and 2007, were the major shocks in the past that have impacted negatively on food production and the poorest communities in the past. Lower levels of resilience and projections of higher incidences of climate shocks (e.g. Khan et al. 2000; Ali 1999) raises the prospect of 'perfect storm' scenarios (Dearing et al. 2012) with a greater likelihood of negative impacts in the future. Dynamic and integrated modelling approaches that can guide the selection of adaptation and sustainable management options for the coming decades will be essential to address these concerns.

Notes

1. available from http://microdata.worldbank.org
2. https://marine.rutgers.edu/globalnews/datasets.htm

References

Ali, A. 1999. Climate change impacts and adaptation assessment in Bangladesh. *Climate Research* 12 (2–3): 109–116. https://doi.org/10.3354/cr012109.

BBS. 2012. *Bangladesh population and housing census 2011 – Socio-economic and demographic report*. Report national series, vol. 4. Dhaka: Bangladesh Bureau of Statistics (BBS) and Statistics and Informatics Division (SID), Ministry of Planning, Government of the People's republic of Bangladesh. http://www.bbs.gov.bd/PageSearchContent.aspx?key=census%202011. Accessed 7 July 2016.

Brammer, H. 2014. Bangladesh's dynamic coastal regions and sea-level rise. *Climate Risk Management* 1: 51–62. https://doi.org/10.1016/j.crm.2013.10.001.

CEGIS. 2007. *Effect of cyclone sidr on the Sundarbans: A preliminary assessment*. Dhaka: Center For Environmental and Geographic Information Services (CEGIS).

Chaffey, D.R., F.R. Miller, and J.H. Sandom. 1985. *A forest inventory of the Sundarbans, Bangladesh*. Project report no 140. Land Resources Development Centre, England: Overseas Development Administration.

Chowdhury, Q. 2001. State of Sundarban. In *Forum of environmental journalist of Bangladesh, ministry of environment and forest, Bangladesh (MoEF)*. Dhaka: United Nations Development Program (UNDP).

Chowdhury, S., L.A. Banu, T.A. Chowdhury, S. Rubayet, and S. Khatoon. 2011. Achieving millennium development goals 4 and 5 in Bangladesh. *Bjog-an International Journal of Obstetrics and Gynaecology* 118: 36–46. https://doi.org/10.1111/j.1471-0528.2011.03111.x.

Chowdhury, A.M.R., A. Bhuiya, M.E. Chowdhury, S. Rasheed, Z. Hussain, and L.C. Chen. 2013. The Bangladesh paradox: Exceptional health achievement despite economic poverty. *The Lancet* 382 (9906): 1734–1745. https://doi.org/10.1016/s0140-6736(13)62148-0.

Dearing, J.A., S. Bullock, R. Costanza, T.P. Dawson, M.E. Edwards, G.M. Poppy, and G.M. Smith. 2012. Navigating the perfect storm: Research strategies for socialecological systems in a rapidly evolving world. *Environmental Management* 49 (4): 767–775. https://doi.org/10.1007/s00267-012-9833-6.

Dey, T.K. 2007. *Deer population in the Bangladesh Sundarban.* Chittagong: The Ad Communication.

Dinda, S. 2004. Environmental Kuznets curve hypothesis: A survey. *Ecological Economics* 49 (4): 431–455. https://doi.org/10.1016/j.ecolecon.2004.02.011.

Ericksen, N.J., Q.K. Ahmad, and A.R. Chowdhury. 1996. Socio-economic implications of climate change for Bangladesh. In *The implications of climate and sea–Level change for Bangladesh*, ed. R.A. Warrick and Q.K. Ahmad, 205–287. Dordrecht: Springer.

FAO. 1999. *Forest resources of Bangladesh.* Country report, working paper no. 15. Rome: Food and Agricultural Organization of the United Nations (FAO).

———. 2007. *Mangroves of Asia 1980–2005.* Country reports. Forest resources assessment programme, working paper 137. Rome: Food and Agriculture Organization of the United Nations (FAO). ftp://ftp.fao.org/docrep/fao/010/ai444e/ai444e00.pdf. Accessed 3 Jan 2013.

———. 2014. *The state of world fisheries and aquaculture: Opportunities and challenges.* Rome: Food and Agriculture Organization of the United Nations (FAO). http://www.fao.org/3/d1eaa9a1-5a71-4e42-86c0-f2111f07de16/i3720e.pdf. Accessed 5 July 2016.

Gittins, S.P. 1980. A survey of the primates of Bangladesh. In *Project report to the forest department of Bangladesh.* London: Flora and Fauna Preservation Society.

Helalsiddiqui, A.S.M. 1998. *Sundarban at a glance.* Khulna: Mangrove Silviculture Division, Bangladesh Forest Research Institute.

Hendrichs, H.H. 1975. The status of the tiger Panthera Tigris (Linné, 1758) in the Sundarbans mangrove forest (Bay of Bengal). *Säugetierekundliche Mitteilungen* 23 (3): 161–199.

Hossain, M.S., M.J. Uddin, and A.N.M. Fakhruddin. 2013. Impacts of shrimp farming on the coastal environment of Bangladesh and approach for management. *Reviews in Environmental Science and Bio-Technology* 12 (3): 313–332. https://doi.org/10.1007/s11157-013-9311-5.

Hossain, M.S., F. Amoako Johnson, J.A. Dearing, and F. Eigenbrod. 2016a. Recent trends of human wellbeing in the Bangladesh delta. *Environmental Development* 17: 21–32. https://doi.org/10.1016/j.envdev.2015.09.008.

Hossain, M.S., J.A. Dearing, M.M. Rahman, and M. Salehin. 2016b. Recent changes in ecosystem services and human well-being in the Bangladesh coastal zone. *Regional Environmental Change* 16 (2): 429–443. https://doi.org/10.1007/s10113-014-0748-z.

IFPRI. 2013. *The status of food security in the feed the future zone and other regions of Bangladesh. Results from the 2011–2012 Bangladesh integrated household survey.* Dhaka: International Food Policy Research Institute. http://www.fao.org/family-farming/detail/en/c/417259/. Accessed 06 Jan 2017.

Iftekhar, M.S., and M.R. Islam. 2004. Degeneration of Bangladesh's Sundarbans mangroves: A management issue. *International Forestry Review* 6 (2): 123–135. https://doi.org/10.1505/ifor.6.2.123.38390.

Iftekhar, M.S., and P. Saenger. 2008. Vegetation dynamics in the Bangladesh Sundarbans mangroves: A review of forest inventories. *Wetlands Ecology and Management* 16 (4): 291–312. https://doi.org/10.1007/s11273-007-9063-5.

Islam, M.R. 2006. Managing diverse land uses in coastal Bangladesh: Institutional approaches. In *Environment and livelihoods in tropical coastal zones: Managing agriculture-fishery-aquaculture conflicts,* ed. C.T. Hoanh, T.P. Tuong, J.W. Gowing, and B. Hardy. Wallingford: CAB International.

Islam, N. 2007. *Exploring the drivers for conversion of agricultural land to shrimp farms in south-west region of Bangladesh.* MSc thesis, Environmental Science Discipline, Khulna University.

Islam, M.S.N. 2008. *Cultural landscape changing due to anthropogenic influences on surface water and threats to mangrove wetland ecosystems: A case study on the Sundarbans.* PhD thesis, Faculty of Environmental Sciences and Process Engineering, Brandenburg University of Technology.

Islam, M.B., M.Y. Ali, M. Amin, and S.M. Zaman. 2011. Climatic variations: Farming systems and livelihoods in the high Barind tract and coastal areas of Bangladesh. In *Climate change and food security in South Asia,* ed. Lal Rattan, V.K. Mannava Sivakumar, S.M.A. Faiz, A.H.M. Mustafizur Rahman, and R. Khandakar Islam, 477–497. Dordrecht: Springer.

IUCN. 1997. *Sundarban wildlife sanctuaries (Bangladesh).* IUCN technical evaluation. Gland: International Union for Conservation of Nature (IUCN). http://whc.unesco.org/archive/advisory_body_evaluation/798.pdf. Accessed 20 Jan 2013.

Khan, M.M.H. 2005. *Project Sundarban tiger: Tiger D=density and tiger-human conflict.* Financial technical report. Washington, DC: Save The Tiger Fund of the National Fish and Wildlife Foundation. http://www.panthera.org/sites/default/files/STF/2005-0013-004.pdf. Accessed 2 Apr 2013.

Khan, T.M.A., O.P. Singh, and M.R. Sazedur. 2000. Recent sea level and sea surface temperature trends along the Bangladesh coast in relation to the frequency of intense cyclones. *Marine Geodesy* 23 (2): 103–116. https://doi.org/10.1080/01490410050030670.

MDG. 2015. *Millennium development goals*. Bangladesh progress report 2015. Dhaka: General Economics Division, Bangladesh Planning Commission, Government of the People's Republic of Bangladesh. http://www.bd.undp. org/content/bangladesh/en/home/library/mdg/mdg-progress-report-2015. html. Accessed 11 Jan 2017.

MEA. 2005. Ecosystems and human well-being: Synthesis. *Millennium ecosystem assessment (MEA)*. Washington, DC: Island Press. http://www.millenniumassessment.org/documents/document.356.aspx.pdf. Accessed 01 Aug 2016.

Mirza, M.M.Q. 1997. Hydrological changes in the Ganges system in Bangladesh in the post-Farakka period. *Hydrological Sciences Journal* 42 (5): 613–631. https://doi.org/10.1080/02626669709492062.

———. 1998. Diversion of the Ganges water at Farakka and its effects on salinity in Bangladesh. *Environmental Management* 22 (5): 711–722. https://doi. org/10.1007/s002679900141.

Mirza, M.Q., and N.J. Ericksen. 1996. Impact of water control projects on fisheries resources in Bangladesh. *Environmental Management* 20 (4): 523–539. https://doi.org/10.1007/bf01474653.

MoEF. 2002. *Survey to determine the relative abundance of tiger wild boar and spotted deer in the Bangladesh Sundarban forest during December 2001–March 2002*. Technical reports-TR no.17. Dhaka: Ministry of Environment and Forest (MoEF), Government of the People's Republic of Bangladesh.

———. 2010. *Fourth national report to the convention on biological diversity. Biodiversity national assessment and programme of action 2020*. Dhaka: Ministry of Environment and Forests (MoEF). www.cbd.int/doc/world/bd/ bd-nr-04-p1-en.pdf. Accessed 5 July 2016.

OECD. 2013. *Guidelines on measuring subjective well-being*. Paris: OECD Publishing. https://doi.org/10.1787/9789264191655-en.

Rahman, A., G. Rabbani, M. Muzammil, M. Alam, S. Thapa, R. Rakshit, and H. Inagaki. 2010. *Scoping assessment on climate change adaptation in Bangladesh*. Report prepared by the Bangladesh center for advanced studies. Bangkok: AIT-UNEP RRC.AP.

Rahman, M.H., T. Lund, and I. Bryceson. 2011. Salinity impacts on agro-biodiversity in three coastal, rural villages of Bangladesh. *Ocean and Coastal Management* 54 (6): 455–468. https://doi.org/10.1016/j. ocecoaman.2011.03.003.

Rahman, M.M., V.R. Giedraitis, L.S. Lieberman, T. Akhtar, and V. Taminskiene. 2013. Shrimp cultivation with water salinity in Bangladesh: The implications of an ecological model. *Universal Journal of Public Health* 1 (3): 131–142. http://dx.doi.org/10.13189/ujph.2013.010313.

Raudsepp-Hearne, C., G.D. Peterson, M. Tengö, E.M. Bennett, T. Holland, K. Benessaiah, G.K. MacDonald, and L. Pfeifer. 2010. Untangling the environmentalist's paradox: Why is human well-being increasing as ecosystem services degrade? *Bioscience* 60 (8): 576–589. https://doi.org/10.1525/bio.2010.60.8.4.

Salim, M., B.U. Maruf, M. Sumsuddoha, A.I. Chowdhury, and A.R. Babul. 2007. Climate change would intensify river erosion in Bangladesh: Climate change impact in Bangladesh. In *Equity and justice working group, campaign brief 6.* Dhaka: COAST Trust.

Swapan, M.S.H., and M. Gavin. 2011. A desert in the delta: Participatory assessment of changing livelihoods induced by commercial shrimp farming in Southwest Bangladesh. *Ocean and Coastal Management* 54 (1): 45–54. https://doi.org/10.1016/j.ocecoaman.2010.10.011.

Uddin, M.N., and A. Haque. 2010. Salinity response in southwest coastal region of Bangladesh due to hydraulic and hydrological parameters. *International Journal of Sustainable Agricultural Technology* 6 (3): 1–7.

UN. 2015. *Sustainable development goals (SDGs).* United Nations (UN). http://www.un.org/sustainabledevelopment/. Accessed 22 Sept 2017.

UNDP. 2014. Human development report 2014. *Sustaining human progress: Reducing vulnerabilities and building resilience.* New York: United Nations Development Program (UNDP). http://hdr.undp.org/en/content/human-development-report-2014. Accessed 5 July 2016.

Wodon, Q.T. 1997. Food energy intake and cost of basic needs: Measuring poverty in Bangladesh. *Journal of Development Studies* 34 (2): 66–101. https://doi.org/10.1080/00220389708422512.

Zmarlicki, C.B. 1994. *Integrated resources development of the Sundarbans Reserved Forest, Bangladesh.* Rome: United Nations Development Programme (UNDP) and Food and Agriculture Organization of the United Nations (FAO).

6

Governance of Ecosystem Services Across Scales in Bangladesh

Andrew Allan and Michelle Lim

6.1 Introduction

Ecosystem services (MEA 2005) are governed and affected by a diverse variety of different legal and governance frameworks (Paavola et al. 2009), although analysis of these legal frameworks has rarely been carried out (Salzman et al. 2001). Ideally, these frameworks are integrated and coordinated with a view to achieving common objectives, but in reality they are often developed and implemented within sectoral boundaries without reference to each other. For example, the provisioning ecosystem service of freshwater is potentially affected by management and policy decisions related to water abstraction, pollution control and agricultural land use and flood risk alleviation, among others. Laws and institutions often fail to address cross-cutting issues that are shared across sectors and approaches to these issues are therefore frequently fragmented and incomplete.

A. Allan (✉)
School of Law, University of Dundee, Dundee, UK

M. Lim
Adelaide Law School, University of Adelaide, Adelaide, SA, Australia

© The Author(s) 2018
R. J. Nicholls et al. (eds.), *Ecosystem Services for Well-Being in Deltas*,
https://doi.org/10.1007/978-3-319-71093-8_6

Lack of policy integration is widely observed, and although it may be explained as a product of the increasing complexity involved in effectively dealing with difficult issues, such as the demands that are imposed by choosing to implement integrated water resources management, it is also in part a function of weaknesses in government planning structures.

An important aspect of the research undertaken in the study area was to understand the ways in which ecosystem services are related to poverty alleviation. One way to begin interrogating the relationship between these two domains is to question whether or not the benefits that might be derived from ecosystem services are actually accessible by those who can most profit from them. The staged approach to ecosystem services proposed by Fisher et al. (2009) is relevant here, the theory positing that intermediate ecosystem services require an additional catalyst in the form of one or other type of capital in order to render their potential benefits actually realisable. Governance frameworks are relevant not only to the quality and extent of the ecosystem services themselves but also to the ability of people to enjoy their benefits (see, e.g. Butler and Oluoch-Kosura 2006). Consequently, the governance research examines the legal, institutional and policy frameworks that both affect the quality of the ecosystem services and those influencing an individual or community's ability to enjoy the derived benefits.

The research therefore focuses on legislative coherence especially across sectors, transposition of international obligations, adherence to international best practice and the extent to which policy objectives are supported by the legal foundation. In the particular context of the Ganges-Brahmaputra-Meghna (GBM) delta, this entails consideration of formal documented policies, regulations and statutes (and case law where appropriate), especially those documents that are legally enforceable, affect multiple sectors, and are at an administrative scale where decisions could be made that would affect ecosystem services and livelihoods. The fact that in some cases decision making is either devolved or decentralised within national boundaries and that the GBM basin is shared by five riparian states (China, India, Bangladesh, Nepal and Bhutan) meant that multiple scales have to be examined. Analysis of the factors that influence the implementation and achievement of policy objectives, and the extent to which legal and institutional frameworks are capable of supporting policy is also necessary (Hill Clarvis et al. 2014). This analysis of barriers was extended to cover informal governance systems as far as possible, in order to understand the degree of coherence between local customary systems and more formal frameworks.

Finally, the governance work investigates ways in which governance quality might be incorporated into the modelling work that integrated the major biophysical elements of ecosystem services with the health and livelihood findings of the household survey. Even an elementary quantification of the influence of governance quality may usefully inform decision-making choices as regards appropriate interventions for alleviating poverty.

6.2   Policy and Legislation

The governance analysis focuses on around 80 pieces of legislation and policy relevant to the sources of ecosystem services (including water and land use management, fisheries, and environmental protection) and to the protection and improvement of livelihoods (e.g. human rights and rural development). Identification of the areas of relevant law is linked primarily to the conceptualisation of the relationship between ecosystem services and poverty alleviation and the factors mediating between these. As seen in Chap. 4, these were identified as access and entitlements, mobility and urban areas, community and informal institutions and health. This wide range of factors is both affected and governed by a vast spectrum of legal and policy frameworks, so limiting the scope to the key elements only is imperative for the preliminary analysis.

Areas of governance of potential relevance are therefore identified. Access and entitlement to resources are governed primarily by the framework for apportioning rights of use and access to water resources, along with property rights and land tenure. These are also influenced by rights to use and access forests for timber and non-timber forest products, with any frameworks allocating rights to fish. In many areas, local customary legal and institutional systems have not yet been supplanted by more formal sources, especially in relation to resource entitlements (e.g. land, water, fisheries—Freestone et al. 1996), so must also be considered.

The quality and extent of resource use entitlements are critically dependent upon the extent to which resources can be protected; understanding the governance context for pollution control and environmental management is consequently required. In relation to broader questions of mobility and health, the existence of appropriate human rights regardless of location is a crucial consideration, in both substantive and procedural form, with access to information rights underpinning the latter, forming part of the foundation for the wider cross-sectoral governance context. Land use planning capacity is also a concern with respect both to access and entitlements, but also in relation to urban areas and disaster risk management. The prevention, management and alleviation of disasters are increasingly addressed through holistic governance frameworks, the quality of which can drastically affect population vulnerability and resilience. Finally, access to justice and remediation (including through local courts) is imperative for the enforcement of rights and obligations under each of the foregoing categories.

6.2.1   The International Context

Not all relevant legislation or policy will be set out in detail here—nor does space allow a detailed analysis of the international legal framework. There are certain key areas of law that are particularly important for both ecosystem services and poverty alleviation, however. The first are those legal frameworks concerning freshwater, which are central to the ecosystem services within the study area (Rieu-Clarke and Spray 2013), addressing both quantity, flow and quality considerations. At the international scale, the GBM system is characterised by a general absence of a pan-watershed agreement (despite some bilateral arrangements further upstream), and none of the basin states have ratified the UN Watercourses Convention (UN 1997) or the UNECE Water Convention (UNECE 1992) although the latter has been open to states outside the UNECE area since 1 March 2016. A draft agreement on the use of the Teesta River has also yet to be signed, despite having been in existence for a number of years, and there is no formal agreement on shared groundwater. This leaves only the Farakka Treaty (1996), though this is limited in its application as it applies only to the dry season, contains no provisions on water quality, applies to only two riparian states and affects one of the basin rivers alone.

In addition, and to some extent, bridging the gap between ecosystem services and livelihoods are legal frameworks relating to human rights. Bangladesh has ratified both of the international human rights covenants, respectively, on Economic, Social and Cultural Rights (ICESCR 1966) and on Civil and Political Rights (ICCPR 1966), some of which have been transposed through the country's 1972 Constitution. The situation as regards human rights for migrants, especially those internally displaced for whatever reason, is less secure however (Allan et al. 2015).

6.2.2 National Governance and Implementation

At the national level, while legislation and policy exist, the general level of inter-sectoral coordination and legislative coherence, with notable exceptions, is rather discouraging. In addition, there are a number of barriers to the effective implementation and the age of legislation in many cases worrying. Examples include the Environment Policy of 1992 (its environmental impact assessment elements remain guidelines only) and the failure to translate the cross-sectoral Coastal Zone Policy of 2005 into practice. There are also a number of examples where primary legislation is in place, but is hobbled by the absence of implementing guidelines, subordinate legislation or rules required for it to have any impact (see, e.g. the Groundwater Management Ordinance[1]).

While the age of legislation is not in itself always problematic, there are areas where policy priorities have moved significantly globally over the past 20 or 30 years, and the pace of change is in fact increasing. Environmental protection and human rights are two such areas. In neither case is there much evidence of significant legislative shifts in Bangladesh over the past 20 years, and, in the context of forest management in particular, much of the land use management regime and compulsory purchase frameworks (where private land is expropriated for public purposes) continue to be heavily influenced by approaches from the British colonial period. Water resource management is something of an exception here, with the recent Water Act of 2013, although the detailed rules needed for implementation are yet to reach the statute book.

An allied issue relates to reporting and monitoring. The Planning Commission of Bangladesh contains a division that is dedicated to the implementation, monitoring and evaluation of public sector development projects, indicating the value that the government puts on ensuring that projects do what they are intended to do, at least initially. However, some of the indicators used by this division are problematic. For example, those relating to environmental sustainability correspond only loosely with what might be regarded as best practice (see, e.g. Bell and Morse 2008): forest coverage, length of waterways that are navigable all year round, number of cyclone shelters, what are described as the number of rural communities with disaster resilient habitats and community assets, and reductions in case backlogs in courts.

Clear criteria are needed to ensure accountability and transparency, but the level of detail necessary for this is not so apparent in some Bangladeshi legislation. For example, with respect to groundwater extraction, the criteria guiding decisions on permits for sinking tube-wells are very vague (Groundwater Management Ordinance 1985, s.5(4)), leaving decision makers almost unfettered discretion. This is compounded by serious restrictions on the ability of disappointed applicants to appeal against these decisions, including with respect to time restrictions. The balance between the length of time available for challenging decisions and the consequences of those decisions may be inappropriate. The Acquisition of Waste Land Act of 1950 provides an example. The authorities have the right under the Act to compulsorily acquire what is defined as 'waste land' for certain public purposes. Objections to such acquisition must be raised within 15 days of notification being made, at a limited number of places and only then by those who may have a right of compensation for the land being taken, taking no account of literacy or access to the appropriate information. Land tenure in Bangladesh is vulnerable to the rapid erosion processes on the GBM, and although legal provision for *khas* land appears to favour the poor, the reality of the application of the State Acquisition and Tenure Act 1950 (as amended) is that the poor are at the mercy of richer farmers (FAO 2010). Land tenure especially is further complicated by the influence of local customary frameworks, through the operation of the *samaj*[2] and adjudication of disputes (including those relating to land) through *shalish*[2] tribunals (Lewis and Hossain 2008).

This latter point is also symptomatic of a more general trend towards asymmetry between the rights, powers and obligations of the government (and its officials) and the public. State authorities are given a significant degree of discretion in their decision making, and this is especially noticeable again when the consequences of a decision are weighed against the degree to which the decision is open to challenge. The Embankment and Drainage Act (1952) endows the engineer with extensive powers to enter into and to acquire land that is relevant to the construction of public embankments. Similarly, the determination of land as being 'waste land' for the purposes of the Acquisition of Waste Land Act, and the very broad definition of 'public purpose', and the lack of clarity in relevant criteria, leave a great deal of latitude to decision makers. Where, as in the case in the latter Act, access to the civil courts as a means of obtaining redress is expressly forbidden, the balance of power is very definitely on the side of the authorities. This compounds the general lack of participation and the marginalisation of the poor that takes place despite the increasing number of references to participatory approaches that appear in policy documents and legislation.

The lack of coordination and cooperation within and between ministries is a recurring theme within the literature and in interviews conducted as part of the project stakeholder engagement, but is also evident in the separation of different policy frameworks. This is compounded by a fragmented legal regime and inconsistencies within laws and regulations (Afroz and Alam 2013). Different aspects relating to the management of the Sundarbans, for example, are governed by separate legal regimes. The Forest Department (part of the Ministry of Environment and Forests—MoEF) is responsible for implementing the Forest Act 1927, and another wing of the same Ministry, the Department of Environment, is responsible for implementing the Environment Conservation Act 1995. The Department of Fisheries (part of the Ministry of Fisheries and Livestock) is responsible for monitoring fish stocks, but the Forest Department (MoEF) issues permits for fish collection inside the mangrove forests. These separate regimes create challenges for management and difficulties for the control of fish stocks (Iftekhar 2010). They also suggest that the quality of governance of ecosystem services and the corresponding ability of people to enjoy the benefits the service provides may vary across social-ecological systems (see Chap. 22).

Frustratingly, however, it is also clear that cross-sectoral coordination is possible and practicable in Bangladesh, with the disaster management framework providing an excellent example of this. As shown in Fig. 6.1, policy and legislation were developed in close chronological proximity, appropriate institutional arrangements put in place rapidly, and the system now appears to work well.

The lack of enforcement financial and technical capacity also has a severe impact on implementation and management in Bangladesh. The effectiveness of legislation is compromised by the lack of enforcement (Afroz and Alam 2013), and this creates a large gap between the *de jure* commitment and the de facto reality.

6.3 Relationship Between Policy and Legislation

As previously noted, the success of the disaster management architecture in Bangladesh may in part be a result of the way in which policy was developed almost in tandem with the required legal framework. What then is the relationship between policy and legal frameworks in this context? While recognising that policy aims may not be directly dependent on the capacity of legal frameworks to support them (e.g. infrastructure development, changes in investment priorities or capacity development), analysis suggests that while there may be policy development in areas that are important for ecosystem services and livelihoods, this may not be followed up with linked improvements in legal frameworks.

This is not to say that there is no connection between the two. The Disaster Management Plan was very rapidly translated into legal effect, for example, through the Standing Orders on Disaster. This is in line with the idea that disasters may precipitate what Pelling and Dill call 'a tipping point in the social contract' (Pelling and Dill 2010), though in this case the development of both the disaster management policy and legislation

Policy	Year	Legislation
National Social Protection Strategy (3rd draft)	2014	
	2013	Water act
	2012	
	2011	
Sixth Five Year Plan Plan for Disaster management National Industrial Polict Perspective Plan Child labour Elimination Policy	2010	Standing Orders on Disasters
National Adaptation Plan of Action National Tiger Action Plan	2009	Right to Information Act
	2008	
	2007	
Coastal Development Strategy National Fisheries Strategy National Food Policy	2006	
Fifth Five Year Plan Coastal Zone Policy	2005	
	2004	
	2003	
Population Policy	2002	
Rural Development Policy	2001	
	2000	Environment Court Act Water Development Board Act
National Water Policy	1999	
National Fisheries Policy	1998	
	1997	Environment Conservation Rules
	1996	Ganges Water Sharing Treaty
	1995	Environment Conservation Act Protection and Conservation of Fisheries (Amendment) Act
National Forest Policy	1994	

Fig. 6.1 Chronology comparing the development of policy and legislation

can be seen partly as a response to Cyclone Aila in 2009 and also as part of an international continuum during the first ten years of this century where many countries adopted similar approaches, driven in part by the Hyogo Framework (Allan 2017) and the earlier Yokohama Strategy (UNISDR 1996).

The factors involved in other legislation and policy development are ordinarily rather different (South Africa is a key exception here, because of its unique circumstances at the end of the apartheid era), and it is therefore more difficult to interpret the relationship between the two elements of governance. A systematic review of policy development and correlative legal frameworks has not been done in relation to ecosystem services, and making firm inferences from such a study would be problematic as there are a multitude of factors that might have influence. Existing legislation may be adequate to support policy innovations or require merely a change in interpretation that can be manifested through operational guidelines (which would be invisible to such a study). Policy on the other hand may not require legislative intervention in order to achieve its objectives. There is also a question of breadth or scope. If principles of human behaviour can be established that are generally adhered to, for example, in national Constitutions (e.g. with respect to human rights in the case of Bangladesh and the adherence to international standards in law at least), it may be that no further change is needed. This is not the case with respect to the management of resources, for instance, where changing management practices are driven in part by increased understanding through science, the application of technology, involvement of broader swathes of stakeholders and potentially also changing resource availability (as a result, for instance, of demographic changes or the impacts of climate change). One consequence of this is that it becomes increasingly difficult for legal practice to keep up with societal change. Furthermore, the development of legislation is in some ways intrinsically more contentious than the development of policy simply because of the power structures inherent in legislatures. Policy puts objectives in place that are generally non-enforceable, while law (at least in theory) puts obligations and rights in place that are supposed to be backed by official sanction.

Figure 6.1 does not unpick these difficulties, and further research is needed to analyse how it compares with other countries. The key elements that stand out include the fact that there is no corresponding legislation to implement the National Fisheries and Forestry policies (1998 and 1994 respectively), especially given the age of the existing forest legislation. The 14-year gap between the development of the National Water Policy and the 2013 Water Act is also noteworthy, especially as the latter is so closely aligned with the former. Finally, there is an unfortunate absence of a legal framework to implement the inter-sectoral 2005 Coastal Zone Policy. It is unrealistic to expect policy objectives to be met when there are no enforceable means of achieving them.

A number of other factors may influence the extent to which policy objectives may be implemented. These may include (i) adequate financial provision within government to support implementation, (ii) institutional capacity and effective coordination between separate agencies or institutions towards a common cross-sectoral objective and (iii) appropriate expectations as regarding the time and budget within which the objective should be fulfilled.

6.4   Linking with the Integrated Modelling

As discussed in Chap. 10, governance issues were clearly identified by interviewees and workshop attendees as being of concern in the study area, and the multiple facets of the concept were elucidated in great detail. In order to begin the process of quantifying the impact of governance on ecosystem services and livelihoods, efforts were made to incorporate governance metrics and indicators into the integrated modelling process in order to try to capture the governance situation in future projections.

The principal objective was to better understand the potential impact of varying qualities of governance regimes on the effectiveness of policy implementation, management, or infrastructural interventions intended to alleviate poverty or enhance the benefits derived from ecosystem services. Of the list of governance issues identified as part of the scenario development process, only around 40 per cent were considered even potentially capable of being represented in the integrated modelling work. This was due in part to limitations on modelling capacity (the integrated model simply cannot model everything) and to the difficulties encountered in trying to reflect governance quality in model-able inputs.

Firstly, identifying appropriate governance datasets that could be directly applicable was difficult. There is no commonly accepted definition of 'governance' across the multitude of indicator systems available (see, e.g. Arndt and Oman 2006), and there is no close sectoral overlap of indicator systems with the areas of governance being studied in the project. In addition, the temporal aspects of governance indicators raise obstacles against their use: even the most regularly updated indicators (such as the World Governance Indicators (Kaufmann et al. 1999) and the Corruption Perceptions Index[3] are revised only once every year, with the more comprehensive assessments (including the World Bank CPIA (Country Policy and Institutional Assessment) measure and the UNDAF CCA (Common Country Assessment) process) conducted only every five years or so. Combined with the general lack of a long-term dataset to aid calibration, demonstrating governance trends over time in Bangladesh and in relation to the relevant governance frameworks is extremely challenging.

Secondly, and more crucially, the establishment of a causal or even an associative relationship between a governance intervention of any sort and a change in an indicator of biophysical or human well-being in such a broad arena as ecosystem services and livelihoods is virtually impossible. There are so many variables that may affect the success or otherwise of policy implementation that drawing a clear link between cause and effect is challenging in the extreme. This was further highlighted in the process of attempting to establish some sort of spatial explicitness with respect to governance indicators (i.e. from national level down to the project case study areas in the south-west of the country). This proved impracticable, not least because existing governance datasets are analysed as regards national state contexts and cannot be magnified to give finer resolution.

Therefore, it became clear that while the existence of certain elements of governance was obviously desirable, direct quantification of exactly how beneficial they might be is extremely challenging. These latter elements included institutional coordination, stakeholder involvement in decision making and forward planning. All are manifestly useful, but quantifying their influence will always involve a degree of speculation.

6.5   Conclusions

The legal and policy infrastructure relating to ecosystem services is immensely complex, crossing multiple sectors and demanding a degree of institutional and policy coordination that is challenging even in the most developed nations. In addition, the capacity of the poor to access and enjoy the benefits produced by those services may be compromised by restrictions imposed by weak enforcement of applicable legal entitlements, difficulty in accessing justice or by the divergent demands of local informal governance frameworks compared to formal state systems.

It appears that governance frameworks in Bangladesh for water resource and land management rely to some extent on rather outdated policy and legislation. Further research is needed on better understanding the relationship between policy development and its translation into legislation (if needed). It is clear however that where the two are closely linked in time, it is possible for the requisite degree of coordination, cross-sectoral coherence and enforceability to be achieved; the disaster risk management process demonstrates that workable solutions can be developed even in the poorest countries.

Understanding the exact relationship between governance frameworks for ecosystem services and poverty alleviation through the incorporation of quantitative indicators in mathematical models requires additional research in order to assess compatibility with existing indicator sets, and for the incorporation of spatial explicitness. This would be bolstered by greater insight into the relationship between governance more generally and the biophysical environment.

Notes

1. All Bangladesh Acts and Laws mentioned in the text are available at http://bdlaws.minlaw.gov.bd/
2. The term '*shalish*' refers to a small-scale local council, distinct from the formal village court, which is convened for the purposes of civil and criminal conflict resolution. '*Samaj*' refers to the community that convenes the *shalish*. (For more information, see Bode and Howes 2002.)
3. See Transparency International www.transparency.org

References

Afroz, T., and S. Alam. 2013. Sustainable shrimp farming in Bangladesh: A quest for an integrated coastal zone management. *Ocean and Coastal Management* 71: 275–283. https://doi.org/10.1016/j.ocecoaman.2012.10.006.

Allan, A. 2017. Legal aspects of flood management. In *Routledge handbook of water law and policy*, ed. A. Rieu-Clarke, A. Allan, and S. Hendry. London: Earthscan.

Allan, A., A. Rieu-Clarke, C. Addoquaye Tagoe, S. Dey, A.K. Ghosh, M.S. Mondal, W. Nelson, M. Salehin, and C.L. Samling. 2015. *Governance analysis.* Working paper. Deltas, Vulnerability and Climate Change: Migration and Adaptation (DECCMA) project. www.deccma.com.

Arndt, C., and C. Oman. 2006. *Uses and abuses of governance indicators.* Paris: Organisation for Economic Co-operation and Development (OECD). http://www.oecd.org/dev/usesandabusesofgovernanceindicators.htm. Accessed 11 Jan 2017.

Bell, S., and S. Morse. 2008. *Sustainability indicators: Measuring the immeasurable?* 2nd ed. London: Earthscan.

Bode, B., and M. Howes. 2002. The north west institutional analysis. In *Go-interfish project report.* Dhaka: CARE Bangladesh. http://www.carebangladesh.org/shouhardoII/publication/Publication_4595642.pdf. Accessed 03 July 2017.

Butler, C.D., and W. Oluoch-Kosura. 2006. Linking future ecosystem services and future human well-being. *Ecology and Society* 11 (1): 30.

FAO. 2010. *On solid ground: Addressing land tenure issues following natural disasters.* Rome: Food and Agriculture Organization of the United Nations (FAO).

Farakka Treaty. 1996. *Treaty between the Government of the people's Republic of Bangladesh and the Government of the Republic of India on sharing of the Ganga/Ganges water at Farakka.* 36 I.L.M 523 (1997).

Fisher, B., R.K. Turner, and P. Morling. 2009. Defining and classifying ecosystem services for decision making. *Ecological Economics* 68 (3): 643–653. https://doi.org/10.1016/j.ecolecon.2008.09.014.

Freestone, D., M. Farooque, and S.R. Jahan. 1996. Legal implications of global climate change for Bangladesh. In *The implications of climate and sea-level change for Bangladesh*, ed. R.A. Warrick and Q.K. Ahmad, 289–334. Dordrecht: Springer.

Hill Clarvis, M., A. Allan, and D.M. Hannah. 2014. Water, resilience and the law: From general concepts and governance design principles to actionable mechanisms. *Environmental Science & Policy* 43: 98–110. https://doi.org/10.1016/j.envsci.2013.10.005.

ICCPR. 1966. *International covenant on civil and political rights.* United Nations. 999 UNTS 171.

ICESCR. 1966. *International covenant on economic social and cultural rights.* United Nations. 993 UNTS 3.

Iftekhar, M.S. 2010. Protecting the Sundarbans: An appraisal of national and international environmental laws. *Asia Pacific Journal of Environmental Law* 13 (2): 249–226.

Kaufmann, D., A. Kraay, and P. Zoido-Lobaton. 1999. *Aggregating governance indicators.* Policy research working paper no. 2195. Washington, DC: World Bank. http://info.worldbank.org/governance/wgi/pdf/govind.pdf. Accessed 31 May 2017.

Lewis, D., and A. Hossain. 2008. *Understanding the local power structure in rural Bangladesh.* SIDA studies no. 22. Stockholm: SIDA. http://personal.lse.ac.uk/lewisd/images/Lewis&H-SidaStudies-22.pdf. Accessed 31 May 2017.

MEA. 2005. Ecosystems and human well-being: Synthesis. In *Millennium ecosystem assessment (MEA).* Washington, DC: Island Press. http://www.millenniumassessment.org/documents/document.356.aspx.pdf. Accessed 01 Aug 2016.

Paavola, J., A. Gouldson, and T. Kluvánková-Oravská. 2009. Interplay of actors, scales, frameworks and regimes in the governance of biodiversity. *Environmental Policy and Governance* 19 (3): 148–158. https://doi.org/10.1002/eet.505.

Pelling, M., and K. Dill. 2010. Disaster politics: Tipping points for change in the adaptation of sociopolitical regimes. *Progress in Human Geography* 34 (1): 21–37. https://doi.org/10.1177/0309132509105004.

Rieu-Clarke, A., and C. Spray. 2013. Ecosystem services and international water law: Towards a more effective determination and implementation of equity? *Potchefstroom Electronic Law Journal* 16 (2): 12–65.

Salzman, J., B.H. Thompson Jr, and G.C. Daily. 2001. Protecting ecosystem services: Science, economics and law. *Stanford Environmental Law Journal* 20: 309–332.

UN. 1997. *Convention on the law of non-navigational uses of international watercourses.* United Nations. May 21 (in force 17 August 2014), reprinted in 36 I.L.M 700.

UNECE. 1992. *Convention on the protection and use of transboundary watercourses and international lakes.* United Nations Economic Commission for Europe. March 17 (in force 6 October 1996) 1936 UNTS 269; 31 ILM 1312.

UNISDR. 1996. *Yokohama strategy and plan of action for a safer world: Guidelines for natural disaster prevention, preparedness and mitigation. World conference on natural disaster reduction.* Geneva: United Nations Office for Disaster Risk Reduction (UNISDR). http://www.unisdr.org/files/8241_doc6841con-tenido1.pdf. Accessed 01 June 2017.

7

Health, Livelihood and Well-Being in the Coastal Delta of Bangladesh

Mohammed Mofizur Rahman and Sate Ahmad

7.1 Introduction

Bangladesh contains one of the largest, most densely populated and heavily farmed deltas of the world which is at great risk of increased flooding and submergence from sea-level rise (Chap. 3; Syvitski et al. 2009). Coastal Bangladesh is a focus for national and international research due to the diverse ecosystem services it provides along with the invaluable natural resources it carries—such as coral reef, fisheries, forest resources—and it includes the world's largest contiguous mangrove forest (Hoque and Datta 2005; Hossain 2001). Approximately 14 million or ten per cent of the total population of Bangladesh live in the south-west coast—the study area. Living in the delta means living on an ever-changing

M. M. Rahman (✉)
International Center for Diarrheal Disease Research, Bangladesh (icddr,b), Dhaka, Bangladesh

S. Ahmad
Faculty of Agricultural and Environmental Sciences, University of Rostock, Rostock, Germany

© The Author(s) 2018
R. J. Nicholls et al. (eds.), *Ecosystem Services for Well-Being in Deltas*,
https://doi.org/10.1007/978-3-319-71093-8_7

131

Fig. 7.1 The ever-changing coast of Bangladesh (Photograph: Mohammed Mofizur Rahman)

balance between land and sea (see Fig. 7.1) and fresh and saline water (van Schendel 2009). To meet the demands of such a large population, high levels of exploitation of natural resources and degradation of the environment have occurred (Hossain 2001).

The social and economical vulnerability of the south-west coast is due to its exposure to severe natural hazards. Amidst existing vulnerabilities, climate change is likely to impact human health directly or indirectly (Shahid 2009). Direct effects on health may be caused by extreme weather events such as cyclones (Shultz et al. 2005) and slow onset hazards such as salinity intrusion in soil, groundwater and surface water (Khan et al. 2011, 2014). Health is also indirectly influenced through changes in air pollution, spread of disease vectors, food and water insecurity, under-nutrition, displacement, changes in livelihoods and through infliction of mental ill-health (Haque et al. 2012, Kabir et al. 2016; Nahar et al. 2014).

7.2   Coastal Livelihoods and Relation to Health

Coastal Bangladesh offers a diverse range of livelihoods including coastal and marine fisheries, aquaculture (e.g. shrimp, crab culture), agriculture and forest resource harvesting. Besides these, other economic activities in the south-west coast include salt production, fish processing (e.g. dry-fish production) (Fig. 7.2), day labour, as well as tourism (mainly in the Sundarbans and Kuakata).

While it has been increasingly recognised that ill-health contributes to poverty by diminishing people's meagre resources (spent on health), it is rarely acknowledged that people often are not able to gain access to health services if they cannot mobilise vital livelihood resources (Obrist et al. 2007).

Fig. 7.2 Thousands of people are engaged in fish drying activities, especially women and children (Photograph: Mohammed Mofizur Rahman)

As such, health and livelihoods are interdependent; change in one is likely to change the status of the other. In the context of coastal regions, physical isolation may make some communities highly dependent on ecosystem services, thereby reducing their access to alternative livelihoods (Pomeroy et al. 2006). This can make them especially vulnerable to environmental shocks or change (e.g. salinity intrusion in coastal waters and soil). In addition coastal areas are also prone to land-use changes such as transformation from rice cultivation to shrimp culture (Ali 2006). Such changes may result in loss of livelihoods of local people (Deb 1998), thereby causing impoverishment, debt and the associated lack of access to healthcare. Moreover, poverty may affect health also by reducing people's *entitlement sets* needed to acquire food and thus resulting in undernourishment (Osmani 1993) (Fig. 7.3). Local environmental change may also result in a change in livelihoods and in some cases may influence migration decision making. Thus, this may have implications for the health and nutrition of such migrants at the destination.

Fig. 7.3 Dry fish is a relatively inexpensive source of protein in coastal Bangladesh (Photograph: Mohammed Mofizur Rahman)

7.3   Food Security and Nutrition in Relation to Health

The tide dominated coastal delta of Bangladesh is exposed to a wide variety of environmental stresses. These include rapid onset hazards such as storm surge, cyclones, river and coastal erosion as well as slow onset hazards such as salinity intrusion in groundwater as well as in surface water. Such hazards not only have direct impacts on mortality and morbidity but also may pose a threat to food security through the loss of arable land and agricultural crops and through loss of livelihoods (Shameem et al. 2014).

Food security plays an important role in the understanding of health and well-being (Jaron and Galal 2009). It is a prerequisite to good health as undernourishment and malnutrition are widely agreed to be the results of hunger and food insecurity. Undernourishment of mothers may result in low birth weight (Kramer 1987), while malnutrition of infants may inhibit brain growth and subsequent intellectual development (Stoch and Smythe 1963). Additionally, food security is a significant determinant of child growth in rural Bangladesh (Saha et al. 2009) as well as throughout the world (Frongillo et al. 1997).

Bangladesh has progressed significantly towards reducing food insecurity. However, despite this progress, domestic food production is increasingly unable to meet the demand of consumers for a more diverse diet which, in turn, has led to increasing food imports since 2006 (HKI and BIGH 2016). Food availability itself may not result in food security, and thus access to food is another important factor to take into account. According to the Food Security Nutritional Surveillance Project (FSNSP), as compared to the rest of the country, the coastal belt shows seasonality in household food insecurity, with highest prevalence (39 per cent of households in 2014) occurring during the monsoon season. Prevalence of food insecurity for the coastal belt is the second highest in comparison to other surveillance zones, second only to the northern charlands of Bangladesh (HKI and BIGH 2016) and significantly higher than the national average level of 23–26 per cent (in 2014) according to the same surveillance project.

7.4   Water and Health

In coastal Bangladesh, drinking water sources mostly include tube-wells of various depth, but also rainwater, rivers, canals and ponds. The community has access to different water sources depending on geographical availability, income and weather conditions (Scheelbeek 2015). These different sources have varying levels of sodium bacterial contamination and arsenic risk (Houghton 2009). Storage containers can also act as breeding grounds for mosquitoes and bacteria, which favour warmer waters, increasing risk of diarrhoeal disease. Saline tolerant mosquitoes have also been discovered which has implications on the risk from malaria and dengue (Yunus et al. 2011).

Salinity is an additional major problem in coastal waters, which has only recently come to light in the literature, with respect to its implications on health outcomes. As most of the population in south-west coastal Bangladesh depend on groundwater as the primary source of drinking water, salinity intrusion into coastal aquifers may have serious consequences on human health. Several studies have found that the levels of sodium in drinking water are much above the WHO standards (Khan et al. 2011). Salt is a well-known trigger of hypertension. Recent studies have raised concern over the increased risk of hypertension from exposure to salinity in coastal Bangladesh (Vineis et al. 2011) and additionally have found an association between drinking water salinity and high blood pressure (Talukder et al. 2017).

Increased salinity exposure through drinking, cooking and bathing has also reported to have potential links with skin diseases, acute respiratory infection, cardio-vascular diseases, strokes and diarrhoeal diseases (Vineis et al. 2011; Khan et al. 2011). Maternal health may also be impacted by increased exposure to salinity as it was found to be significantly associated with increased risk of (pre)eclampsia and gestational hypertension in south-west coastal Bangladesh (Khan et al. 2014).

A major cause for concern is that a positive feedback mechanism may exist whereby saline water consumption increases thirst leading to higher fluid intake, which if saline may result in even higher levels of thirst (Islam et al. 2011).

7.5 Health Risks from Natural Hazards

Bangladesh ranks fifth in the WorldRiskIndex for disaster risk related to natural disasters in 2016 (Comes et al. 2016). Around 26 per cent of the population are affected by cyclones, while 70 per cent live in flood-prone regions (Cash et al. 2013). The coastal delta is particularly vulnerable to cyclones and tidal surges owing to its low elevation (Chap. 8). Other hazards that the coastal area is exposed to include coastal and river erosion and water logging from polders.

Since the 1950s changes in extreme weather events have been observed (IPCC 2014). In Bangladesh weather and extreme weather were found to be associated with mortality (Lindeboom et al. 2012). In Abhoynagar, a rural coastal sub-district of Bangladesh, it was found that heavy rainfall was associated with mortality (Lindeboom et al. 2012).

Extreme precipitation events and higher temperatures are likely to cause widespread food-borne and water-borne disease outbreaks including cholera and other diarrheal diseases. Cholera (*Vibrio cholerae*) outbreaks have been shown to be linked to rising sea surface temperatures which trigger the release of cholera into the environment. Cholera risk may be exacerbated by coastal and estuarine water warming or by local flooding (Rodó et al. 2002; Lipp et al. 2002). Thus it was found that cholera risk in Bangladesh increases with the increase in sea surface temperature (Shahid 2009). Increased precipitation in the country has been also linked to increased non-cholera diarrhoea (Hashizume et al. 2007).

Bangladesh is highly prone to flooding because of its location at the confluence of the Ganges, Brahmaputra and Meghna (GBM) rivers and because of the hydro-meteorological and topographical characteristics of the GBM delta (Mirza 2002). On average, annual floods inundate around 20 per cent of the country and can reach up to 70 per cent during an extreme flood event (Mirza 2002). As a result about two-thirds of the tube-wells and all toilets have become unusable in Bangladesh (Annya et al. 2010). During floods water supply and sanitation condition becomes severely disrupted and allows the spread of various water-borne diseases.

Low-lying coastal regions across the world are prone to salt water intrusion due to sea-level rise (Chap. 17 and 18). In Bangladesh, the south-west coast is subject to additional drivers of salinity ingress such as brackish shrimp cultivation (Deb 1998) and upstream withdrawal and diversion of freshwater (Mirza 1998; Gain and Giupponi 2014). Such changes have implications for the population of the delta.

7.6   Gender and Health

Women, due to their role in the gendered division of labour, face more imminent health risks than men (Denton 2002). As mentioned earlier, coastal Bangladesh is particularly vulnerable to natural disasters and impacts of climate change. However, women and children are disproportionately affected by such impacts in rural coastal Bangladesh (Kapoor 2011; MoEF 2009). The gender-specific levels of control over assets, as well as roles and responsibilities in the household and society, imply distinct effects of shocks from climate and health events among men and women (Brody et al. 2008).

Globally, women and children are 14 times more likely to die than men as a result of a natural disaster (SIA 2006). In the 1991 cyclone of Bangladesh, female deaths were higher than those of men. Restrictions on women's spatial movement (social restrictions, gender-specific clothes, gendered role of caring for the young), fewer opportunities to access information on risk levels and ways to minimise risk and lack of agency with regard to decisions to evacuate have all been credited with contributing to these gender differences in mortality (Ikeda 1995). Similarly during the East Bengal cyclone of 1970, more deaths occurred in women, elderly people and children in Bangladesh (Sommer and Mosley 1972). For temperature-related mortality, gender and age-based differences have been observed across the whole country (Lindeboom et al. 2012).

Violation of women's rights becomes more prominent during disasters (Nasreen 2008). According to the World Disaster Report, women and girls are at higher risk of sexual violence, sexual exploitation and abuse, trafficking and domestic violence during natural disasters (International Federation for Red Cross and Red Crescent Societies [IFRC] 2007). Adolescent girls report especially high levels of sexual harassment and abuse in the aftermath of disasters and complain of the lack of privacy in emergency shelters (Bartlett 2008). Natural disasters also exacerbate child marriage in the coastal areas of Bangladesh; frequent flooding and river erosion means many families live with constant insecurity and variable poverty levels, which impact decisions about schooling and marriage for girls (HRW 2015).

Traditionally, the burden of coping with food shortages largely falls on the shoulders of women (Nasreen 2008). Both women and girls suffer more from shortages of food and economic resources in the aftermath of disasters (Neumayer and Plumper 2007). In Bangladesh, due to unequal food distribution in the family, a girl child receives 20 per cent fewer calories than a boy child (Ikeda 1995). The recent FSNSP study showed that of those who had to sleep hungry, 64 per cent were female children (up to four years old) compared to 24 per cent of male children (HKI and BIGH 2016).

Changes in seasons and climatic conditions imply that women and children in coastal areas may have to use water supplies from unsafe sources such as dirty ponds, which tend to dry up during the dry season, increasing their exposure to water-borne diseases such as diarrhoea. Furthermore, the increased time taken to fetch water may entail that young female household members are additionally required to help with household duties, increasing the likelihood of their missing school (Denton 2002). Moreover, women face respiratory problems caused by indoor pollution due to their direct contact with traditional fuels. In the event of a disaster, male children are likely to receive preferential treatment in rescue efforts (Neumayer and Plumper 2007). Vulnerability of women to certain diseases may also be high compared to men.

7.7 Conclusion

Bangladesh has progressed significantly in terms of family planning, food security, economic development, disaster preparedness and overall health (increased life expectancy, decreased maternal and infant mortality, etc.). However, considerable challenges remain for the coastal area which is highly exposed to natural disasters due its geophysical setting and social context. Cyclones, tidal flooding, coastal erosion, salt water intrusion, change in precipitation patterns, may directly or indirectly lead to adverse health outcomes, some of which are already being observed. The coastal areas have undergone a transition of livelihoods in recent decades accompanied by land-use changes. The health and livelihoods of the delta people are intertwined and impact on one may lead to impact on the other.

Therefore any loss or degradation of ecosystem services due to natural hazards or human induced land-use change may result in loss of livelihoods and income which may ultimately lead to worsened health and well-being. Food, water and nutritional security are other areas of concern, and thus more research needs to be carried out and interventions should be implemented. Health risks from certain non-communicable and infectious diseases are more likely to increase and thus demands extra attention from the government and health planners.

References

Ali, A.M.S. 2006. Rice to shrimp: Land use/land cover changes and soil degradation in Southwestern Bangladesh. *Land Use Policy* 23 (4): 421–435. https://doi.org/10.1016/j.landusepol.2005.02.001.

Annya, C.S., A.P. Gulsan, B. Chaitee, and S. Rajib. 2010. Impact and adaptation to flood: A focus on water supply, sanitation and health problems of rural community in Bangladesh. *Disaster Prevention and Management: An International Journal* 19 (3): 298–313. https://doi.org/10.1108/0965 3561011052484.

Bartlett, S. 2008. Climate change and urban children: Impacts and implications for adaptation in low-and middle-income countries. *Environment and Urbanization* 20 (2): 501–519.

Brody, A., J. Demetriades, and E. Esplen. 2008. Gender and climate change: Mapping the linkages; a scoping study on knowledge and gaps. Brighton: Institute of Development Studies. http://siteresources.worldbank.org/EXTSOCIALDEVELOPMENT/Resources/DFID_Gender_Climate_Change.pdf. Accessed 20 Apr 2017.

Cash, R.A., S.R. Halder, M. Husain, M.S. Islam, F.H. Mallick, M.A. May, M. Rahman, and M.A. Rahman. 2013. Reducing the health effect of natural hazards in Bangladesh. *The Lancet* 382 (9910): 2094–2103. https://doi.org/10.1016/s0140-6736(13)61948-0.

Comes, M., M. Dubbert, M. Garschagen, M. Hagenlocher, R. Sabelfeld, Y.J. Lee, L. Grunewald, M. Lanzendörfer, P. Mucke, O. Neuschäfer, S. Pott, J. Post, S. Schramm, D. Schumann-Bölsche, B. Vandemeulebroecke, T. Welle, and J. Birkmann. 2016. *World risk report 2016*. Germany: Bündnis Entwicklung Hilft, United Nations University – Institute for Environment and Human Security. http://weltrisikobericht.de/wp-content/uploads/2016/08/WorldRiskReport2016.pdf. Accessed 20 Apr 2017.

Deb, A.K. 1998. Fake blue revolution: Environmental and socio-economic impacts of shrimp culture in the coastal areas of Bangladesh. *Ocean and Coastal Management* 41 (1): 63–88. https://doi.org/10.1016/S0964-5691(98)00074-X.

Denton, F. 2002. Climate change vulnerability, impacts, and adaptation: Why does gender matter? *Gender and Development* 10 (2): 10–20. https://doi.org/10.1080/13552070215903.

Edenhofer, O., R. Pichs-Madruga, Y. Sokona, E. Farahani, S. Kadner, K. Seyboth, ... and B. Kriemann. (2014). *IPCC, 2014: Summary for policymakers*. Climate Change.

Frongillo, E.A., M. de Onis, and K.M.P. Hanson. 1997. Socioeconomic and demographic factors are associated with worldwide patterns of stunting and wasting of children. *Journal of Nutrition* 127 (12): 2302–2309.

Gain, A.K., and C. Giupponi. 2014. Impact of the Farakka Dam on thresholds of the hydrologic flow regime in the Lower Ganges River Basin (Bangladesh). *Water* 6 (8): 2501–2518. https://doi.org/10.3390/w6082501.

Haque, M.A., S.S. Yamamoto, A.A. Malik, and R. Sauerborn. 2012. Households' perception of climate change and human health risks: A community perspective. *Environmental Health* 11 (1): 1. https://doi.org/10.1186/1476-069X-11-1.

Hashizume, M., B. Armstrong, S. Hajat, Y. Wagatsuma, A.S.G. Faruque, T. Hayashi, and D.A. Sack. 2007. Association between climate variability and hospital visits for non-cholera diarrhoea in Bangladesh: Effects and vulnerable groups. *International Journal of Epidemiology* 36 (5): 1030–1037. https://doi.org/10.1093/ije/dym148.

HKI, and BIGH. 2016. *State of food security and nutrition in Bangladesh: 2014*. Dhaka: Helen Keller International (HKI) and BRAC Institute of Global Health (BIGH).

Hoque, A.K.F., and D.K. Datta. 2005. The mangroves of Bangladesh. *International Journal of Ecology and Environmental Sciences* 31 (3): 245–253.

Hossain, M.S. 2001. Biological aspects of the coastal and marine environment of Bangladesh. *Ocean and Coastal Management* 44 (3): 261–282. https://doi.org/10.1016/S0964-5691(01)00049-7.

Houghton, J.T. 2009. *Global warming: The complete briefing*. 4th ed. Cambridge, UK: Cambridge University Press.

HRW. 2015. *Marry before your house is swept away: Child marriage in Bangladesh*. New York: Human Rights Watch (HRW). https://www.hrw.org/sites/default/files/report_pdf/bangladesh0615_web.pdf. Accessed 20 Apr 2017.

Ikeda, K. 1995. Gender differences in human loss and vulnerability in natural disasters: A case study from Bangladesh. *Bulletin (Centre for Women's Development Studies)* 2 (2): 171–193.

International Federation for Red Cross and Red Crescent Societies [IFRC]. 2007. *World disaster report 2007: Focus on discrimination,* Geneva. http://www.ifrc.org/Global/Publications/disasters/WDR/WDR2007-English.pdf.

Islam, M., H. Sakakibara, M. Karim, and M. Sekine. 2011. *Rural water consumption behavior: A case study in southwest coastal area, Bangladesh.* Proceedings of the 2011 World Environmental and Water Resources Congress, May 22–26, Palm Springs.

Jaron, D., and O. Galal. 2009. Food security and population health and well being. *Asia Pacific Journal of Clinical Nutrition* 18 (4): 684–687.

Kabir, M.I., M.B. Rahman, W. Smith, M.A.F. Lusha, and A.H. Milton. 2016. Climate change and health in Bangladesh: A baseline cross-sectional survey. *Global Health Action* 9. https://doi.org/10.3402/Gha.V9.29609.

Kapoor, A. 2011. *Engendering the climate for change: Policies and practices for gender-just adaptation.* New Delhi: Alternative Futures.

Khan, A.E., A. Ireson, S. Kovats, S.K. Mojumder, A. Khusru, A. Rahman, and P. Vineis. 2011. Drinking water salinity and maternal health in Coastal Bangladesh: Implications of climate change. *Environmental Health Perspectives.* https://doi.org/10.1289/ehp.1002804.

Khan, A.E., P.F. Scheelbeek, A.B. Shilpi, Q. Chan, S.K. Mojumder, A. Rahman, A. Haines, and P. Vineis. 2014. Salinity in drinking water and the risk of (pre)eclampsia and gestational hypertension in coastal Bangladesh: A case-control study. *PLoS One* 9 (9): e108715. https://doi.org/10.1371/journal.pone.0108715.

Kramer, M.S. 1987. Determinants of low birth-weight – Methodological assessment and meta-analysis. *Bulletin of the World Health Organization* 65 (5): 663–737.

Lindeboom, W., N. Alam, D. Begum, and P.K. Streatfield. 2012. The association of meteorological factors and mortality in rural Bangladesh, 1983–2009. *Global Health Action* 5 (s1): 61–73. https://doi.org/10.3402/Gha.V5i0.19063.

Lipp, E.K., A. Huq, and R.R. Colwell. 2002. Effects of global climate on infectious disease: The cholera model. *Clinical Microbiology Reviews* 15 (4): 757–770. https://doi.org/10.1128/CMR.15.4.757-770.2002.

Mirza, M.M.Q. 1998. Diversion of the Ganges water at Farakka and its effects on salinity in Bangladesh. *Environmental Management* 22 (5): 711–722. https://doi.org/10.1007/s002679900141.

————. 2002. Global warming and changes in the probability of occurrence of floods in Bangladesh and implications. *Global Environmental Change* 12 (2): 127–138. https://doi.org/10.1016/S0959-3780(02)00002-X.

MoEF. 2009. *National Adaptation Programme of Action (NAPA)*. Update of 2005 report. Ministry of Environment and Forests (MoEF), Government of the People's Republic of Bangladesh. http://www.climatechangecell.org.bd/Documents/NAPA%20october%202009.pdf. Accessed 01 Aug 2016.

Nahar, N., Y. Blomstedt, B. Wu, I. Kandarina, L. Trisnantoro, and J. Kinsman. 2014. Increasing the provision of mental health care for vulnerable, disaster-affected people in Bangladesh. *BMC Public Health* 14 (1): 708. https://doi.org/10.1186/1471-2458-14-708.

Nasreen, M. 2008. *Impact of climate change on food security in Bangladesh: Gender and disaster perspectives*. International symposium on climate change and food cecurity in South Asia, Dhaka.

Neumayer, E., and T. Plumper. 2007. The gendered nature of natural disasters: The impact of catastrophic events on the gender gap in life expectancy, 1981–2002. *Annals of the Association of American Geographers* 97 (3): 551–566. https://doi.org/10.1111/j.1467-8306.2007.00563.x.

Obrist, B., N. Iteba, C. Lengeler, A. Makemba, C. Mshana, R. Nathan, S. Alba, A. Dillip, M.W. Hetzel, and I. Mayumana. 2007. Access to health care in contexts of livelihood insecurity: A framework for analysis and action. *PLoS Medicine* 4 (10): e308. https://doi.org/10.1371/journal.pmed.0040308.

Osmani, S.R. 1993. *The entitlement approach to famine: An assessment*. Working paper 107. Helsinki: UNU World Institute for Development Economics Research. https://www.wider.unu.edu/sites/default/files/WP107.pdf.Licence 107. Accessed 20 Apr 2017.

Pomeroy, R.S., B.D. Ratner, S.J. Hall, J. Pimoljinda, and V. Vivekanandan. 2006. Coping with disaster: Rehabilitating coastal livelihoods and communities. *Marine Policy* 30 (6): 786–793. https://doi.org/10.1016/j.marpol.2006.02.003.

Rodó, X., M. Pascual, G. Fuchs, and A. Faruque. 2002. ENSO and cholera: A nonstationary link related to climate change? *Proceedings of the National Academy of Sciences* 99 (20): 12901–12906. https://doi.org/10.1073/pnas.182203999.

Saha, K.K., E.A. Frongillo, D.S. Alam, S.E. Arifeen, L.A. Persson, and K.M. Rasmussen. 2009. Household food security is associated with growth of infants and young children in rural Bangladesh. *Public Health Nutrition* 12 (9): 1556–1562. https://doi.org/10.1017/S1368980009004765.

Scheelbeek, P. 2015. *Hypertensive diseases in salinity prone coastal areas*. PhD thesis, Imperial College, London.

Shahid, S. 2009. Probable impacts of climate change on public health in Bangladesh. *Asia-Pacific Journal of Public Health* 22 (3): 310–319. https://doi.org/10.1177/1010539509335499.

Shameem, M.I.M., S. Momtaz, and R. Rauscher. 2014. Vulnerability of rural livelihoods to multiple stressors: A case study from the southwest coastal region of Bangladesh. *Ocean and Coastal Management* 102: 79–87. https://doi.org/10.1016/j.ocecoaman.2014.09.002.

Shultz, J.M., J. Russell, and Z. Espinel. 2005. Epidemiology of tropical cyclones: The dynamics of disaster, disease, and development. *Epidemiologic Reviews* 27 (1): 21–35.

SIA. 2006. *Reaching out to women when disaster strikes.* White paper – Updated May 2011. Philadelphia: Soroptimist International of the Americas. http://www.soroptimist.org/whitepapers/whitepaperdocs/wpreachingwomendisaster.pdf. Accessed 20 Apr 2017.

Sommer, A., and W. Mosley. 1972. East Bengal cyclone of November, 1970: Epidemiological approach to disaster assessment. *The Lancet* 299 (7759): 1030–1036.

Stoch, M.B., and P.M. Smythe. 1963. Does undernutrition during infancy inhibit brain growth and subsequent intellectual development. *Archives of Disease in Childhood* 38 (202): 546.

Syvitski, J.P.M., A.J. Kettner, I. Overeem, E.W.H. Hutton, M.T. Hannon, G.R. Brakenridge, J. Day, C. Vorosmarty, Y. Saito, L. Giosan, and R.J. Nicholls. 2009. Sinking deltas due to human activities. *Nature Geoscience* 2 (10): 681–686. https://doi.org/10.1038/ngeo629.

Talukder, M.R.R., S. Rutherford, C. Huang, D. Phung, M.Z. Islam, and C. Chu. 2017. Drinking water salinity and risk of hypertension: A systematic review and meta-analysis. *Archives of Environmental and Occupational Health* 72 (3): 1–13. https://doi.org/10.1080/19338244.2016.1175413.

van Schendel, W. 2009. *A history of Bangladesh.* Cambridge, UK: Cambridge University Press.

Vineis, P., Q. Chan, and A. Khan. 2011. Climate change impacts on water salinity and health. *Journal of Epidemiology and Global Health* 1 (1): 5–10. https://doi.org/10.1016/j.jegh.2011.09.001.

Yunus, M., N. Sohel, S.K. Hore, and M. Rahman. 2011. Arsenic exposure and adverse health effects: A review of recent findings from arsenic and health studies in Matlab, Bangladesh. *The Kaohsiung Journal of Medical Sciences* 27 (9): 371–376. https://doi.org/10.1016/j.kjms.2011.05.012.

8

Floods and the Ganges-Brahmaputra-Meghna Delta

Anisul Haque and Robert J. Nicholls

8.1 Introduction

Bangladesh is a highly flood prone country, reflecting the strongly seasonal regional climate and monsoon run-off of three large rivers from the Himalayas (Brammer 1990; Hofer and Messerli 2006; Brammer 2014), heavy local precipitation during the monsoon and tropical cyclones in the Bay of Bengal (Nicholls 2006). As a result, flooding can occur for multiple reasons, posing a threat to life and damage to economic assets. Of relevance to this analysis, floods can cause damage to ecosystem services, particularly agriculture and associated livelihoods.

In coastal Bangladesh, floods are related to a number of inter-related physical processes, which can be classified according to the main driver: (i) fluvial

A. Haque (✉)
Institute of Water and Flood Management, Bangladesh University of
Engineering and Technology, Dhaka, Bangladesh

R. J. Nicholls
Faculty of Engineering and the Environment and Tyndall Centre for Climate
Change Research, University of Southampton, Southampton, UK

© The Author(s) 2018
R. J. Nicholls et al. (eds.), *Ecosystem Services for Well-Being in Deltas*,
https://doi.org/10.1007/978-3-319-71093-8_8

floods, (ii) tidal floods, (iii) fluvio-tidal floods and (iv) storm surge floods. While all types of floods can cause damage and disruption, coastal Bangladesh is best known for storm surge flooding due to large historic death toll associated with some of these events most notably between 1970 and 1991 (Nicholls 2006; Alam and Dominey-Howes 2015; Lumbroso et al. 2017). Fluvial and fluvio-tidal flooding is largely dictated by the flow in the major rivers of the delta due to precipitation and run-off upstream. The study area in this research (see Chap. 4, Fig. 4.2) has a large tidal range (three to six metres) which can cause tidal flooding in unprotected (un-poldered) areas. The magnitude of storm surge flooding due to tropical cyclones is also related to the tide; landfall of a cyclone during high tide can cause more extensive storm surge flooding, while cyclone landfall at low tide may not be noticed in terms of water level. The consequences for the resident population are therefore varied.

Coastal Bangladesh has an extensive system of coastal embankments and polders built since the 1960s with the goal to reduce flooding, manage water levels and enhance agriculture. While reducing the extent and frequency of coastal flooding, these have greatly modified the flood characteristics and associated coastal morphodynamics and rates of subsidence (e.g. Auerbach et al. 2015). Moreover, there are plans for substantial upgrade to some embankments as part of the Bangladesh Delta Plan 2100 (BanDuDeltAS 2014). In this chapter, a general overview of the four types of floods defined above is provided, followed by a summary and refection on the need for further research on flooding in coastal Bangladesh. More details of the flood modelling conducted in this research are given in Chap. 16.

8.2   Fluvial Floods

The Ganges, Brahmaputra and Meghna (GBM) basin and numerous minor rivers provide the main freshwater flow into the northern and central part of Bangladesh (see Fig. 8.1). During the monsoon period (June–October), the volume of water often exceeds the carrying capacity of these rivers, and fluvial flooding occurs.

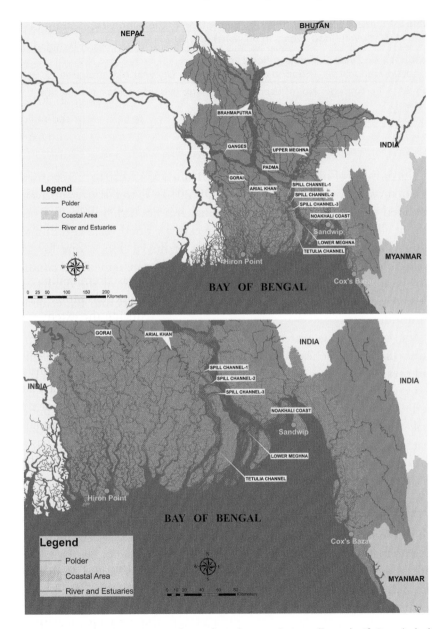

Fig. 8.1 Major river systems (*upper*) and coastal area (*lower*) of Bangladesh showing polders, river systems, main fluvial inflow inlets and tide gauge locations

Table 8.1 Fluvial flooding extents, as percentage of the national area, from 1954 to 2014 (60 fluvial flood events in 60 years) (BWDB 2015)

Extreme flood year (flooded area >24%)		Average flood year (flooded area 20–24%)		Dry year (flooded area <20%)	
No. of events	Percentage	No. of events	Percentage	No. of events	Percentage
15	25	9	15	36	60

The number of historical flood events from 1954 to 2014, categorised by extent, are shown in Table 8.1 (BWDB 2015). Extreme, dry and average flood years are defined by the national area flooded. Twenty-five per cent of all flood events occurred during extreme flood years and caused significant damage. Two typical examples of fluvial floods, one for an average flood condition (year 2000) and one for an extreme flood condition (year 1998), are shown in Fig. 8.2. The location of the discharge measurement stations are the entry points of the rivers into Bangladesh from India.

The Brahmaputra water level starts rising from the early monsoon (June–July) and reaches its first peak in the third week of July. It then falls and rises again and, in an average flood year, attains its second peak in the first week of August. The Ganges has a single peak of flood level occurring in the second week of September. For the Upper Meghna, the first flood peak occurs in the second or third week of May, and, in an average flood year, a second peak occurs close to the second peak of the Brahmaputra. The effect of flow from the Upper Meghna on overall flooding is small as its discharge only represents ten per cent of the combined discharges of Ganges and Brahmaputra Rivers (Islam et al. 2010). The different timing of the flood peaks of the major rivers are mainly due to variations in rainfall in the upper catchments and the travel time to reach the discharge measurement points considered. Synchronisation of the peaks across the three rivers is rare, but has occurred: the floods of 1988 and 1998 (see Fig. 8.2) being examples (Islam and Chowdhury 2002). During peak synchronisation, the second peak of the Brahmaputra is delayed or may occur as a third peak coinciding with the single peak of the Ganges. This triggers significant fluvial flooding in Bangladesh.

Average Flood Year

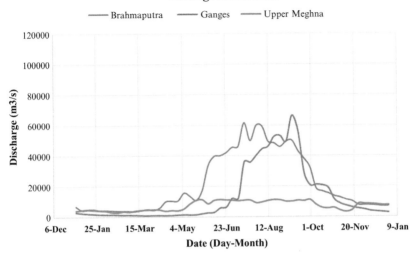

Extreme Flood Year

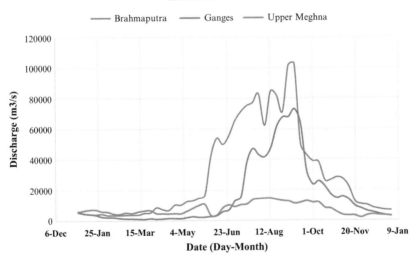

Fig. 8.2 Examples of typical flood peaks for the major rivers of Bangladesh in (*upper*) an average flood year (year 2000) and (*lower*) an extreme flood year (year 1998) (Data source: Bangladesh Water Development Board (BWDB))

Most fluvial flow adjacent to the study area occurs through the Lower Meghna estuary (MoWR 2005; Haque et al. 2016). This can cause inundation directly through the Lower Meghna, or via the estuarine networks that ultimately drains the combined flows of the three major rivers (Fig. 8.1). The drainage rate of fluvial flood water is largely dependent on sea levels. An elevated sea level due to either spring tides or a sustained monsoon may slow drainage prolonging flood duration, as happened during the 1998 flood (Haque et al. 2002). Over the longer term, subsidence and sea-level rise will increase mean sea levels (see Chaps. 14 and 16), influencing the fluvial flooding pattern in the coastal region.

There are few channels to take fluvial flow from the upper part of the country into the study area (Fig. 8.1), and most of these channels are restricted due to sedimentation. This limits flow and hence fluvial flooding. In addition, when polder embankments are considered (see Fig. 8.1 (lower)), only a small part of the study area is susceptible to fluvial flooding as they significantly determine flood extents. The maximum, minimum and average polder heights in the study region are 5.75, 4.50 and 4.79 m, respectively. Areas within the embankments are generally therefore not flooded during average fluvial events by overtopping (because flood water levels rarely exceed the embankment height), but there are incidents when areas within the embankments are flooded due to polder breaching (because embankments are not strong enough to resist the thrust of the fluvial flood). During extreme fluvial events (e.g. floods of 1998), both overtopping and breaching of polders happens.

8.3   Tidal Floods

Tides along the Bangladesh coast are semi-diurnal. The 18.6-year lunar nodal cycle has no influence along the Bangladesh coast, but the 4.4-year lunar perigean cycle modulates the tide by 4 cm (Sumaiya 2017). Tidal range is greatest along the Noakhali coast, immediately east of the Lower Meghna estuary (see Fig. 8.1), and declines to the east and west (Ahmed and Louters 1997). The mean tidal range at Hiron Point near the Sundarbans (west coast) is around 3 m, increases up to 6 m near Sandwip (west coast) and then decreases to 3.6 m further east (Cox's Bazar).

Tides propagate up to 100 km inland along coastal Bangladesh's estuaries (Choudhury and Haque 1990). Coastal flooding due to tides is more pronounced in the central and eastern parts of the coast compared to west coast, reflecting the tidal range.

The construction of polders since the 1960s has eliminated significant tidal flooding in these areas. However, the polders also significantly reduce the area available for sediment deposition. The entire sediment load coming from the catchments of GBM basins drains into the Bay of Bengal through the different estuaries of the coastal region (see Chap. 15). Sedimentation of the river bed has reduced the capacity of the channels (Haque et al. 2016). At high tide, water levels can easily overtop the estuary bank and flood any unprotected land. During extreme fluvial flood conditions, high tides can cause overtopping and/or breaching of the polders. Once overtopped, the flood water inside the polder is unable to drain due to the difference in land elevation caused by confined sedimentation. This creates water logging inside the polder with negative implications for agriculture and other land uses. However, water logging inside polders also happens due to flooding from internal canals (without polder overtopping and/or breaching) and drainage congestion due to unplanned road networks and confined sedimentation.

Monsoon winds need to be considered when assessing tidal flooding in coastal regions. Along the Bangladesh coast, south-westerly and south-easterly winds during the monsoon season (June–October) are termed the monsoon wind. If they exceed 10 m/s, this raises the sea surface at the coast by 0.45 m in the west to 1.65 m in the east (Sumaiya 2017). This slows the fall of the tide and prolongs high water levels (Haque et al. 2002). Hence a combination of the highest astronomical tides and strong sustained monsoon winds give the highest potential for severe tidal flooding (Sumaiya 2017).

8.4   Fluvio-Tidal Floods

Fluvio-tidal flooding is compound flooding caused by the combination of fluvial flows and high tides. This is important in coastal Bangladesh, but has not been described previously. A typical hydrograph of fluvio-tidal

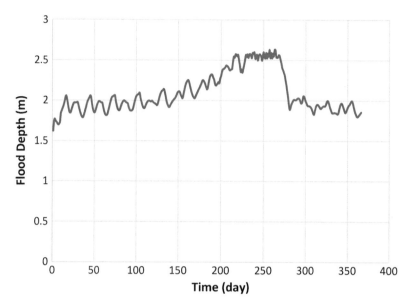

Fig. 8.3 A typical flood hydrograph showing the fluvio-tidal characteristic of floods (Source: model simulation)

flooding shows both rising and falling due to fluvial effects from the monsoon combined with tidal variation due to tidal effects (see Fig. 8.3). Where the fluvial flows are highest at the end of the monsoon, the tidal range is reduced and tidal flooding is reduced as around day 250 in Fig. 8.3.

Figure 8.1 indicates the main channels controlling fluvial flow. During the monsoon, there are large freshwater flows along the Lower Meghna. These enter the study area along three spill channels and then flow south along the Tetulia channel, which is situated to the west of Lower Meghna. The other potential channels from the Meghna into the study area are the Gorai and Arial Khan, but currently these have limited conveyance restricting freshwater flow. The coastal area does not therefore receive any significant fluvial flow during the monsoon except adjacent to the Lower Meghna. Hence, tidal fluctuation plays a dominant role in determining the pattern of flooding across coastal Bangladesh. Due to the dynamic interaction between the fluvial flow and tides, and considering the flood versus ebb dominance of the estuaries (see Choudhury and Haque 1990), the landward part of coastal Bangladesh is characterised by fluvial flood, the middle part is characterised by fluvio-tidal floods and the seaward part is characterised by tidal flood. The exception is around the Lower Meghna estuary. Here flooding is always fluvially dominated by the large Meghna flow as it dominates tidal action during the monsoon.

8.5   Storm Surge Floods

The coastal region of Bangladesh is well known for storm surge flooding with major events going back several centuries or more (Alam and Dominey-Howes 2015); it dominates the global death toll due to storm surges over the last century (Nicholls 2006). The primary cause of storm surge flooding is the landfall of tropical cyclones—the magnitude, location and extent of flooding depend largely on cyclone intensity, its landfall location and time of landfall relative to the tide. The event has two inter-related components. One is the high wind speed of the cyclone leading to storm surge, and the other is the resulting land inundation from the sea and rivers/estuaries. Note that an intense cyclone does not always result in significant flood event. For example, in May 1997, a cyclone with wind speeds of 275 km/hr made landfall near the Noakhali-Chittagong coastline. This is the highest wind speed ever recorded in Bangladesh, and the landfall location is one of the most vulnerable locations for cyclone landfall. However, landfall occurred during low tide, and there was no inundation and only 155 people died. In 1991, when a cyclone with wind speeds 50 km/hr lower (225 km/hr) made landfall almost at the same location as the 1997 cyclone, there were about 138,000 deaths (Ali 1999; Dube et al. 2004); one of the most devastating cyclones in terms of human consequences in Bangladesh's history. The variable that distinguished these two cyclones was landfall timing. The 1997 cyclone occurred at low tide, while the 1991 cyclone occurred at high tide, generating extreme water levels exceeding 8 m in a few places. For the Bangladesh coast, when a cyclone makes landfall during high tide, it has more potential for generating extensive storm surge flooding and associated damage.

Since 1991, major efforts have been made to mitigate these surge events (Lumbroso et al. 2017). These include flood forecasts and warning systems and the construction of robust surge shelters where the resident population can take refuge during these events. As a result fatalities have been greatly reduced by two orders of magnitude during recent major cyclones (e.g. Cyclone Sidr in 2007) compared to earlier events.

8.6   Summary

As coastal Bangladesh is the drainage route of the fluvial flows of the Ganges-Brahmaputra-Meghna River systems, the nature of fluvial floods in the region depends on the flooding patterns from these major rivers. The tidal range is large (between 3 and 6 m) and increases from west to east. Due to high tides, the unprotected land along the coast (outside the polders) is regularly flooded at high tide. Extreme astronomical tides combined with strong monsoons intensify the flooding situation; historical data shows that sea levels generated by tropical cyclones making landfall at high tide have the most impact on flood extents and associated damage. Polders and coastal embankments have stopped most fluvial and tidal floods in coastal Bangladesh. In the case of surges, forecasts, warnings and shelters have collectively great reduced fatalities during cyclones.

This chapter has focussed on describing the sources of flooding in coastal Bangladesh individually. As indicated in this chapter, it is increasingly recognised that floods occur due to multiple causes—so-called compound flooding (Leonard et al. 2014). Generally, it may therefore be more beneficial to consider flooding for Bangladesh in a more systemic manner: considering all sources, pathways, receptors and consequences. This allows analysis of the changing flood system, as illustrated for the United Kingdom by Evans et al. (2004a, b), to support national policy. Some examples of compound events have been shown here in terms of fluvial, tidal and surge events interacting. In the future, it would be useful to use this type of approach to analyse sources of coastal flooding in Bangladesh. The effects of polders on tides, morphodynamics and land elevation (subsidence) also deserve more attention.

References

Ahmed, S., and T. Louters. 1997. Residual flow volume and sediment transport patterns in the Lower Meghna Estuary during pre-monsoon and post-monsoon: An analysis of available LRP data collected during 1986–1994. *Meghna Estuary study technical note*. Dhaka: Bangladesh Water Development Board (BWDB).

Alam, E., and D. Dominey-Howes. 2015. A new catalogue of tropical cyclones of the Northern Bay of Bengal and the distribution and effects of selected landfalling events in Bangladesh. *International Journal of Climatology* 35 (6): 801–835. https://doi.org/10.1002/joc.4035.

Ali, A. 1999. Climate change impacts and adaptation assessment in Bangladesh. *Climate Research* 12 (2–3): 109–116. https://doi.org/10.3354/cr012109.

Auerbach, L.W., S.L. Goodbred, D.R. Mondal, C.A. Wilson, K.R. Ahmed, K. Roy, M.S. Steckler, C. Small, J.M. Gilligan, and B.A. Ackerly. 2015. Flood risk of natural and embanked landscapes on the Ganges-Brahmaputra tidal delta plain. *Nature Climate Change* 5 (2): 153–157. https://doi.org/10.1038/nclimate2472.

BanDuDeltAS. 2014. *Inception report. Bangladesh delta plan 2100 formulation project*. Dhaka: General Economics Division (GED), Planning Commission, Government of the People's Republic of Bangladesh.

Brammer, H. 1990. Floods in Bangladesh: Geographical background to the 1987 and 1988 floods. *The Geographical Journal* 156 (1): 12–22. https://doi.org/10.2307/635431.

———. 2014. Bangladesh's dynamic coastal regions and sea-level rise. *Climate Risk Management* 1: 51–62. https://doi.org/10.1016/j.crm.2013.10.001.

BWDB. 2015. *Annual flood report 2015*. Dhaka: Bangladesh Water Development Board (BWDB). http://www.ffwc.gov.bd/index.php/reports/annual-flood-reports. Accessed 9 May 2016.

Choudhury, J.U., and A. Haque. 1990. Permissible water withdrawal based upon prediction of salt-water intrusion in the Meghna delta. *The hydrological basis for water resources management proceedings of the Beijing symposium*. Publication no. 197. Wallingford: International Association of Hydrological Sciences (IAHS).

Dube, S.K., P. Chittibabu, P.C. Sinha, A.D. Rao, and T.S. Murty. 2004. Numerical modelling of storm surge in the head Bay of Bengal using location specific model. *Natural Hazards* 31 (2): 437–453. https://doi.org/10.1023/B:NHAZ.0000023361.94609.4a.

Evans, E.P., R.M. Ashley, J. Hall, E. Penning-Rowsell, A. Saul, P. Sayers, C. Thorne, and A. Watkinson. 2004a. *Foresight; future flooding. Scientific summary, volume I: Future risks and their drivers.* London: Office of Science and Technology.

Evans, E.P., R.M. Ashley, J. Hall, E. Penning-Rowsell, P. Sayers, C. Thorne, and W. Watkinson. 2004b. *Foresight; future flooding. Scientific summary, volume II: Managing future risks.* London: Office of Science and Technology.

Haque, A., M. Salehin, and J.U. Chowdhury. 2002. Effects of coastal phenomena on the 1998 flood. In *Engineering concerns of flood: A 1998 perspective*, ed. M.A. Ali, S.M. Seraj, and S. Ahmad. Dhaka: Directorate of Advisory, Extension and Research Services, Bangladesh University of Engineering and Technology.

Haque, A., S. Sumaiya, and M. Rahman. 2016. Flow distribution and sediment transport mechanism in the estuarine systems of Ganges-Brahmaputra-Meghna Delta. *International Journal of Environmental Science and Development* 7 (1): 22–30. https://doi.org/10.7763/IJESD.2016.V7.735.

Hofer, T., and B. Messerli. 2006. *Floods in Bangladesh: History, dynamics and rethinking the role of the Himalayas.* Paris: United Nations University Press. http://archive.unu.edu/unupress/sample-chapters/floods_in_Bangladesh_web.pdf. Accessed 09 Aug 2017.

Islam, A.K.M.S., and J.U. Chowdhury. 2002. Hydrologic characteristics of the 1998 flood in major rivers. In *Engineering concerns of flood: A 1998 perspective*, ed. M.A. Ali, S.M. Seraj, and S. Ahmad. Dhaka: Directorate of Advisory, Extension and Research Services, Bangladesh University of Engineering and Technology.

Islam, A.S., A. Haque, and S.K. Bala. 2010. Hydrologic characteristics of floods in Ganges–Brahmaputra–Meghna (GBM) Delta. *Natural Hazards* 54 (3): 797–811. https://doi.org/10.1007/s11069-010-9504-y. Licence journal article.

Leonard, M., S. Westra, A. Phatak, M. Lambert, B. van den Hurk, K. McInnes, J. Risbey, S. Schuster, D. Jakob, and M. Stafford-Smith. 2014. A compound event framework for understanding extreme impacts. *Wiley Interdisciplinary Reviews: Climate Change* 5 (1): 113–128. https://doi.org/10.1002/wcc.252.

Lumbroso, D.M., N.R. Suckall, R.J. Nicholls, and K.D. White. 2017. Enhancing resilience to coastal flooding from severe storms in the USA: International lessons. *Natural Hazards and Earth Systems Sciences* 17: 1357–1373. https://doi.org/10.5194/nhess-17-1-2017.

MoWR. 2005. *Coastal zone policy*. Dhaka: Ministry of Water Resources (MoWR), Government of the People's Republic of Bangladesh. http://lib.pmo.gov.bd/legalms/pdf/Costal-Zone-Policy-2005.pdf. Accessed 20 Apr 2017.

Nicholls, R.J. 2006. Storm surges in coastal areas. Natural disaster hotspots: Case studies. In *Disaster risk management 6*, ed. M. Arnold, R.S. Chen, U. Deichmann, M. Dilley, A.L. Lerner-Lam, R.E. Pullen, and Z. Trohanis, 79–108. Washington, DC: World Bank.

Sumaiya. 2017. *Impacts of dynamic interaction between astronomical tides and monsoon wind on coastal flooding in Bangladesh*. M.Sc thesis, Institute of Water and Flood Management (IWFM), Bangladesh University of Engineering and Technology (BUET), Dhaka.

Part 3

Scenarios for Policy Analysis

9

Integrating Science and Policy Through Stakeholder-Engaged Scenarios

Emily J. Barbour, Andrew Allan, Mashfiqus Salehin, John Caesar, Robert J. Nicholls, and Craig W. Hutton

9.1 Introduction

Scenarios are widely used to explore plausible future alternatives given the high degree of uncertainty in future projections (see Mahmoud et al. 2009; Rounsevell and Metzger 2010 for reviews). They can be effective in guiding planning strategies through identifying shared visions of future outcomes, as well as assessing the effectiveness of different interventions

E. J. Barbour (✉)
School of Geography and the Environment, University of Oxford, Oxford, UK

A. Allan
School of Law, University of Dundee, Dundee, UK

M. Salehin
Institute of Water and Flood Management, Bangladesh University of Engineering and Technology, Dhaka, Bangladesh

J. Caesar
Met Office Hadley Centre for Climate Science and Services, Exeter, Devon, UK

© The Author(s) 2018
R. J. Nicholls et al. (eds.), *Ecosystem Services for Well-Being in Deltas*,
https://doi.org/10.1007/978-3-319-71093-8_9

in terms of performance against future uncertainties. Scenarios already form a key part of Bangladesh's planning process through the development of Vision 2021 and the associated Perspective Plan, which map out a desired scenario for Bangladesh in 2021 (GED 2012). Further, scenarios are a core element of the developing Bangladesh Delta Plan 2100, which provides a long-term adaptive and integrative planning framework up to 2050 and 2100 (BanDuDeltAS 2014; GED 2015).

This research adopts a scenario process which draws upon the latest global climate change community scenarios which consider climate emissions and socio-economic change (Moss et al. 2010; Kriegler et al. 2012; IPCC 2014). These global scenarios are modified to reflect local issues which are of specific relevance to stakeholders and decision makers. The project combines three climate change scenarios with three socio-economic scenarios to cover a range of plausible futures for coastal Bangladesh. The combination of these nine scenarios was aimed at identifying a range of possible future change, as well as investigating the effectiveness of different management interventions. Most importantly, the development of these scenarios was aimed at generating a dialogue across institutions and sectors to create a shared vision of the future and address existing challenges in a holistic and integrated way.

A critical component of the scenario development process has been stakeholder involvement as part of an iterative learning loop (Fig. 9.1). As described in Chap. 4, the iterative learning loop involved co-development of qualitative and quantitative scenarios with stakeholders using an iterative process. Around 60 institutions (including government authorities, multi- and bi-lateral donors, local and international non-governmental organisations (NGOs), academic institutions and individual experts) have been involved in identifying key issues, developing baseline scenarios and identifying management interventions over a series of stakeholder

R. J. Nicholls
Faculty of Engineering and the Environment and Tyndall Centre for Climate Change Research, University of Southampton, Southampton, UK

C. W. Hutton
Geodata Institute, Geography and Environment, University of Southampton, Southampton, UK

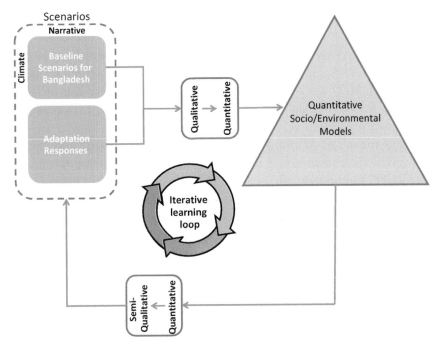

Fig. 9.1 Developing qualitative and quantitative scenarios through an iterative learning loop with project partners and stakeholders (Adapted from Nicholls et al. 2016)

workshops. This process has created three plausible future storylines considering governance, institutional change, economics, social values and the natural environment. It has enabled the quantification of these scenarios in a range of integrated detailed biophysical and socio-economic models which can be used to identify pathways out of poverty and improved well-being. Furthermore, it has facilitated the cross-institutional and cross-sectoral discussions needed to address and manage such complex systems (Holling 1978; Liu et al. 2007).

The following sections provide further details of the scenario process, beginning with an overview of the scenario framework, followed by an outline of the climate and socio-economic scenarios, and finishing with key outcomes. The results of the scenario analysis using the individual and integrated models are described in Chaps. 13, 14, 15, 16, 17, 18, 19, 20, 21, 22, 23, 24, 25, 26, 27, and 28, whilst further detail regarding stakeholder engagement is provided in Chap. 10.

9.2   Scenario Framework

The scenarios adopted combine three socio-economic development pathways up to 2050 and three alternative future climates up to 2099 (Table 9.1). 2050 was selected as the limit of the socio-economic projections due to both the high level of uncertainty in projecting such changes to longer time frames and to maximise consistency with local strategic planning timescales. Whilst there is also significant uncertainty in changes in climate, for known inputs the existing models allow such projections up to 2099 (and potentially longer). The climate projections represent a range of plausible changes in temperature and precipitation over the Ganges-Brahmaputra-Meghna (GBM) region, incorporating uncertainty in model parameter values. The three socio-economic development scenarios are based on the shared socio-economic pathways (SSPs) (Kriegler et al. 2012; O'Neill et al. 2012; IPCC 2014) as explained below. Each scenario was downscaled in a stakeholder-led process of discussion in workshop rather than as a quantitative exercise. The result was three locally specific socio-economic scenarios with a high level of detail, providing a distinct approach to identifying locally relevant scenarios for coastal Bangladesh.

The climate and socio-economic scenarios are summarised in Sects. 9.3 and 9.4, whilst further information on the climate analysis can be found in Chap. 11 and Caesar et al. (2015). Specific economic scenarios are described in Chap. 12.

Table 9.1 Combination of climate and socio-economic scenarios with nomenclature

		Socio-economic scenario		
Climate scenario		Less Sustainable	Business As Usual	More Sustainable
Q0	Warmer (+2.2 °C) and wetter (+8%)	Q0-LS	Q0-BAU	Q0-MS
Q8	Warmer (+2.5 °C) and drier (−1%)	Q8-LS	Q8-BAU	Q8-MS
Q16	Warmer (+2.7 °C) and much wetter (+10%)	Q16-LS	Q16-BAU	Q16-MS

9.3   Climate

A coupled global (HadCM3) (Gordon et al. 2000; Pope et al. 2000; Collins et al. 2001) and regional (HadRM3P) (Massey et al. 2015) climate model developed by the Met Office Hadley Centre was used to project temperature and precipitation across the GBM region up to 2099 (Collins et al. 2011; Caesar et al. 2015). Given uncertainty in climate model parameter values, 17 different combinations of parameter values were used to identify ranges in possible temperature and precipitation changes (see Chap. 11, Fig. 11.3). Despite substantial variability between the 17 iterations, all projections indicate a warmer and wetter climate by 2099.

From these 17 scenarios, three were selected for application in the models to cover a range of possible change (Q0, Q8 and Q16). Ensemble members were selected to provide a central baseline (Q0), a warmer and drier scenario during the mid-century (Q8), and a much warmer and wetter scenario for both 2050 and 2099 (Q16). Sea-level rise is also a critical issue for Bangladesh. Global sea-level rise scenarios are based directly on the Fifth Assessment Report of the Intergovernmental Panel on Climate Change (Church et al. 2013). These are not explicitly coupled to the Q0, Q8 and Q16 scenarios, and their application is explained in the relevant chapter (see Chaps. 14 and 16). Land subsidence is taken from observations (Chap. 15), while example cyclones are used to illustrate their possible effects (Chap. 16).

9.4   Socio-economic Change

Three socio-economic scenarios were developed to cover a range of plausible patterns of future development based on locally relevant issues identified by stakeholders. A Business As Usual (BAU) scenario was used to represent stakeholder expectations of the future assuming development patterns continue as they have in the recent past and is similar to the SSP2 Middle of the Road scenario. A More Sustainable (MS) scenario was used to represent an 'improved' (but still realistic) future, drawing upon elements of the SSP1 (Sustainability) scenario, whilst a Less Sustainable (LS) scenario provided a less desirable outcome and combines components of the SSP3 (Fragmentation) and SSP4 (Inequality) scenarios (O'Neill et al. 2012). No objective measure of sustainability was used in the three resulting scenarios, with More and Less Sustainable being determined solely with reference to BAU—there is therefore no suggestion that the More Sustainable future would actually achieve sustainability.

Scenario development focused on adopting an interdisciplinary approach covering key elements of the biophysical environment as well as changes in livelihoods, education, economics and governance both locally and internationally. The approach adopted involved close collaboration with stakeholders and the project team with a view to developing both qualitative narratives and quantitative scenarios for the evaluation of management interventions. As such, scenario development involved four main stages: (i) identification of key issues of relevance to stakeholders, (ii) qualitative narratives for the three baseline scenarios to 2050, (iii) quantification of the narratives for baseline scenarios and (iv) identification and evaluation of management interventions (Fig. 9.2).

The four stages of scenario development were conducted as part of an iterative process with six national level stakeholder workshops held over the period from October 2013 to May 2016, as outlined in Chap. 10. Identification of key issues and downscaling of the SSPs into the three socio-economic baseline scenarios (stage 1) and the development of the qualitative narratives (stage 2) are outlined with further description in Chap. 10. Stages 3 and 4 are described in 9.4.1 and 9.4.2.

A series of interviews were held during 2012 and 2013 with relevant stakeholders, primarily at the national level, with a view to determining

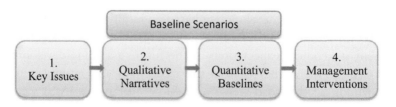

Fig. 9.2 Four stages in scenario development

the key issues of concern in relation to longer-term livelihood and environmental protection in the south-western part of the GBM delta. This produced a list of around 15 broad issues that were broken down by stakeholders at the first workshop into 105 constituent elements, categorised into natural resource management, food security, health/livelihoods/poverty and governance. Having deconstructed each of these key issues, participants at this workshop then agreed on the extent to which each element would change by 2050 in a BAU environment, using a seven-point scale, making assumptions regarding the extent to which current trends were likely to continue or not, and taking account of existing government policy initiatives and strategic direction.

Outcomes of these interviews and workshop were used to produce a detailed and internally consistent narrative by the project team that encompassed all of the elements and their anticipated status in 2050, effectively providing a downscaled qualitative narrative for the BAU scenario. This was augmented by project members through the preparation of draft narratives describing the Less and More Sustainable scenarios, and all three narratives were closely examined by stakeholders at a subsequent workshop for plausibility and for internal and cross-scenario consistency, with further revisions made in line with stakeholder recommendations. The result was three detailed co-produced descriptions of how the socio-economic situation in Bangladesh might look in 2050, along with what stakeholders believed are the consequences on the biophysical environment (included in Chap. 10 Appendix 1). The latter consequences form the basis of the next stage, which entailed more detailed quantification.

9.4.1 Quantification of the Qualitative Narratives

Representation of the baseline scenarios in a modelling framework enables further investigation to understand potential changes in Bangladesh's natural resources and the dynamics between the natural environment and human well-being. The use of an integrated modelling approach can facilitate interdisciplinary discussions regarding such complex dynamics between the biophysical and socio-economic systems, to identify key uncertainties, and explore the effectiveness of management interventions (Voinov and Gaddis 2008). However, this requires the quantification of the qualitative narratives such that the three socio-economic scenarios could be modelled.

This qualitative to quantitative process (referred to here as Q2Q) presents a challenge given the uncertainty in projecting changes to 2050 (and to support the biophysical modelling in some cases to 2100). As such, a five-step approach was adopted: (i) estimation of quantitative values by project members based on published data and expert knowledge where possible, (ii) identification amongst the project team of estimates which are most uncertain or unknown, and where stakeholder input would be of most value, (iii) individual stakeholder questionnaires to explore the range of responses across stakeholders and to avoid responses being influenced by dominant group members, (iv) stakeholder group discussions and consensus regarding key assumptions, and (v) iterative testing of assumptions in the modelling framework and presentation to stakeholders for feedback and modification.

Thirteen categories of model assumptions were used for consultation with stakeholders (Table 9.2). These were divided into a biophysical and socio-economic questionnaire which was emailed to participants with follow-up phone calls. Participants were primarily identified by local partners as well as through previous connections formed as part of earlier stakeholder interviews and workshops.

Following distribution of the individual questionnaires, a stakeholder workshop was held to facilitate group discussions. The workshop was attended by twenty participants from twelve external organisations and a further ten participants from partner organisations who were not directly involved in the research work. Given limited individual response to the questionnaires, the workshop began with independent completion of the questionnaires followed by sectoral group discussions, finishing with a cross-sectoral discussion.

The outcome of the Q2Q process was largely successful in terms of engaging representatives from different institutions and disciplines to discuss future changes across a range of key issues. Informal participant feedback indicated the process was interesting, useful and informative, although a number of participants found the questions challenging.

Table 9.2 Scenario categories for quantitative biophysical and socio-economic model assumptions

Biophysical	Socio-economic
Water resources	**Migration**
• Dam construction	• Changes in migration type
• Major water transfers	• Factors influencing migration
• Drought indices	• Influence of policy and policy
• Water extractions	makers
• Effluent discharge	**Employment**
• Sewage treatment plants	• Change in employment
• Deep/shallow groundwater	**Literacy**
extraction	• Change in national literacy
• Subsidy programmes for	• Change in rural literacy
groundwater	**Subsidies and loans**
• Flood management: polder height	• Current and planned subsidies
and maintenance	• Changes in loan provisions in
	rural areas
Land use	**Poverty metrics**
• Sundarbans encroachment	• Poverty dimensions
• Crop yields and salinity tolerance	• Commonly used poverty metrics
• Aquaculture area and technology	• Advantages/disadvantages of
• Change in land cover types	different metrics
• Land zoning programme and	
incentives	
• Access—rail/road/bridge construction	
Fisheries	
• Fishing effort	
• Fishing subsidies	

Fourteen participants completed a formal feedback form, of which the large majority indicated that the workshop had contributed to their wider understanding of ecosystem services at least to some extent, through the quantification of real conditions and assumptions, the use of narratives, assumptions and scenarios and discussion with economists about economic valuation of ecosystem services.

The process was also valuable in providing quantitative input for the modelling framework which is hoped to have improved the acceptance and validity of the modelling outputs. However, there were varying levels of response given the wide range of subjects covered by the questionnaire relative to the number of participants. Some components received much greater attention than others based on the expertise and interest within the groups. Time was also a limiting factor, particularly in the water resources group, where questions regarding future infrastructure development consumed most of the time with other questions on groundwater, water quality and polder management receiving little attention.

Where participants were asked to comment on values previously estimated within the project team, in general there was reasonable agreement with what had been proposed, or small suggested variations. Where participants were asked to provide new values for different assumptions, in general there was reasonable agreement between respondents in terms of overall direction and magnitude of change, but with the specific value of change varying between responses. The greatest disagreement for proposed assumptions concerned water transfer volumes and timing. Despite requesting individual responses, it was evident from some questionnaire responses that there was likely to have been discussion between participants sitting near one another. Whilst this is a limitation in not showing a full spread of individual perceptions prior to the group session, the generation of discussion about such topics is still a positive outcome.

In general, group responses reflected some elements of the individual responses, whilst in a few cases the group discussion introduced additional perspectives or changed the majority view of individuals. This change may also have been influenced by some participants joining group discussions on topics which were different from the sections they completed for the individual questionnaires. Results from the Q2Q process were disseminated to the modelling teams and incorporated into the boundary conditions for the component and integrated modelling runs.

9.4.2 Management Interventions and Adaptation Responses

A key strength of the adopted approach is the use of a modelling framework to allow management interventions to be tested against a range of uncertain futures, considering both climate and socio-economic change. Alongside the testing of baseline model assumptions, the main management interventions of relevance to stakeholders were also identified.

A critical element for management interventions was providing stakeholders with sufficient guidance as to what approaches could be modelled whilst not limiting or influencing input through leading questions. Initial efforts to have stakeholders identify interventions, irrespective of modelling capacity, were unsuccessful as stakeholders defaulted to the policies they already knew and understood. For example, stakeholder recommendations included mainstreaming poverty reduction in all development projects, proper implementation of the national social protection scheme and capacity development of local government. As these could not be analysed in the models, a series of suggestions that could be analysed were subsequently provided to stakeholders in advance of their technical discussions. These proposals included renegotiation of the Farakka Treaty, construction of the Ganges Barrage, changing polder height and other structural interventions to manage flooding and sea-level rise, improved drainage to reduce waterlogging, groundwater use policies, land zoning policies, new potential crops and subsidies for farming.

Unfortunately however, this appeared to drive group discussions and frame their proposed interventions, instead of providing indicative suggestions. It remains a challenge to have stakeholders identify interventions that are outside of existing sectoral approaches or beyond research-driven suggestions. It is for this reason that the iterative learning loop and integrated model approach (Fig. 9.1) are particularly valuable, as stakeholders have multiple opportunities to engage. After developing the inputs, they see the modelled results of their sector-driven interventions in terms of broader impacts across ecosystem services and livelihoods. They are then able to propose alternative interventions that are able to respond more directly to the integrated model results considering cross-sectoral impacts. With the application of the integrated model to test specific interventions in the Bangladesh Delta Plan 2100, this process is ongoing. The continuation of this work demonstrates the success of the approach.

9.5   Key Insights

Development of both qualitative and quantitative scenarios across a diverse range of biophysical and socio-economic issues facilitates cross-disciplinary discussion and learning. The scenarios present a range of plausible futures which are used to evaluate the impact of future change and the effectiveness of different management interventions. Adopting a systems-based approach of this scale is challenging but essential to the effective management of coastal Bangladesh and can be used to support the development of existing and future government plans including the Five-Year Plans, Vision 2021 and the Delta Plan 2100.

Involving stakeholders is critical to the success of the scenario process, as well as improving the acceptance and validity of the model and creating ownership of the process. As such, the scenario workshops are a key component of the stakeholder engagement within the research. It is intended that the scenarios are adaptive and continue to be iterated as new data and knowledge becomes available.

The scenario development processes highlighted a number of future challenges. Identifying trends in socio-economic processes with multiple interactions and dependencies is severely limited by the current capacity to understand and represent these processes, particularly in a quantitative way (Berkhout et al. 2002; Swart et al. 2004). This is further compounded by the range of scales considered, from international cooperation and macroeconomic issues through to individual and household behaviour. Despite an increasing number of studies adopting interdisciplinary scenario development down to the regional scale, the majority of these remain focused on a sub-set of future changes (e.g. flood risk—Hall et al. 2005), water resources (Soboll et al. 2011) and land use change (Baker et al. 2004; Rounsevell et al. 2005; Audsley et al. 2006), with few addressing the extent of biophysical and socio-economic changes considered in this research. In particular, there is limited evaluation of socio-economic scenarios focusing on human well-being and poverty (Lázár et al. 2015). This complexity creates a challenge for effective stakeholder engagement, requiring participants from multiple sectors with sufficient time to engage in the scenario process. The workshops identified that there was generally inadequate time in a single workshop to both sufficiently explain the overall context, as well as to allow stakeholders to discuss and respond to questions and invited feedback. Hence, multiple workshops and repeated engagement are critical to better engagement.

Despite these challenges, continued engagement with stakeholders throughout the scenario process is successful in developing both qualitative and quantitative plausible futures. The process has assisted in promoting dialogue about the complex dynamics influencing changes in the natural and human environment and breaking down silos between those with different expertise.

References

Audsley, E., K.R. Pearn, C. Simota, G. Cojocaru, E. Koutsidou, M.D.A. Rounsevell, M. Trnka, and V. Alexandrov. 2006. What can scenario modelling tell us about future European scale agricultural land use, and what not? *Environmental Science and Policy* 9 (2): 148–162. https://doi.org/10.1016/j.envsci.2005.11.008.

Baker, J.P., D.W. Hulse, S.V. Gregory, D. White, J. Van Sickle, P.A. Berger, D. Dole, and N.H. Schumaker. 2004. Alternative futures for the Willamette River basin, Oregon. *Ecological Applications* 14 (2): 313–324. https://doi.org/10.1890/02-5011.

BanDuDeltAS. 2014. *Inception report. Bangladesh delta plan 2100 formulation project.* Dhaka: General Economics Division (GED), Planning Commission, Government of the People's Republic of Bangladesh.

Berkhout, F., J. Hertin, and A. Jordan. 2002. Socio-economic futures in climate change impact assessment: Using scenarios as 'learning machines'. *Global Environmental Change-Human and Policy Dimensions* 12 (2): 83–95. https://doi.org/10.1016/s0959-3780(02)00006-7.

Caesar, J., T. Janes, A. Lindsay, and B. Bhaskaran. 2015. Temperature and precipitation projections over Bangladesh and the upstream Ganges, Brahmaputra and Meghna systems. *Environmental Science-Processes and Impacts* 17 (6): 1047–1056. https://doi.org/10.1039/c4em00650j.

Church, J.A., P.U. Clark, A. Cazenave, J.M. Gregory, S. Jevrejeva, A. Levermann, M.A. Merrifield, G.A. Milne, R.S. Nerem, P.D. Nunn, A.J. Payne, W.T. Pfeffer, D. Stammer, and A.S. Unnikrishnan. 2013. Sea level change. In *Climate change 2013: The physical science basis. Contribution of working group I to the fifth assessment report of the intergovernmental panel on climate change,* ed. T.F. Stocker, D. Qin, G.-K. Plattner, M. Tignor, S.K. Allen, J. Boschung, A. Nauels, Y. Xia, V. Bex, and P.M. Midgley. Cambridge, UK/New York: Cambridge University Press.

Collins, M., S.F.B. Tett, and C. Cooper. 2001. The internal climate variability of HadCM3, a version of the Hadley Centre coupled model without flux adjustments. *Climate Dynamics* 17 (1): 61–81. https://doi.org/10.1007/s003820000094.

Collins, M., B.B.B. Booth, B. Bhaskaran, G.R. Harris, J.M. Murphy, D.M.H. Sexton, and M.J. Webb. 2011. Climate model errors, feedbacks and forcings: A comparison of perturbed physics and multi-model ensembles. *Climate Dynamics* 36 (9–10): 1737–1766. https://doi.org/10.1007/s00382-010-0808-0.

GED. 2012. *Perspective plan of Bangladesh 2010–2021: Making vision 2021 a reality*. Dhaka: General Economics Division (GED), Planning Commission, Government of the People's Republic of Bangladesh. http://www.plancomm.gov.bd/perspective-plan/. Accessed 20 July 2016.

———. 2015. *Seventh five year plan FY2016 – FY2020: Accelerating growth, empowering citizens*. Final draft. Dhaka: General Economics Division (GED), Planning Commission, Government of the People's Republic of Bangladesh. http://plancomm.gov.bd/wp-content/uploads/2015/11/7FYP_after-NEC_11_11_2015.pdf. Accessed 6 Jan 2017.

Gordon, C., C. Cooper, C.A. Senior, H. Banks, J.M. Gregory, T.C. Johns, J.F.B. Mitchell, and R.A. Wood. 2000. The simulation of SST, sea ice extents and ocean heat transports in a version of the Hadley Centre coupled model without flux adjustments. *Climate Dynamics* 16 (2–3): 147–168. https://doi.org/10.1007/s003820050010.

Hall, J.W., P.B. Sayers, and R.J. Dawson. 2005. National-scale assessment of current and future flood risk in England and Wales. *Natural Hazards* 36 (1–2): 147–164. https://doi.org/10.1007/s11069-004-4546-7.

Holling, C.S., ed. 1978. *Adaptive environmental assessment and management, IIASA international series on applied systems analysis*. Chichester: Wiley.

IPCC. 2014. *Climate change 2014: Impacts, adaptation, and vulnerability. Part A: Global and sectoral aspects. Contribution of working group II to the fifth assessment report of the intergovernmental panel on climate change*. Cambridge/New York: Cambridge University Press.

Kriegler, E., B.C. O'Neill, S. Hallegatte, T. Kram, R.J. Lempert, R.H. Moss, and T. Wilbanks. 2012. The need for and use of socio-economic scenarios for climate change analysis: A new approach based on shared socio-economic pathways. *Global Environmental Change-Human and Policy Dimensions* 22 (4): 807–822. https://doi.org/10.1016/j.gloenvcha.2012.05.005.

Lázár, A.N., D. Clarke, H. Adams, A.R. Akanda, S. Szabo, R.J. Nicholls, Z. Matthews, D. Begum, A.F.M. Saleh, M.A. Abedin, A. Payo, P.K. Streatfield, C. Hutton, M.S. Mondal, and A.Z.M. Moslehuddin. 2015. Agricultural livelihoods in coastal Bangladesh under climate and environmental change – A model framework. *Environmental Science-Processes and Impacts* 17 (6): 1018–1031. https://doi.org/10.1039/c4em00600c.

Liu, J.G., T. Dietz, S.R. Carpenter, M. Alberti, C. Folke, E. Moran, A.N. Pell, P. Deadman, T. Kratz, J. Lubchenco, E. Ostrom, Z. Ouyang, W. Provencher, C.L. Redman, S.H. Schneider, and W.W. Taylor. 2007. Complexity of coupled human and natural systems. *Science* 317 (5844): 1513–1516. https://doi.org/10.1126/science.1144004.

Mahmoud, M., Y.Q. Liu, H. Hartmann, S. Stewart, T. Wagener, D. Semmens, R. Stewart, H. Gupta, D. Dominguez, F. Dominguez, D. Hulse, R. Letcher, B. Rashleigh, C. Smith, R. Street, J. Ticehurst, M. Twery, H. van Delden, R. Waldick, D. White, and L. Winter. 2009. A formal framework for scenario development in support of environmental decision-making. *Environmental Modelling and Software* 24 (7): 798–808. https://doi.org/10.1016/j.envsoft.2008.11.010.

Massey, N., R. Jones, F.E.L. Otto, T. Aina, S. Wilson, J.M. Murphy, D. Hassell, Y.H. Yamazaki, and M.R. Allen. 2015. weather@home—Development and validation of a very large ensemble modelling system for probabilistic event attribution. *Quarterly Journal of the Royal Meteorological Society* 141 (690): 1528–1545. https://doi.org/10.1002/qj.2455.

Moss, R.H., J.A. Edmonds, K.A. Hibbard, M.R. Manning, S.K. Rose, D.P. van Vuuren, T.R. Carter, S. Emori, M. Kainuma, T. Kram, G.A. Meehl, J.F.B. Mitchell, N. Nakicenovic, K. Riahi, S.J. Smith, R.J. Stouffer, A.M. Thomson, J.P. Weyant, and T.J. Wilbanks. 2010. The next generation of scenarios for climate change research and assessment. *Nature* 463 (7282): 747–756. https://doi.org/10.1038/nature08823.

Nicholls, R.J., C.W. Hutton, A.N. Lázár, A. Allan, W.N. Adger, H. Adams, J. Wolf, M. Rahman, and M. Salehin. 2016. Integrated assessment of social and environmental sustainability dynamics in the Ganges-Brahmaputra-Meghna delta, Bangladesh. *Estuarine, Coastal and Shelf Science* 183 (Part B): 370–381. https://doi.org/10.1016/j.ecss.2016.08.017

O'Neill, B.C., T.R. Carter, K.L. Ebi, J. Edmonds, S. Hallegatte, E. Kemp-Benedict, E. Kriegler, L.O. Mearns, R.H. Moss, K. Riahi, B. Van Ruijven, and D. Van Vuuren. 2012. *Meeting report of the workshop on the nature and use of new socio-economic pathways for climate change research, November 2–4, 2011.* Boulder: National Center for Atmospheric Research (NCAR). http://www2.cgd.ucar.edu/research/iconics/events/20111102. Accessed 20 July 2016.

Pope, V.D., M.L. Gallani, P.R. Rowntree, and R.A. Stratton. 2000. The impact of new physical parametrizations in the Hadley Centre climate model: HadAM3. *Climate Dynamics* 16 (2–3): 123–146. https://doi.org/10.1007/s003820050009.

Rounsevell, M.D.A., and M.J. Metzger. 2010. Developing qualitative scenario storylines for environmental change assessment. *Wiley Interdisciplinary Reviews-Climate Change* 1 (4): 606–619. https://doi.org/10.1002/wcc.63.

Rounsevell, M.D.A., F. Ewert, I. Reginster, R. Leemans, and T.R. Carter. 2005. Future scenarios of European agricultural land use II. Projecting changes in cropland and grassland. *Agriculture Ecosystems and Environment* 107 (2–3): 117–135. https://doi.org/10.1016/j.agee.2004.12.002.

Soboll, A., M. Elbers, R. Barthel, J. Schmude, A. Ernst, and R. Ziller. 2011. Integrated regional modelling and scenario development to evaluate future water demand under global change conditions. *Mitigation and Adaptation Strategies for Global Change* 16 (4): 477–498. https://doi.org/10.1007/s11027-010-9274-6.

Swart, R.J., P. Raskin, and J. Robinson. 2004. The problem of the future: Sustainability science and scenario analysis. *Global Environmental Change-Human and Policy Dimensions* 14 (2): 137–146. https://doi.org/10.1016/j.gloenvcha.2003.10.002.

Voinov, A., and E.J.B. Gaddis. 2008. Lessons for successful participatory watershed modeling: A perspective from modeling practitioners. *Ecological Modelling* 216 (2): 197–207. https://doi.org/10.1016/j.ecolmodel.2008.03.010.

10

Incorporating Stakeholder Perspectives in Scenario Development

Andrew Allan, Michelle Lim, and Emily J. Barbour

10.1 Introduction

The incorporation of stakeholder views is integral to the research from beginning to end and served a number of critical functions. Creating a clear link between research outputs and stakeholder needs was identified as an early priority giving stakeholders tangible input to and ownership of modelling and scenario processes. This component of the research places particular emphasis on stakeholder-identified issues and priorities to develop intervention recommendations that are relevant to both local

A. Allan (✉)
School of Law, University of Dundee, Dundee, UK

M. Lim
Adelaide Law School, University of Adelaide, Adelaide, SA, Australia

E. J. Barbour
School of Geography and the Environment, University of Oxford, Oxford, UK

© The Author(s) 2018
R. J. Nicholls et al. (eds.), *Ecosystem Services for Well-Being in Deltas*,
https://doi.org/10.1007/978-3-319-71093-8_10

users of ecosystem services and national decision-makers. For example, efforts to understand the reality of how legal, institutional and policy frameworks can mediate the translation of ecosystem services to benefits (see Chaps. 2 and 4) were driven in large part by stakeholders identifying key issues of interest during the early stages of the research.

A highly structured approach was accordingly adopted to ensure the ability to respond to stakeholder priorities and knowledge and ensure that stakeholder expectation of findings were realistic. In addition to the manifest need to match stakeholder needs with research capacity, the views of stakeholders are also integral to the scenario development process described in Chap. 9. This chapter describes the stakeholder engagement process as well as the first two stages of the four-stage scenario development process (see Fig. 10.1), identification of key issues and development of scenario narratives.

10.2   Stakeholder Mapping

Before elaborating further on the scenario development process, some background should be given regarding the rationale and methods used for incorporating stakeholder views. The increased use of stakeholder analysis in natural resource management reflects recognition of the extent to which stakeholders can influence the decision-making process (Prell et al. 2009). The identification, analysis and engagement of stakeholders are therefore central components of the research. Networks of stakeholders

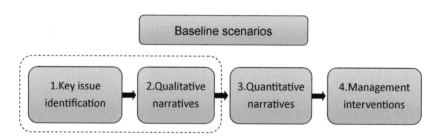

Fig. 10.1 The stages in scenario development highlighting the focus of this chapter

were identified through interviews and stakeholder workshops involving government officials, local project partners, experts (e.g. academics) and Non-Governmental Organisations (NGOs).

Stakeholders were identified by way of an extensive mapping process (principally based on a snowball approach (Reed et al. 2009)). Additional stakeholders were identified at the more localised level (i.e. in the project case study areas in the Khulna and Barisal divisions). This process was bolstered through enhanced engagement with a small number of key stakeholders whose interests aligned most closely with the project from the perspective of use and uptake, data provision and cross-sectoral relevance. It became clear that for the mutual benefit of both the project and a small number of these key stakeholders, a more formal relationship would be preferable, and this resulted in the Water Resources Planning Organisation (WARPO) becoming a formal project partner and the establishment of a strategic alliance with the General Economics Division of the Planning Commission of the Government of Bangladesh.

Representatives from round 60 institutions were actively involved in the engagement between the project and relevant institutions across multiple scales in Bangladesh. Initially, this engagement primarily took the form of formal one-to-one interviews and latterly through more widely attended workshops.[1] The range of stakeholders who were contacted by project partners and who chose to participate included the following:

- National government officials in a range of Ministries, the portfolios of which relate to ecosystem services and human well-being (e.g. Planning Commission; Ministry of Agriculture, WARPO)
- Relevant non- and inter-governmental organisations at the international level (e.g. International Union for Conservation of Nature (IUCN), International Organisation for Migration, Global Water Partnership, CARE)
- UN organisations, multi- and bi-lateral donor agencies (e.g. United Nations Development Programme (UNDP), World Health Organisation (WHO), World Food Programme (WFP), World Bank, Asian Development Bank)

- National NGOs, research groups and subject experts, including representatives from academic institutions (e.g. BRAC and BRAC University, WildTeam, Bangladesh Agricultural Research Institute (BARI), Bangladesh Rice Research Institute (BRRI))
- Key informants, that is, those whose social/institutional positions give them specialist knowledge about other people, processes or happenings that is more extensive, detailed or privileged than ordinary people (Payne and Payne 2004)

The extent of engagement varied according to the relevance of individual stakeholders to the research: some agreed to individual interviews with a view to identifying the main issues of concern and others to attending the series of workshops that were held during the latter half of the research period.

10.3   Identification of Key Issues

An iterative approach was adopted to identify key issues of concern for stakeholders. This includes a review of relevant literature (i.e. academic papers, government legal and policy documents and available grey literature) and a series of workshops and unstructured interviews with national level stakeholders and other key informants. As shown in Table 10.1 (and further discussed in Allan et al. 2013), this process revealed the breadth of issues that needed to be addressed to effectively assess ecosystem service provision and human well-being. These represent a range of interlinked issues, challenges and uncertainties across the social, economic and environmental landscape of the study area.

Table 10.1 summarises the key issues, based on a synthesis and categorisation of issues identified in the literature, which also overlap those of stakeholder opinion. Issues identified in the literature that were not seen as priorities by stakeholders were therefore omitted from the list. Many stakeholder concerns mirror issues identified in the literature and underline the challenge of managing complex social-ecological systems. At this stage, the issues reflect the views of stakeholders and should not be taken to be a comprehensive and exhaustive categorisation of all the issues that may relate ecosystem services to livelihoods.

Table 10.1 Key biophysical and socio-economic issues of importance as identified by stakeholders

Socio-economic	Biophysical
Food security	Salinisation
Human-wildlife conflict	Riverbank erosion and sedimentation
Human-induced challenges to flow/ freshwater availability	Arsenic
Changes in livelihoods	Freshwater availability
Barriers to accessing ecosystem services in the Sundarbans	Unpredictability of weather
Migration	Extreme weather events
Shrimp vs. crop	Location of biggest embankments (sea dykes)/coastal defence
Upstream/international issues	
Availability of land	

10.4 Scenario Development Process

The rationale for adopting a scenario-based narrative approach of possible (and plausible) futures is that this allows responses to environmental and social changes over time to be explored in a way that addresses the significant levels of uncertainty. It also helps facilitate the integration of the views of stakeholders with the scientific findings, as well as their engagement in the assessment process.

In order to fully enable the connection between stakeholder priorities and the available modelling capacity, two separate stages are required: (i) to qualitatively describe what the future might look like at the scenario time horizon and (ii) to translate these qualitative descriptions into the quantitative form required by the numerical models. This section outlines the process adopted with respect to the first of these (the later stages are discussed in Chap. 9). The approach adopted was inspired by the shared socio-economic reference pathways (SSPs) approach (Arnell et al. 2011; O'Neill et al. 2012). Based on this, three socio-economic scenarios were developed for Bangladesh and coastal Bangladesh as devices for engaging with stakeholders: (i) Less Sustainable (LS), (ii) Business As Usual (BAU) and (iii) More Sustainable (MS). BAU was defined as the situation that might exist if existing policies were to continue and development trajectories proceed along similar lines to the previous 30 years or so, irrespective of whether or not this in itself is sustainable. LS and MS are alternative trajectories that are broadly less or more sustainable than BAU. This scenario elaboration approach effectively produces what Arnell et al. (2011) call 'extended SSPs' as it takes what is a global approach unsuited to direct application at the national level and, through the addition of more locally relevant characteristics, facilitates the downscaling of the SSPs. It also allowed the stakeholder issues of concern to be projected up to 2050, on the basis of the climate projections that used in the research (see Chap. 11).

In order to better facilitate the integration of stakeholder concerns with the development of scenario narratives, the issues were categorised into broad groups and further elaborated in accordance with stakeholder views as expressed at the first stakeholder workshop. This produced a consolidated list (see Table 10.2) that was used as the basis for downscaling.

Each of the resulting issues was categorised and, during a workshop held on 22–23 October 2013, broken down by participants into more than 100 separate elements. The meeting was attended by a total of around 35 people, representing donor organisations, government ministries and academic or sectoral experts across a wide spectrum of disciplines. Once agreement (consensus or majority agreement) was reached on the breakdown of each of the issues and categories, the groups were asked to assess the extent of the expected improvement or deterioration over time, using a scale from '+' to '+++', with '+' being slight and '+++' being strong (and with 'no change' comprising a seventh element of the scale). The result was a detailed and internally consistent matrix of how the participants believed society in the Ganges Brahmaputra Meghna (GBM) delta would look like in 2050 considering (i) existing policy direction within Bangladesh; (ii) trends over the previous 35 years; (iii) factors influencing the likelihood of these trends continuing for the next 35 years; (iv) externally imposed boundary conditions with respect to

Table 10.2 Factors considered in the downscaling of socio-economic pathways for the development of local scenario narratives

Well-being	Livelihood issues	Physical change
Nutrition	Land use	Extreme weather events
Food security	Crop types/diversification	Seasonality
Access and availability	Excessive and unplanned use of fertilisers	Frequency of natural disaster
Household equity	Pollution/sanitation	Arsenic
Food prices	Natural resource management	Salinity
Migration	Changes in livelihoods (e.g. crop to shrimp)	Riverbank erosion and sedimentation
Remoteness	Local elite	Coastal defence
Coordination (sectoral and geographical)	Barriers to accessing ecosystem services in the Sundarbans	
Capacity	Corruption/governance	
Lack of participation and marginalisation of the poor	Implementation and enforcement of regulations	
Disease		

freshwater, temperature, storminess and sea-level rise; and (v) relevant international and global influences.

The list of issues was elaborated in far greater detail than had been expected and effectively downscaled the BAU scenario to the study area context. In effect therefore, the considerable effort required to elucidate each of the 100 or so elements constituted a downscaling of the SSP approach to a national context in a way that was considered credible by the cross-sectoral group of stakeholders present and maintained internal consistency. Unfortunately, it was not possible to undergo the same process for the other two scenarios due to time limitations, and adjustments had to be made, as discussed in the next section, during the next (narrative) stage in the scenario development. The completed matrix for BAU is shown in Table 10.3.

Table 10.3 Completed matrix for the Business As Usual (BAU) scenario from the workshop on 22–23 October 2013

Natural resource management	Food security
Salinity/freshwater	**Availability and access**
Freshwater ↓ +++	Rice (area) ↓ +
Ingress salinity ↑	Rice (yield) ↑ +
Mangrove ↓ +	Others (area) ↑ +
Flow dynamics/riverbank erosion and	Others (yield) ↑ +
sedimentation	Storage ↑ ++
Mech: Accretion ↑ +	Household storage↑ +
Erosion ↑ +	Market access ↑ +
Water logging ↑ ++ and flooding ↑ ++	Farmer knowledge↑ +
Land use	**Water security**
Land use change rate ↑ ++	Freshwater:
Rice production ↓ +	Quality ↓ ++
Shrimp production↑ +	Quantity ↓ ++
Floodplain fisheries ↓ +++	Predictability ↓ +++
Coastal defence	Accessibility ↑ +
Infrastructure ↑ +	**Nutrition**
Maintenance/rehabilitation ↑ +	Food habit ↑ +
Mangrove/forest ↓ +	Pricing (% income)↓+
Impact of extreme weather events	Protein ↑
Asset damage ↑ ++	**Agriculture production systems/R&D**
Loss of life ↓ ++	Efficient Fertiliser use ↑ +
Conservation effort ↑ +	R&D/ technology↑ ++
Biodiversity ↓ +	Crop diversification ↑+
Management (local involvement) ↑ +	Subsidies ↑ +
	Wheat production ↑ +
	Household equity
	Intra- ↑ +
	Inter-↓ +
	Market dynamics
	Role of intermediaries ↓ +
	Information technology (price information
	e.g. mobile phones) ↑ ++
	Seasonality
	Shift in traditional practices

Key: ↓, expected deterioration; ↑, expected improvement; +, slight; ++, stronger; +++, strong

Health/livelihoods/poverty	Governance
Migration Net migration (urban: rural ratio) ↑++ Outmigration from project area ↑ ++ Push ↑++ Pull ↑+++ **Remoteness/communication/infrastructure** Infrastructure↑ + Communication ↑++ **W.A.S.H** Community ↑+ Urban (formal)↑++ Urban (informal/slum) ↑+ Water: Sanitation ↑+ **Changes in livelihoods** Diversification ↑ ++ **Utilisation of ecosystem services** Availability/Access ↑ Private sector: Community↓++ (access ratio) Private/community ↓++ **Disease** Non-communicable↑+ Water borne ↑+ Vector borne↑+ Zoonotic ↑+ **Frequency and intensity of disasters** **Gender** InFl DM ↑+ Disaster risk reduction +climate change adaptation↑++ Access to natural resources/ecosystem services ↑+	**Coordination and collaboration (sectoral and geographical) NRM benefits the most, (2 livelihoods 3) food security** Sectoral: ↑+ Geographical: Transboundary ↔ Bangladesh ↑+ **Power structure/conflict** Conflict ↓ Inter-sectoral (e.g. fisherman vs. farmers) ↓+ Intra-sectoral ↓++ Power structure ↔ **Human and financial capacity/awareness/extension agents** Human and financial capacity ↑ + (likely to have most impact on pollution, NRM ↑+) Awareness ↑ ++ **Local government empowerment ↑+** **Implementation and enforcement ↑+** **Law and Order/security (dacoits/pirates)** Fisheries ↑++ Unauthorised inputs (pesticides, fertiliser, etc.) ↓+ Piracy ↔ **Lack of participation and marginalisation of the poor** Participation ↑ ++ Marginalisation ↓++ **Role of NGOs/civil society/private sector/farmers' assn., public organisations** NGOs/CSO ↑+ Private/corporate/entrepreneurs ↑++ **Transparency/access to information/accountability** Transparency ↑+ Access to information ↑++ Accountability ↑+ **Land management/zoning and distribution** Land management ↑+ Zoning ↑+ Distribution ↔ **Transboundary (India, China)** Water ↓++ Trade ↓+ **Planning** Central ↑+ Local ↑+ **Maintenance of existing infrastructure ↑+** **Rules and regulations ↑ + and local level policy ↑ +, local courts ↔** **Service delivery efficiency ↑+**

10.5   Scenario Narratives

In order to translate the rough categories shown in Table 10.3 into the modelling efforts, they first had to be converted into a credible and representative narrative that could draw each element together in a more digestible format. A detailed narrative description was prepared to represent the detailed elaboration of the BAU scenario, with additional provisional narratives prepared by the project team for LS and MS, based on reasonable extrapolations of the BAU exposition. These three narratives were then presented at a further, larger workshop in May 2014. The main objective of the workshop, which was attended by 100 people, was to critically assess these detailed scenario narratives. The narratives needed to be stress-tested for credibility, internal consistency and for consistency between themselves—especially as only the BAU narrative was based on stakeholder-derived information. The narratives re-framed the multitude of elements that were detailed in the first workshop (Table 10.3) into six categories: (i) land use, (ii) water, (iii) international cooperation, (iv) disaster management, (v) environmental management and (vi) quality of life and livelihoods. A coherent story was then developed by combining local, regional and global drivers and highlighting their impact for Bangladesh, and more particularly the study area in the south-west of the delta. This produces a greater alignment between the breadth of the matrix and the individual elements of the modelling and survey frameworks within the project.

Attendees were presented with a copy of the draft consolidated scenario narratives, and then split into three representative groups, mixing the diverse sectoral interests present. Each group was allocated one of the scenarios and given instructions (and some background) on how they should interpret the document and what they should do with it. Reports from the groups were made in plenary, consisting of identifying problems in their respective scenarios, highlighting potential policy or management interventions, and identifying barriers to policy implementation.

The scenario narratives stood up well to the sustained critical assessment of 100 experts. Many comments were made, and these were incorporated into the revised (and final) version of the narratives (see Appendix to this chapter). There was a generally lower level of consensus at the second stakeholder workshop than the first, although groups were still able to produce critical evaluation that was broadly agreed to by their members, and this level of disagreement could be explained simply by the fact that there were almost three times as many people present at the second meeting.

10.6   Conclusions

There was a great deal of value in conducting the scenario development in the manner described. Focusing on the elaboration of the BAU scenario during the initial stages allowed for better understanding of stakeholder views on the baseline situation current in Bangladesh, through reflection on both existing and recent historical trends. Stakeholders were often pleasantly surprised that they were able to maintain their involvement from the interview stage and then on to the workshops. This continuity provided evidence that the project was serious in taking their views into account. There was a general level of acceptance on the part of those attending workshops that the approach being taken was credible and was addressing the correct issues, even though there might be a strong element of disagreement over potential solutions or the magnitude of the problem. Over the duration of the project, it became clear that the credibility of research outputs was increased significantly by the fact that stakeholder views and inputs had been integral to each successive stage from the identification of the key issues right through to the modelling.

What also became clear was that availability of time was an issue for stakeholders, both from the perspective of project members trying to achieve too much within the limited time available during workshops but also because of the length of time that stakeholders were involved. In retrospect, the amount of information that stakeholders were expected to read and absorb was unrealistic: the narratives are complex documents of around 1,500 words each. This requires a considerable commitment on the part of stakeholders, who derive no other benefit from the process than the opportunity to discuss issues in a forum with others from outside their immediate sphere of contact, and the hope that they might acquire a greater understanding through the research outputs. Other alternative approaches may perform better—for example, by establishing a standing stakeholder expert group who could comment on technical detail, perhaps in return for a fee reflecting the degree of commitment needed—but it was not possible to follow these through in this project.

Appendix: Socio-economic Scenario Narratives for Bangladesh

For future policy analysis, three socio-economic scenarios for the study area within Bangladesh were developed using the process described in Chaps. 9 and 10: Business As Usual (BAU), More Sustainable (MS) than BAU and Less Sustainable (LS) than BAU. The key aspects of each scenario are described in the following sections.

Business as Usual

Land Use

While the rate of change in land use has risen, there has been a gradual move to increased diversification of crops, for example to include more wheat and more vegetables, with continuing increases in shrimp production. Due to improvements in cultivation techniques (following decent hikes in the level of investment in research and development (R&D)), more efficient use of fertilisers and pesticides, more targeted subsidy programmes and the use of high-yield varieties, yield per hectare for all crops has increased. Consequently, although cultivated areas given over to rice have decreased, overall production has risen.

Reductions in the level of resource conflict, between farmers and fish-farmers for example, along with the enhanced role of agricultural extension officers and more integrated rice/fish farming, provide positive contributions to increasing farm yield, along with higher levels of understanding of appropriate techniques on the part of farmers. Overall, these have the effect of cancelling out the detrimental impact of the changes in seasonality that have been experienced. Less helpfully, the combined effect of more intensive land use and patchy environmental management compliance has been an increase in land degradation.

The extent of coastal defence infrastructure has been enhanced, and natural flood barriers, such as the mangrove forest, have been slightly reduced in extent. Regulation of land use, including for floodplain and sectoral use zoning, has improved, as have levels of Central and devolved planning capacity.

Water

Improvements to the technology used for irrigation have been driven in part by a reduction in the amount of water coming down from India, with some reductions in predictability of availability and water quality. Predictability and availability are affected in part by increased river regulation in Nepal, India and China, with water pollution levels being driven by a combination of lower flows and higher levels of upstream industrial pollution. These improvements in irrigation have been to some extent offset by a significant overall increase in the use of water for agriculture.

Reduced freshwater flow and greater use of water for agriculture coupled with sea-level rise have heightened problems associated with saline intrusion in coastal areas. Despite this, provision of water to households, even in informal settlement areas, has improved to some extent, with better service delivery efficiency and infrastructure maintenance, following investment in water and sanitation service provision pursuant to achievement of development goals.

As a result of the decreasing flow in cross-border rivers, accretion is increasing, with erosion also increasing in the upper reaches of the delta.

Cooperation between water users across and within sectors has improved as a result of the relative scarcity of water and amplified levels of demand.

International Cooperation

Maintaining these levels of cooperation has not been aided by a deterioration in the extent to which basin states on the Ganges and Brahmaputra Rivers are cooperating, both with respect to water and in relation to trade. This is one of the most significant drivers of the reduction in transboundary flows. China has retained its observer status with the SAARC, and efforts to accord it full membership have not yet succeeded.

Disaster Management

Along with increases in the extent of coastal defence and emergency infrastructure (such as cyclone shelters), efforts have been made to better maintain these constructions. Storage of harvested crops is substantially better than in 2013, through initiatives such as cyclone-resistant households. One of the benefits from these improvements has been a drastic reduction in the loss of life as a result of cyclones but there have also been relative increases in the level of economic damage caused.

Environmental Management

After decades of reasonably stable forest cover, the mangrove forest in the case area has suffered a small degree of encroachment. With reduced levels of water flow and increasing use of agricultural fertilisers across the country, for example, water quality has deteriorated to a certain extent, with governance capacity having improved to some degree but not sufficiently to control diffuse pollution.

Improvements to reticulated water supplies have not been quite adequate to compensate for this, and consequently levels of water-borne diseases have risen slightly. Protection of biodiversity has been detrimentally affected by a government focus on economic development, though efforts by civil society groups to remedy this have been stepped up. Coastal fisheries have dwindled due to the use of illegal and destructive gear, defiance of the ban period by the fishers and catching of undersized fish. Despite this, over-fishing continues as enforcement is weak.

Quality of Life and Livelihoods

The means by which households in the case areas maintain themselves have diversified significantly since 2013, in addition to incorporating changes in cropping patterns. This includes substantial outward migration from the case areas, driven in part by rural pressures but more so by the economic attractions of urban areas such as Khulna, Chittagong and Dhaka. Population levels have remained largely static in coastal regions, though the population is ageing and the fertility rate has decreased.

Long-term upward trends in literacy rates have continued, with education levels much improved on their 2014 levels.

Income levels are affected positively by a downturn in the importance of intermediaries in production processes, driven in part by rapid developments in mobile information technology and communications, price transparency and market access. Household storage of food has also increased, alleviating periods of scarcity somewhat. As a proportion of income, food is cheaper than it was in previous decades, with better eating habits and protein intake. However, this is offset by a slight increase in the incidence of non-communicable diseases and conditions, such as hypertension, with vector-borne (and zoonotic) diseases also rising, mainly as a result of rising temperatures and climatic conditions.

Increased household income coupled with continuing problems with significant disparities in income has resulted in a drop in inter-household equity, although this is complicated by broad advances in the participation of marginalised groups in society. Community power structures of patronage still govern much of rural society, but increasing involvement of the private sector and of Non-Governmental Orgaisations (NGOs) in local economic activities is changing the dynamic. Progress in the availability of mobile communications has enhanced awareness of legal rights and obligations, and improved access to information to a great degree. Enforcement of these rights has improved slightly, in line with some advancements in local enforcement capacity (through better local government empowerment), though these are somewhat restricted by a lack of progress in the capacity of local courts to process claims. This is highlighted by the disturbing lack of progress in tackling dakoits, which continues to blight the lives and economies of those who rely on fishing in particular.

More Sustainable (MS) than BAU

Land Use

Cultivated areas continued to be dominated by rice, but diversification of crops, especially the more intense cultivation of cash crops, driven by better access to markets (local and international) and effective agricultural extension and educational outreach, has flourished. The environmental impact of shrimp cultivation has decreased substantially in extent due to the adoption of more sustainable techniques. Investment in agricultural research and development, along with adoption of more climate-smart agricultural techniques, has bolstered the use of high-yield varieties and more salt-tolerant varieties because of the need to reduce the area under crops, in the interests of environmental protection and natural flood defence.

This pressure to reduce or at least maintain no more than existing levels of agricultural land has been helped by the general stabilisation in population numbers and continuing (if slightly reduced) rural-urban migration. The proportion of urban against rural populations has risen steadily, thereby increasing the need for greater intensification of agriculture, a process that has not been alleviated by the global market place.

Greater intensification of agriculture has led to a slight deterioration in soil quality parameters. This has been offset by special development programmes that have produced new crop varieties that are suitable for coastal areas and less hazardous to soil health. The proportion of chemical fertilisers and pesticides used has declined compared to organic manure and integrated pest management.

Coastal protection has been extended, mainly through the efforts of the Delta plan, using a mixture of structural and non-structural options. Better zoning and monitoring of land use change has been beneficial, and the quality of land use management is now one of the key factors in the management of water use. Conflict over land use, including over ownership rights, has been very much reduced, due mainly to improvements in transparency and accountability through the land ownership cadastre and significant improvements to the local judicial hierarchy.

Water

Surface water flow patterns in the Ganges and Brahmaputra Rivers have varied over time, the arrival of the monsoon has become less predictable, and periods of drought have become extended due to the impacts of climate change. With better coordination between the states riparian to these rivers, however, management of water resources in Bangladesh has been able to make progress. The application of efficient land and water management practices and effective enforcement processes in India have enhanced predictability and availability of flow into Bangladesh and reduced levels of industrial and nutrient pollution.

Similar progress has taken place in Bangladesh: advances in communications technology provide regulators with detailed knowledge of river flow, level and quality in real time, with sophisticated modelling ability aiding the regulation of water use management. Legal frameworks allow water use to be varied in response to changes in resource availability, social and environmental priorities, and the better balancing of periods of flood and inter-annual scarcity.

In line with the stronger economic situation in Bangladesh, water and sewage service provision have been extended, and the careful planning of urban expansion has greatly restricted water pollution and reduced the incidence of water-borne disease. This has increased riverine fish stocks and rural engagement, with cultured and floodplain fish production also growing. Subsistence and artisanal fisheries have decreased but commercial fisheries have conversely increased, though the impact of this has been reduced through improved national and international governance of fisheries, which is now focused on sustainable coastal fishing. The successful achievement of the Millennium Development Goals, and subsequent iterations, has created a society where the vast majority of the population have access to piped water in their homes and improved sanitation facilities. This has been aided through excavation of ponds and tanks for conservation of water and the use of local technology for water treatment, such as pond sand filtering.

This improvement in drinking water availability, combined with the use of deeper aquifers in many places, has helped people avoid the problems associated with consumption of saline- and arsenic-contaminated water. Steps have been taken to ensure sustainable use and management of groundwater. There has been a major focus on conjunctive management of surface water and groundwater. Better monitoring of water table levels and groundwater / surface water interactions, and the ability to amend water use rights, is progressively improving the situation, although alternative supplies may still be difficult to apply.

Adequate upland flow has been ensured in water channels through the construction of the Ganges Barrage, which has helped preserve the coastal estuary ecosystem threatened by seawater intrusion. With the rapid development of upstream energy generation facilities, sediment transport downstream has been curtailed. This remains a major issue for the health of the delta, but basin states are working together to formulate a solution under the terms of existing water use treaties. The increased focus on sediment has resulted in improved tidal basin management and increased navigation potential.

As part of the general improvement in the management of water resources, principles of subsidiarity have been applied such that local management of water takes better account of upstream and downstream needs. Cooperation between these has therefore improved, helped by the cross-sectoral management of water resources as a whole and effective compliance monitoring. Levels of conflict between users and sectors and justiciable disagreements have consequently fallen.

International Cooperation

Relations have greatly improved between Bangladesh and India, and between India and China, a process driven partly by the regionalisation of energy markets and the critical importance of hydropower as a fossil fuel replacement. Coordination of electricity generation at basin level, taking account of downstream impacts in terms of flood alleviation, augmentation of dry season flow, improved scarcity management and the sediment requirements of the delta, has sprung from a regional realisation that the benefits of cooperation can be spread equitably and strategically throughout the basin. Improved transport links between Chittagong and both Kunming in China and the north-eastern states of India, coupled with investments in the delta area by both upstream countries, have resulted in greater trade links between the three nations and more effective abstraction and pollution control in the upper reaches of the Brahmaputra and Ganges Rivers.

Detailed multi- and bi-lateral treaties have been agreed by GBM basin states addressing water issues, closely linked to agreements on trade and energy distribution. Independent management authorities are in place, with detailed compliance and reporting requirements, and national legal and policy frameworks work to effect these agreements.

International fisheries agreements relevant to the Bay of Bengal have led to greater food security for coastal fishermen, and improved enforcement has reduced levels of sea piracy.

Disaster Management

With the gradual decentralisation of Bangladesh, drawing populations from Dhaka to regional hubs, disaster management has also been further devolved, with disaster risk reduction being linked closely with adaptation. Disaster forecasting and preparedness is of world-standard quality, benefiting from advances in communication technology. The network of cyclone shelters, financed primarily by local and regional authorities and through private sector initiatives, has evolved such that the impact of increasing storm surges has been largely negated, with loss of life being maintained at relatively minimal levels.

Adaptive agri- and aqua-culture systems have also helped to substantially reduce production losses during and post disaster and aided post-disaster resilience. Storage of local crops and livestock has been significantly improved, with effective local insurance schemes in place to ameliorate livelihood losses. Improved transport networks between urban centres has also had a positive effect on the response times of emergency and remediation teams. The successful and ongoing implementation of the Delta Plan has been advantageous for disaster impact reduction.

Environmental Management

Mangrove forest cover has been maintained in the Sundarbans at the levels seen earlier in the century, augmented by active planting programmes that have taken place as part of the Delta Plan. The result has been an increase in terrestrial and aquatic biodiversity as the mangrove belt has expanded along the coast. The forest has benefited from improvements in water quality, but the balancing of livelihood maintenance for those living in the vicinity, and protection of biodiversity, remains problematic.

Improvements in the economic situation for those living in the case areas has reduced the need to use the Sundarbans directly for their livelihood maintenance, but a significant increase in 'eco-tourism', some of it still unregulated, continues to complicate matters.

Soil and water health has increased overall, driven by improvements in water quality and the use of state-of-the-art agricultural techniques. Although salt water intrusion remains problematic, better surface/groundwater management and improved polder maintenance have helped to keep this in check.

Quality of Life and Livelihoods

Standards of education in the countryside have leapt exponentially, especially for females. This, coupled with agricultural intensification and the managed expansion of decentralised urban hubs, has perpetuated general levels of migration away from the countryside. The gradual erosion of the traditional village and regional hierarchies and power structures has opened up a wide variety of possible livelihood alternatives for those in the case areas. The principal agents of this erosion have been the astonishingly rapid development of mobile technology (providing greater visibility for those working against the law), more effective enforcement mechanisms resulting from economic development, and improvements in educational ability stemming from enforced mandatory standards.

As regards population structure, fertility and mortality rates have been declining for some time now, and, critically, levels of out-migration to regional urban hubs have gone down slightly. The consequence of this is that population levels have dropped very slightly from their 2014 levels, but the structure has changed since then such that there are proportionately significantly more people aged over 65 and substantially fewer aged under 14.

The availability of credit has improved significantly, through a profusion of public and private providers, with reliance on local moneylenders non-existent. The availability of insurance for all has had significant impacts on the resilience of those in the case areas, reducing vulnerability to flood events, for example. Better access to local markets especially, combined with the diversification of crops, has improved the health of the population, although meat is very expensive and protein intake remains problematic for some. Incidence of hypertension has risen alarmingly as populations have grown more sedentary, with higher temperatures discouraging physical activity still further.

Levels of inter-household inequity have fallen in the case areas, as local remittances have increased, the gap between the richest elites and those on average incomes has narrowed with the crumbling of traditional social structures, and income levels for females have gone up (a process that has been mirrored at regional level, reducing income disparities more generally). This has also limited intra-household inequity, with male family members finding it progressively more difficult to maintain economic hegemony over others in their families. The number of NGOs has gone down over time, but their effectiveness has risen, in part because they are more coordinated, and in part because they are better positioned to take advantage of mobile technology.

Less Sustainable (LS) than BAU

Land Use

Areas that were formerly cultivated have been given over to a mixture of brackish shrimp and, to a lesser extent, rice, respectively serving the export market and local consumption needs of subsistence farmers.

Freshwater prawn production has decreased. Brackish shrimp production has taken increasingly large shares of cultivable land, pushing subsistence farm land into areas more vulnerable to inundation and less protected by coastal engineering infrastructure. More intensive rice cultivation is characterised by high levels of fertiliser use, although yields per hectare have not risen as fast as they might because R&D priorities have focused on producing shrimp for the richest nations.

Inter-sectoral cooperation (e.g. between fishermen and farmers) is on the decrease, and intra-sectoral conflict between the owners of industrial farming concerns (and their tenant farmers) and subsistence farmers is growing. Scarcity of available secure land and the difficulty in obtaining clean water for irrigation from reduced water resources exacerbates disagreements. Agricultural extension officers prioritise the production of exportable crops, leaving subsistence farmers struggling to take advantage of new techniques and subsidies, and subject to heightened levels of insecurity as seasonal cropping patterns change with the climate.

In addition to the encroachment of brackish shrimp production, mangrove forests have been slowly sacrificed to commercial agriculture, salt pans and unplanned urban spread, as a result of a combination of the government need for hard currency, increasing soil and surface water salinity, and population migration from rural poverty. Vulnerability to flooding has therefore increased as natural barriers have been removed and existing embankments are poorly managed due to lack of financial resources and sectoral conflicts. While floodplain and land use zoning is in place, implementation levels are low because of a lack of enforcement.

Water

Water resources have decreased significantly as a result of a combination of a number of factors: the rapid development of dams and barrages constructed upstream for the purposes of energy production; flood alleviation and irrigation schemes; the impact of the now fully-implemented Inter-linking Rivers Project; and large-scale transfers from the Brahmaputra River in China to provide water for northern irrigation schemes and domestic consumers in Beijing. The efficiency of industrial agricultural irrigation is high, but this is heavily reliant on the unregulated use of groundwater (driven in part by energy subsidies that fuel pumping), necessary because of the lack of surface water flow and the need to access higher-quality water untainted by polluted surface water.

The unfettered use of groundwater from the less saline shallow aquifers in the northern part of the southwest coastal zone, coupled with the rise in sea level, has hastened saline intrusion of aquifers. The spread of unplanned urban settlements, especially in Dhaka, driven by population growth in the country as a whole and by out-migration from coastal areas, has adversely affected water quality downstream as a result of a lack of sewage treatment works. Early advances in achieving development goals have been undermined by this population growth. Although economic gains have to a certain extent continued, they have not been sufficient to counteract changes in population patterns and location.

Levels of cooperation between upstream and downstream districts have decreased within Bangladesh, mirroring the rise in inter-sectoral conflict between land and water users. As land use ownership patterns have moved to a greater proportion of tenant farmers, local water management institutions have found themselves toothless and ineffective, with longer-term management decisions being compromised by short-term priorities.

International Cooperation

Cooperation in terms of access to global markets has increased in some ways, although exports are very much higher than imports. Cooperation at the more regional level has, however, deteriorated, with basin co-riparians in direct competition with each other, especially with respect to agricultural commodities. This has destroyed efforts to manage regional watercourses at the basin level, with corresponding impacts on the amount of freshwater flowing into Bangladesh. Remaining basin-level governance efforts are focused on maintaining flows needed for commercial agriculture and aquaculture.

Disaster Management

Although there has been some increase in the extent of coastal defence and emergency infrastructure (such as cyclone shelters), maintenance efforts have concentrated on protecting agricultural investments. This has resulted in a creeping process of polderisation in downstream areas, although storage of harvested subsistence crops has increased at village level. These are seldom strong enough to withstand the pressures from cyclones and storm surges, however. Loss of life as a result of these pressures remains low, but the disproportionately high numbers of female deaths mean that impacts on livelihoods are drastic.

Environmental Management

Water quality has been detrimentally affected by the relatively low surface water flows coming into Bangladesh and diffuse pollution as a consequence of the liberal use of fertilisers both upstream and in Bangladesh itself. This has been compounded by the effluent resulting from the expansion of unplanned informal settlements. Encroachment in areas previously covered by mangrove has continued, with commensurate effects on biodiversity and the capacity of supporting ecosystem services. Civil society efforts to combat loss of biodiversity have been dissipated by a lack of inter- and intra-sectoral coherence, although the incidence of poverty has been responsible for an increase in the numbers of Civil Society Organisation (CSOs). Fish stocks in coastal rivers are under severe pressure, as are coastal fisheries, partly as a result of irresponsible shrimp farming methods and partly because of poor regulation and enforcement.

Levels of water-borne diseases have risen because poorer families have few alternatives to using contaminated surface water for domestic use: groundwater levels have fallen below the limits of cheap pumps, and salt-water intrusion is common.

Quality of Life and Livelihoods

The embedded power structures characteristic of rural Bangladesh at the beginning of the twenty-first century have become even more entrenched as local elites take advantage of the economic gains to be made through the production of brackish shrimp and the low cost of labour. Outward migration to urban centres within Bangladesh, particularly Dhaka, has risen as populations have grown and commercialisation of agriculture has reduced still further labour needs in rural areas. Ever-expanding urban areas and low employment opportunities in cities mean that monetary transfers back to rural areas by migrant workers have reduced markedly, and migration out of Bangladesh to traditional remittance-generating regions has become more challenging as the traditional international migration destinations are now much more selective about immigration because of the sheer volume of immigrant labour sources globally.

Population levels in the case areas have not changed drastically in recent decades, but this is only because higher fertility levels have been offset by stubbornly high mortality rates and the marked increase in outward migration. Livelihood sources also have not changed greatly, though the number of older tenant farmers has risen, as people of working age have moved to industrial farms for employment, leaving the young and old behind. Remittances from family members who have moved abroad or to urban centres have diminished, but the capacity of the land to support the growing population, coupled with climate-driven changes in cropping cycles, has meant that such migration has become a necessity.

This is exacerbated by the outward movement of those whose livelihoods have been destroyed by storms. Those living in the largely unplanned informal urban settlements are often forced to live in a hand-to-mouth way, with only the luckiest progressing on to secure jobs. Family structures are weaker than they were 30 or 40 years ago, although family networks are of great importance in maintaining remittance levels at even their current position.

Those working in industrial agricultural operations enjoy greater security of income, although salaries are kept low by the constant need to keep Bangladesh competitive in a very difficult market. Subsistence farmers remain almost completely outside national and international markets, and are unable to take advantage of technological advances in mobile telecommunications. The main developing market for those engaged in business outside the major agricultural conglomerates lies in West Bengal, with cross-border trade in the area between Kolkata and Khulna growing rapidly, a process aided by the gradual destruction of the Sundarban mangroves. Electricity distribution networks are unreliable especially in coastal areas, an ongoing problem caused mainly by the poverty in the area and the high frequency of damage by storms. Food and protein scarcity in subsistence areas have become a problem, leading to an increase in open water fishing by residents, despite the risks. The incidence of vector-borne (and zoonotic) diseases has risen, mainly as a result of rising temperatures and climatic conditions.

The erosion of family structures has, surprisingly, raised levels of intra-household equity as earners of any kind have become more important, but inter-household equity has dropped as the split between subsistence and tenant farmers has deepened. More urgent efforts by NGOs and CSOs to help the very poorest have been beneficial in terms of encouraging broader civic participation, but the power differential between largely locally-focused groups and the large-scale farming concerns has rendered the work of the former largely irrelevant. Earlier weaknesses in local dispute resolution and access to rights have multiplied with the involvement of local elites in wealth development activities. Creeping centralisation over a period of decades has left an emboldened local governance framework characterised by lack of accountability and transparency, and an absence of central oversight. Backlogs in local courts have fallen, but this is the result not of greater efficiency but of an increasing fatalism on the part of the aggrieved population.

The ability of the poorest to access lending facilities is very restricted as formal institutions are reluctant to lend. The poor remain reliant on lending at usurious rates by local lenders. The increased incidence of piracy further affects livelihoods, especially those of fishers, whose numbers are dwindling as stocks collapse and migration becomes more attractive.

Note

1. A full list of the stakeholders, along with details of their attendance at project meetings, is available at www.espadeltas.net

References

Allan, A.A., M. Lim, N. Islam, and H. Huq. 2013. *Livelihoods and ecosystem service provision in the southwest coastal zone of Bangladesh: An analysis of legal, governance and management issues.* ESPA Deltas Working Paper #1. University of Dundee. http://espadelta.net/resources_/working_papers/. Accessed 3 Oct 2016.

Arnell, N.W., T. Kram, T.R. Carter, K.L. Ebi, J. Edmonds, S. Hallegatte, E. Kriegler, R. Mathur, B. O'Neill, K. Riahi, H. Winkler, D. Van Vuuren, and T. Zwickel. 2011. A framework for a new generation of socioeconomic scenarios for climate change impact, adaptation, vulnerability and mitigation research. Working Paper. https://www2.cgd.ucar.edu/sites/default/files/icon-ics/Scenario_FrameworkPaper_15aug11.pdf. Accessed 20 July 2016.

O'Neill, B.C., T.R. Carter, K.L. Ebi, J. Edmonds, S. Hallegatte, E. Kemp-Benedict, E. Kriegler, L.O. Mearns, R.H. Moss, K. Riahi, B. Van Ruijven, and D. Van Vuuren. 2012. *Meeting report of the workshop on the nature and use of new socio-economic pathways for climate change research*, November 2–4, 2011. Boulder: National Center for Atmospheric Research (NCAR). http://www2.cgd.ucar.edu/research/iconics/events/20111102. Accessed 20 July 2016.

Payne, G., and J. Payne. 2004. *Key concepts in social research*. London: Sage.

Prell, C., K. Hubacek, and M. Reed. 2009. Stakeholder analysis and social network analysis in natural resource management. *Society and Natural Resources* 22 (6): 501–518. https://doi.org/10.1080/08941920802199202.

Reed, M.S., A. Graves, N. Dandy, H. Posthumus, K. Hubacek, J. Morris, C. Prell, C.H. Quinn, and L.C. Stringer. 2009. Who's in and why? A typology of stakeholder analysis methods for natural resource management. *Journal of Environmental Management* 90 (5): 1933–1949. https://doi.org/10.1016/j.jenvman.2009.01.001.

11

Regional Climate Change over South Asia

John Caesar and Tamara Janes

11.1 Introduction

The combination of a highly variable climate, a large and increasing population and a high reliance on water dependent sectors such as agriculture means that South Asia could be particularly at risk to future climate change and variability. Changes in the climate occur in a variety of ways, including changes in mean temperature, changes in extreme daily maximum temperatures and changes in the length, frequency or magnitude of heatwaves. Similarly, increases or decreases in the frequency and magnitude of precipitation could affect water availability or change the characteristics of floods or droughts. For a monsoon climate, such as South Asia, changes in seasonality could also have a large impact on the quantities and timings of water availability, a critical factor in agrarian societies. Climate change and variability is therefore a key underpinning input into many of the biophysical models which form part of this research. The focus in this chapter is on climate projections

J. Caesar (✉) • T. Janes
Met Office Hadley Centre for Climate Science and Services,
Exeter, Devon, UK

© The Author(s) 2018
R. J. Nicholls et al. (eds.), *Ecosystem Services for Well-Being in Deltas*,
https://doi.org/10.1007/978-3-319-71093-8_11

207

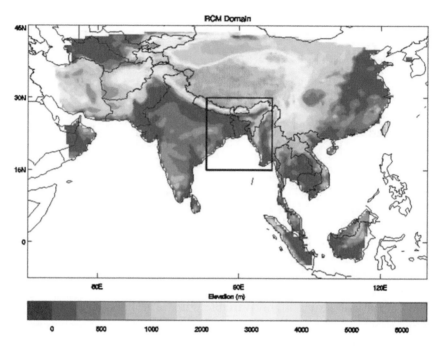

Fig. 11.1 Regional model domain for South Asia used in this research showing land surface elevation. The solid box (83–97°E, 15–30°N) shows the analysis region used throughout this chapter. Area averaged values used throughout this chapter are based upon all land grid points within this box (Caesar et al. 2015—Reproduced by permission of The Royal Society of Chemistry)

over the land regions of Bangladesh and the wider Ganges-Brahmaputra-Meghna (GBM) basins (Fig. 11.1). Climate change over the marine regions of the Bay of Bengal is considered in more detail in Chap. 14.

This chapter provides an overview of the climate change projections used and a consideration of the uncertainty in those projections. The basis for the projections is a widely used medium-high scenario (SRES A1B; Nakićenović et al. 2000) of global increases in greenhouse gas emissions. An important component of this research is the use and integration of a range of biophysical models where climate model data (both for the current period and future projections) is a key input. Whilst this chapter focuses upon temperature and rainfall projections, a wide range of other climate variables (atmospheric, land surface and ocean) were provided from the model simulations to input into other biophysical models.

11.2   Climate of Bangladesh

Bangladesh is widely recognised as being vulnerable to extreme weather events and climate change. A large proportion of the country's topography is low-lying and forms the delta of the GBM rivers (Karim and Mimura 2008). The country is densely populated and socio-economically challenged with agriculture playing a significant role in livelihoods. Bangladesh has a tropical monsoon climate, and this brings high temperature, heavy rainfall, high humidity and strong seasonal variations. Changes in precipitation can lead to enhanced flooding or drought as well as dislocation between the timing of rainfall and the agricultural calendar. Seasons can be separated into four groups, differentiated by the percentage of the mean annual rainfall falling in each season (Islam and Uyeda 2005):

- Pre-monsoon (March–May) ~20 per cent
- Monsoon (June–September) ~62.5 per cent
- Post-monsoon (October–November) ~15.5 per cent
- Winter (December–February) ~2 per cent

These estimates vary between studies, and post-monsoon rainfall in particular can be affected by the impacts of cyclones, but the bulk of annual rainfall occurs during the summer monsoon, with some studies reporting higher proportions of up to 70–80 per cent of the annual total (Shahid 2010a). Winter generally sees much cooler and drier conditions and accounts for only two to four per cent of annual rainfall totals. Tropical cyclones are a particular weather-related hazard for Bangladesh, and can bring periods of intense rainfall, particularly during the post-monsoon season.

Studies that have examined observed climate changes over Bangladesh show that during the twentieth century annual mean temperatures increased by around 0.5 °C (Immerzeel 2008) and were accompanied by increasing rainfall (Shahid 2010b).

Sea-level rise and storm surges associated with tropical cyclones are key risks for this region. These factors are addressed further in Chaps. 14 and 16, including future projections of extreme storm surges. This assessment uses output from the regional climate simulations (which are atmosphere-only models) presented in this chapter, alongside sea-level rise projections from the associated global models, which simulate both the atmosphere and oceans, albeit at a coarser spatial resolution.

11.3   Climate Model Simulations

Simulations of the global climate system are made using computer models, known as general circulation models (GCMs), and these are often used to simulate the impact of different greenhouse gas concentrations in the atmosphere in the future. GCMs typically have coarse spatial resolutions and simulate the global atmosphere split into large 'boxes' of several hundred kilometres in size. They are not necessarily able to provide the high-resolution climate change information that is now often required by climate impacts and adaptation studies. To enable a high-resolution representation of climate change at regional and local levels, regional climate models (RCMs) can be used to provide higher-resolution grids (typically 50 km or finer) and provide better representations of features such as coastlines and mountains and their effects on the climate (Giorgi et al. 2009). RCMs take output of the global scale GCM and use it to drive a simulation over a specific region at a much higher spatial resolution. In this work, the Met Office Hadley Centre HadRM3P model with a 25 km × 25 km spatial grid resolution has been used, downscaling the output of the HadCM3 GCM (Caesar et al. 2015). The focus in the past has tended to be on climate modelling over the wider South Asia region, and there had been few studies using high-resolution climate models over the Bangladesh region, though this number is beginning to increase (e.g. Caesar et al. 2015; Kumar et al. 2013; Mathison et al. 2013).

Previous climate model studies have projected increases in temperature over Bangladesh by the end of the twenty-first century (e.g. May 2004; Agrawala et al. 2003). Projections also indicate an increase in annual rainfall totals in the future (Immerzeel 2008; Islam et al. 2008) and an increase in the intensity of rainfall events (May 2004; Agrawala et al. 2003). The uncertainty associated with these model projections is large, partly as a result of uncertainty in the possible range of future greenhouse gas emissions, and also in part due to differences between the characteristics of different climate models (Immerzeel 2008; Islam et al. 2008; IPCC 2013). The climate projections used for this research make use of an approach called a 'perturbed physics ensemble' (Collins et al. 2006). This involves running the GCM multiple times but each time making small changes to the representation of certain physical processes which may be uncertain and difficult to constrain using observations (Murphy et al. 2004). One example of such a variable is critical relative humidity, which relates the humidity of the atmosphere in the model grid box with the amount of cloud simulated. Model errors are assessed against observational data spanning a range of variables. This approach is one method of representing and assessing uncertainty in the climate models, and is an alternative approach to using projections from a range of different climate models (Collins et al. 2011).

Currently available climate models developed at different institutions represent the Asian monsoon with variable skill (Turner and Slingo 2009) and present conflicting outcomes; in some cases future strengthening of the summer monsoon circulation are projected (May 2004) and in some cases a weakening (Sabade et al. 2011). For validation, it is therefore important that climate models simulate the present-day rainfall during the summer monsoon period reasonably well in order to produce user-relevant climate projections into the future. However, it should be noted that a model which performs well in a present-day evaluation may not necessarily perform as well under different future forcing scenarios (Knutti 2008).

To gain an understanding of how well the RCM used in this research simulates present-day climate, a range of observational datasets of temperature and rainfall were compared with the model output for the period of 1981–2000. The comparison for the region shows that the simulations provide a good representation of the spatial patterns of temperature and rainfall for 1981–2000. Considering the annual cycles, the simulations tend to overestimate temperatures by around 1 °C during March to June (Fig. 11.2, upper). In terms of representing the timing and magnitudes of monsoon rainfall, the simulations tend to be reasonably effective, but show a slightly early peak in monsoon rainfall, and an overestimation of

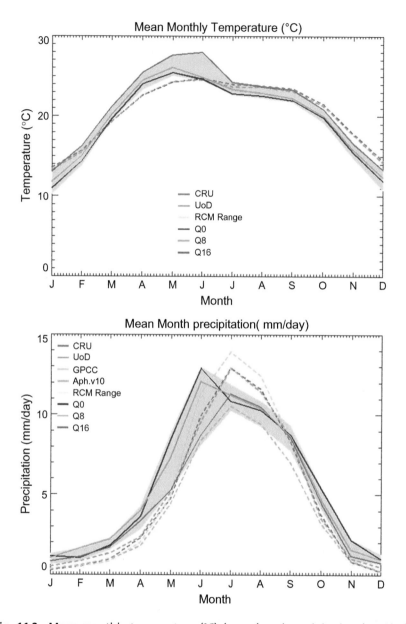

Fig. 11.2 Mean monthly temperature (°C) (upper) and precipitation (mm/day) (lower) over the GBM sub-region (shown in Fig. 11.1) for the period of 1981–2000 for the RCM 17 member ensemble range (shaded), three selected RCM ensemble members (Q0, Q8 and Q16), and the relevant observational datasets. Dashed lines indicate observations (Caesar et al. 2015—Reproduced by permission of The Royal Society of Chemistry)

rainfall compared with observations through most of the year, except during the monsoon season itself when simulated rainfall amounts fall within the observational range (Fig. 11.2, lower).

11.4   Future Climate Change Projections for Bangladesh

To simulate future climate change, a number of inputs to the model need to be defined. These include the concentrations of greenhouse gasses such as carbon dioxide and methane. A range of future scenarios exist which represent a spectrum from low to high future greenhouse gas emissions. These are determined based upon detailed scenarios which represent a storyline of how the world could develop and includes factors such as potential economic and technological changes (Nakićenović et al. 2000).

The climate model simulations use the Special Report on Emissions Scenarios (SRES) A1B scenario, which was generated for the IPCC Third Assessment Report (2001), but is still used widely in climate change studies (IPCC 2014). The A1B scenario assumes a global future of strong economic growth and an increase in the rate of greenhouse gas emissions over the twenty-first century. The IPCC Fifth Assessment Report (IPCC 2013) uses a new set of scenarios called Representative Concentration Pathways (RCPs) (Moss et al. 2010). RCPs differ from the SRES scenarios in that SRES scenarios used socio-economic storylines as the basis to derive climate change pathways, whereas RCPs start from different levels of radiative forcing at the year 2100, which are then used to derive greenhouse gas concentration pathways for the twenty-first century. Whilst the methodologies are different, which can make comparisons between the scenario sets challenging, it is possible to identify analogues between the two sets of scenarios (Rogelj et al. 2012). SRES A1B is situated between the two high-end RCPs (RCP6.0 and RCP8.5) in terms of the projected global temperature increase by the year 2100 and atmospheric CO_2 concentration levels. The projections from the RCM used in this study show a consistent signal of increasing temperatures over the region during the twenty-first century (Fig. 11.3). By the 2060s, the

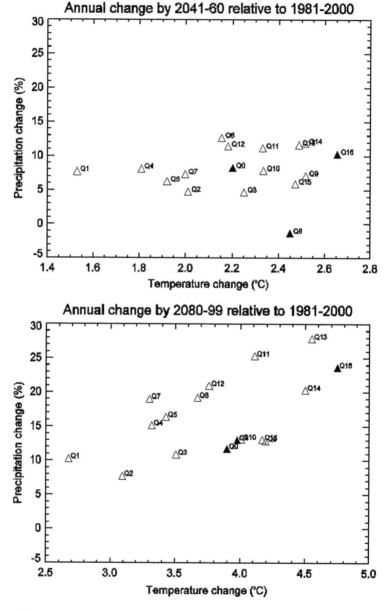

Fig. 11.3 Annual mean temperature and precipitation responses in the 17 regional climate model simulations over the GBM region for 2041–2060 (upper) and 2080–2099 (lower). Filled triangles indicate the three climate projections selected (Caesar et al. 2015—Reproduced by permission of The Royal Society of Chemistry)

temperature, based upon the A1B emissions scenario, is projected to increase by between 1.5 °C and 2.7 °C, and by the 2090s it is in the range of 2.6–4.8 °C.

There is also a consistent signal, using this particular climate model, towards increasing annual rainfall by the end of the twenty-first century (Fig. 11.2). The range of projections for the 2080s is between +8 per cent and +28 per cent relative to 1981–2000. However, it should be noted that year to year variability is large. By 2050 the range is between −1.4 per cent and around +13 per cent compared to present day. The projections also show a decrease in light to moderate rainfall events, a slight increase in heavy events and a large increase in very heavy events (Caesar et al. 2015). So, despite a decrease in the overall number of days with rainfall, the intensity on the days when rainfall occurs is projected to increase. This could result in flooding and have an impact upon the availability of fresh water during the agricultural calendar.

From the total of 17 RCM simulations, three were selected for application within the other biophysical models (Table 11.1, Figs. 11.4 and 11.5). One was chosen based upon the model variant that represents the 'standard' model parameter configuration (known as Q0). Two additional simulations were chosen for their ability to span a wide range of uncertainty across the full collection of 17 model simulations. The first of these, Q8 indicates warmer but drier conditions compared to the present day in the 2060s, but warmer and wetter by 2090. The other simulation (Q16) projects the largest temperature increase by both 2050 and 2090, and also large precipitation increases at both 2050 and 2090.

Table 11.1 Annual mean temperature and precipitation changes relative to 1981–2000 in three selected regional climate model projections at the middle and end of the twenty-first century over the GBM region

Model simulation	Temperature (°C)		Precipitation (%)	
	2041–2060	2080–2099	2041–2060	2080–2099
Q0	+2.20	+3.90	+8.26	+11.75
Q8	+2.45	+3.98	−1.35	+13.01
Q16	+2.65	+4.75	+10.28	+23.66

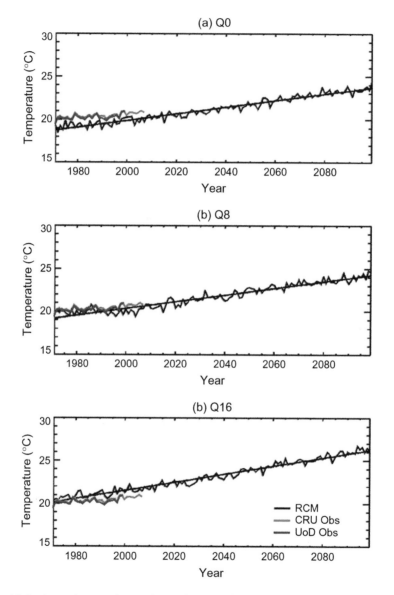

Fig. 11.4 Annual mean time series and trend of air temperature at 1.5 m for each of the selected climate projections from 1971–2099 inclusive for the GBM region. Coloured lines indicate two observational datasets (Caesar et al. 2015— Reproduced by permission of The Royal Society of Chemistry)

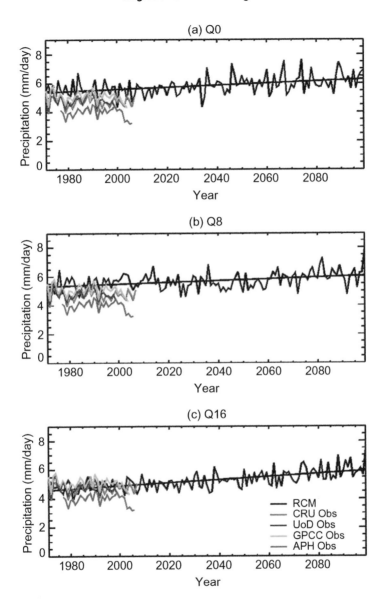

Fig. 11.5 Annual mean time series and trend of precipitation for each of the three selected climate projections from 1971–2099 inclusive for the GBM region. Coloured lines indicate four observational datasets (Caesar et al. 2015— Reproduced by permission of The Royal Society of Chemistry)

11.5   Conclusions

A total of 17 high-resolution regional climate model simulations were used to asses future climate change over Bangladesh, with three being selected to drive a range of biophysical impact models. The simulations used the SRES A1B greenhouse gas emissions scenario, which is a mid- to high-range representation of potential future emissions. Three simulations, which represent a range of future temperature and rainfall changes, were selected for use in the more detailed impact sector models.

The projections show an increase in temperature over the region through the twenty-first century, of between around 2.6 °C and 4.8 °C by 2100. Rainfall changes are projected to be more variable, but generally show increases ranging from 8 to 28 per cent by the end of the century. Additionally, the frequency of heavy rainfall events is projected to increase in the future, with decreases in the frequency of lighter rainfall. This indicates a shift towards a lower number of wet days, but an increase in intensity on those days when rain falls. This implies an increased risk of flash flooding and a potential change in the freshwater availability across Bangladesh.

As further high-resolution climate projections become available for the Bangladesh region, it will be possible to better assess the differences between different climate models, and also a wider range of greenhouse gas emissions scenarios, and therefore provide improved assessments of the ranges and uncertainties associated with future climate change over Bangladesh.

References

Agrawala, S., T. Ota, A.U. Ahmed, J. Smith, and M.V. Aalst. 2003. *Development and climate change in Bangladesh: Focus on coastal flooding and the Sundarbans.* Paris: Organisation for Economic Co-operation and Development. www.oecd.org/env/cc/21055658.pdf. Accessed 28 June 2016.

Caesar, J., T. Janes, A. Lindsay, and B. Bhaskaran. 2015. Temperature and precipitation projections over Bangladesh and the upstream Ganges, Brahmaputra and Meghna systems. *Environmental Science-Processes and Impacts* 17 (6): 1047–1056. https://doi.org/10.1039/c4em00650j.

Collins, M., B.B.B. Booth, G.R. Harris, J.M. Murphy, D.M.H. Sexton, and M.J. Webb. 2006. Towards quantifying uncertainty in transient climate change. *Climate Dynamics* 27 (2–3): 127–147. https://doi.org/10.1007/s00382-006-0121-0.

Collins, M., B.B.B. Booth, B. Bhaskaran, G.R. Harris, J.M. Murphy, D.M.H. Sexton, and M.J. Webb. 2011. Climate model errors, feedbacks and forcings: A comparison of perturbed physics and multi-model ensembles. *Climate Dynamics* 36 (9–10): 1737–1766. https://doi.org/10.1007/s00382-010-0808-0.

Giorgi, F., C. Jones, and G.R. Asrar. 2009. Addressing climate change needs at the regional level: The CORDEX framework. *World Meteorological Organisation Bulletin* 58 (3): 175–183.

Immerzeel, W. 2008. Historical trends and future predictions of climate variability in the Brahmaputra basin. *International Journal of Climatology* 28 (2): 243–254. https://doi.org/10.1002/joc.1528.

IPCC. 2013. *Climate change 2013: The physical science basis. Contribution of working group I to the fifth assessment report of the intergovernmental panel on climate change.* Cambridge, UK/New York: Cambridge University Press.

———. 2014. *Climate change 2014: Impacts, adaptation, and vulnerability. Part A: Global and sectoral aspects.* Contribution of working group II to the fifth assessment report of the intergovernmental panel on climate change. Cambridge, UK/New York: Cambridge University Press.

Islam, M.N., and H. Uyeda. 2005. Comparison of TRMM 3B42 products with surface rainfall over Bangladesh. In *Geoscience and remote sensing symposium (IGARSS) 2005*, 25–29 Seoul.

Islam, M.N., M. Rafiuddin, A.U. Ahmed, and R.K. Kolli. 2008. Calibration of PRECIS in employing future scenarios in Bangladesh. *International Journal of Climatology* 28 (5): 617–628. https://doi.org/10.1002/joc.1559.

Karim, M.F., and N. Mimura. 2008. Impacts of climate change and sea-level rise on cyclonic storm surge floods in Bangladesh. *Global Environmental Change-Human and Policy Dimensions* 18 (3): 490–500. https://doi.org/10.1016/j.gloenvcha.2008.05.002.

Knutti, R. 2008. Should we believe model predictions of future climate change? *Philosophical Transactions of the Royal Society A: Mathematical, Physical and Engineering Sciences* 366 (1885): 4647. https://doi.org/10.1098/rsta.2008.0169, Licence https://doi.org/10.1098/rsta.2008.0169.

Kumar, P., A. Wiltshire, C. Mathison, S. Asharaf, B. Ahrens, P. Lucas-Picher, J.H. Christensen, A. Gobiet, F. Saeed, S. Hagemann, and D. Jacob. 2013. Downscaled climate change projections with uncertainty assessment over India using a high resolution multi-model approach. *Science of the Total Environment* 468: S18–S30. https://doi.org/10.1016/j.scitotenv.2013.01.051.

Mathison, C., A. Wiltshire, A.P. Dimri, P. Falloon, D. Jacob, P. Kumar, E. Moors, J. Ridley, C. Siderius, M. Stoffel, and T. Yasunari. 2013. Regional projections of North Indian climate for adaptation studies. *Science of the Total Environment* 468: S4–S17. https://doi.org/10.1016/j.scitotenv.2012.04.066.

May, W. 2004. Potential future changes in the Indian summer monsoon due to greenhouse warming: Analysis of mechanisms in a global time-slice experiment. *Climate Dynamics* 22 (4): 389–414. https://doi.org/10.1007/s00382-003-0389-2.

Moss, R.H., J.A. Edmonds, K.A. Hibbard, M.R. Manning, S.K. Rose, D.P. van Vuuren, T.R. Carter, S. Emori, M. Kainuma, T. Kram, G.A. Meehl, J.F.B. Mitchell, N. Nakicenovic, K. Riahi, S.J. Smith, R.J. Stouffer, A.M. Thomson, J.P. Weyant, and T.J. Wilbanks. 2010. The next generation of scenarios for climate change research and assessment. *Nature* 463 (7282): 747–756. https://doi.org/10.1038/nature08823.

Murphy, J.M., D.M.H. Sexton, D.N. Barnett, G.S. Jones, M.J. Webb, M. Collins, and D.A. Stainforth. 2004. Quantification of modelling uncertainties in a large ensemble of climate change simulations. *Nature* 430 (7001): 768–772. https://doi.org/10.1038/nature02771.

Nakićenović, N., J. Alcamo, G. Davis, B. Devries, J. Fenhann, S. Gaffin, K. Gregory, A. Gruebler, T.Y. Jung, T. Kram, E. Lebre Larovere, L. Michaelis, S. Mori, T. Morita, W. Pepper, H. Pitcher, L. Price, K. Riahi, A. Roehrl, H.-H. Rogner, A. Sankovski, M. Schlesinger, P. Shukla, S. Smith, R. Swart, S. Vanrooijen, N. Victor, and Z. Dadi. 2000. *Special report on emissions scenarios, a special report of working group III of the intergovernmental panel on climate change.* Cambridge, UK: Cambridge University Press.

Rogelj, J., M. Meinshausen, and R. Knutti. 2012. Global warming under old and new scenarios using IPCC climate sensitivity range estimates. *Nature Climate Change* 2 (4): 248–253. https://doi.org/10.1038/nclimate1385.

Sabade, S.S., A. Kulkarni, and R.H. Kripalani. 2011. Projected changes in South Asian summer monsoon by multi-model global warming experiments. *Theoretical and Applied Climatology* 103 (3): 543–565. https://doi.org/10.1007/s00704-010-0296-5.

Shahid, S. 2010a. Rainfall variability and the trends of wet and dry periods in Bangladesh. *International Journal of Climatology* 30 (15): 2299–2313. https://doi.org/10.1002/joc.2053.

———. 2010b. Recent trends in the climate of Bangladesh. *Climate Research* 42 (3): 185–193.

Turner, A.G., and J.M. Slingo. 2009. Uncertainties in future projections of extreme precipitation in the Indian monsoon region. *Atmospheric Science Letters* 10 (3): 152–158. https://doi.org/10.1002/asl.223.

12

Future Scenarios of Economic Development

Alistair Hunt

12.1 Introduction

The research reported in this book has, as its overarching focus, environmental change and its interaction with natural resource management and poverty alleviation in the Bangladesh delta region. It therefore undertakes bio-physical and socio-environmental modelling of change in the region to 2100 and 2050, respectively, framed by three socio-economic scenario narratives that describe alternative policy contexts to 2050. These narratives (see Appendix to Chap. 10) outline possible trends in land use, water management, international cooperation, disaster management, environmental management and quality of life and livelihoods. In order to enhance the usefulness of these scenarios for national stakeholder groups, including the Government of Bangladesh, this chapter provides further descriptive detail about the economic development paths implied by these scenarios. The policy levers available to national and regional administrations to influence the likelihood of

A. Hunt (✉)
Department of Economics, University of Bath, Bath, UK

© The Author(s) 2018
R. J. Nicholls et al. (eds.), *Ecosystem Services for Well-Being in Deltas*,
https://doi.org/10.1007/978-3-319-71093-8_12

223

individual development paths are also highlighted. Thus, this chapter serves to emphasise the high degree to which the direction of economic development and the extent to which environmental risks threaten this remains in the hands of policy makers in Bangladesh.

12.2 Principal Components of Macro-economic Change: Bangladesh and Delta

Figure 12.1 identifies the critical components of macro-economic performance in the delta and in Bangladesh that are included in the scenario analysis. The outer ring identifies components of the enabling conditions for macro-economic performance, whilst the inner ring identifies the direct aspects of macro-economic management. Macro-economic indicators, along with equity and poverty, are the key outcomes of the evolution of these enabling conditions over time. However, the arrows serve to highlight that the linkages to poverty and equity objectives may be

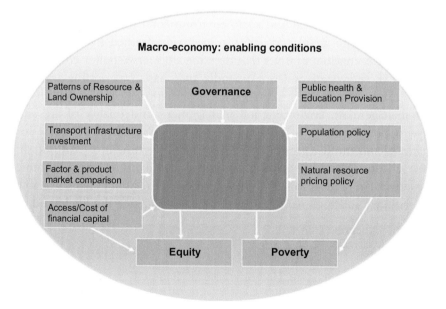

Fig. 12.1 Components of macro-economic development

influenced directly by the macro-economic enabling conditions, as well as through macro-economic performance. For example, implementation of natural resource pricing policy may result in resource ownership rights being distributed more equitably than currently, thereby reducing poverty, but without having immediate macro-economic consequences.

In the following subsections, a brief description of the ways in which these individual components help to determine patterns of macro-economic development is provided. How they relate specifically to the delta region is identified, and how in broad terms their development may evolve under the three descriptive scenarios to 2030 can be characterised. This period is adopted in accordance with the long-term planning horizons adopted by the Government of Bangladesh and development partners. Within these scenario characterisations, an indication of how the bio-physical changes identified in the integrated model (see Chap. 28), and the research more broadly, may interact with the components of macro-economic development is included. The characterisations of macro-economic components are developed on the basis of the experience and judgement and are informed primarily by the data compiled in the series of reports prepared as background input to the preparation of the Seventh Five-Year Plan for Bangladesh (GED 2015). Thus, whilst they are indicative, it is hoped that the wider stakeholder community would be able to provide further validation in the event that they are used to inform policy development.

Data is reviewed and derived from (i) existing short-medium-term projections made by the General Economics Division (GED), World Bank, Asian Development Bank and (ii) longer-term projections— Shared Socio-economic Pathways (SSPs)—created by the climate change research community. It is then compiled data to provide coherent outlines of three economic development paths consistent with the scenario narratives.

12.2.1 Current Macro-economic Context

Trends in gross national product (GNP) and GNP per capita both have a sharply rising profile since 2000. Economic growth has been a result of an expansion in the export sector, driven by clothing and overseas remittances, together with the growth of non-farm employment that has raised agricultural productivity. Until the recent outbreak of political instability, further rapid growth seemed guaranteed, as low-cost manufacturing, as well as information technology (IT) and service industries, began rapid development. Though these trends will not be halted, uncertainties in the political environment may, for example, deter inward foreign direct investment in these sectors. The following subsections outline potential developments in the individual macro-economic enabling components over the longer term, to 2030.

12.2.2   Patterns of Resource and Land Ownership

The extent to which land has legal security and enforceability of ownership is important in determining whether and how an efficient market for land can operate. Security of ownership influences the form and extent of investment that an individual or business makes in their use of land. In general, it is expected that if ownership rights of land exist the individual/business will be willing to invest more resources in the land, on the assumption that they will be able to earn an income over a longer time period. Similarly, if a farmer owns rather than rents land they are likely to invest more resources in machinery that raise the productivity of the land, which increases income and over time allows for the repayment of the investment.

The role of insecure land ownership and access to land in determining the productivity of agriculture, aquaculture and mangroves is significant in the delta region. At present, unequal access to land and its ownership means that a large proportion of the rural population in the area have no incentive, or means, to invest in equipment that could increase yields and introduce more environmentally sustainable production practices. Hossain (2015) comments that "the diversity of ways by which land records is updated and the problems associated with each, give rise to numerous disputes in which the rich and powerful inevitably enjoy the upper hand."

12.2.2.1   Future Scenarios: To 2030

Business As Usual The frequency of land ownership disputes slowly declines as the legal rules surrounding ownerships are made more transparent. Consequently, economic productivity rises—exacerbated by the continuation of the trend towards urbanisation and industrial production. Responses to climate change risks are generally made on a piecemeal, reactive basis.

More Sustainable The institutional framework is rationalised so that the system for recording or registration of property rights is streamlined and so reduces the potential for land conflicts. Security of land-based assets for landless and other low-income groups is increased, raising the returns they can make. Pro-active responses to anticipated climate change risks to land availability and quality are introduced as a result of effective co-ordinated action by the state and other stakeholders.

Less Sustainable Existing vested interests continue to prevent significant reform of land ownership, resulting in ongoing time-consuming legal conflicts and the maintenance of the current inequalities in land ownership patterns. Degradation of productive land mass due to cyclones in the delta region results in exacerbated land conflicts.

12.2.3   Levels of Public Health and Education Provision

Additional to its social and cultural importance, the economic role of education is to increase the availability of human capital, including analytical and technical skills, knowledge and other expertise, to the economic productive process. Good nutrition, supported by good health-care, facilitates the benefits of education by allowing the population to wish to invest more in future prosperity without being concerned or limited by current health problems. Nutritional constraints in the delta region, as in Bangladesh more generally, place limits on the long-term health and longevity of the region, as agricultural production fails to keep pace with population growth. The two tables below provide an overview of recent trends in educational achievement levels and nutritional deficiency in Bangladesh. In both cases, over the decade 2000–2010, the data shows a gradual improvement. For example, in Table 12.1, the percentage of the population with no formal education is shown to fall from 40.3 to 38.7 per cent over the period. In Table 12.2, the percentage of the population with moderate

Table 12.1 Educational achievement levels in Bangladesh, 2000–2010 (percentage) (based on data from Sen and Rahman 2015)

Education level	Year		
	2000	2005	2010
No education	40.33	36.7	38.7
Classes i–v	21.72	21.84	29.93
Classes vi–viii	12.74	13.15	13.9
Classes ix–x	7.68	9.57	7.35
SSC/HSC/equivalent	12.08	14.45	8.36
Diploma	0.5	–	–
Bachelor	3.52	2.85	0.85
Masters	1.01	1.16	0.75
Agriculture	0.02	–	–
Engineering	0.18	0.12	0.07
Medical	0.14	0.15	0.09

Table 12.2 Bangladesh population with moderate and severe deficiency in calorie intake (percentage) (Joliffe et al. 2013)

Year	Moderate deficiency (<2122 kcal/person/day)			Severe deficiency (<1805 kcal/person/day)		
	Rural	Urban	National	Rural	Urban	National
2000	42.3	52.5	44.3	18.7	25.0	20.0
2005	39.5	43.2	40.4	17.9	24.4	19.5
2010	36.8	42.7	38.4	14.9	19.7	16.1

calorie deficiency falls from 44.3 to 38.4 per cent over the period, whilst the percentage with severe deficiency declines from 20 to 16.1 per cent.

12.2.3.1   Future Scenarios: To 2030

Business As Usual The percentage of the population with no education declines to 35 per cent by 2030, whilst those with Bachelor or Master's Degrees increase to five per cent. The percentage of the population with severe calorific deficiency falls to eight per cent. Negative climate change-induced health impacts are restricted by reactive adaptation.

More Sustainable The percentage of the population with no education declines to 30 per cent by 2030, whilst those with Bachelor or Master's Degrees increase to seven per cent. The percentage of the population with severe calorific deficiency falls to five per cent. Pro-active adaptation significantly limits any negative health effects of climate change in the delta region.

Less Sustainable The percentage of the population with no education declines to 36 per cent by 2030, whilst those with Bachelor or Master's Degrees increases to three per cent. The percentage of the population with severe calorific deficiency falls to 12 per cent. Health is negatively impacted by the effects of sea-level rise and increased storm frequency, unconstrained by sufficient adaptation.

12.2.4   Level of Transport Infrastructure Investment

The provision of transport infrastructure is a supply-side measure that helps to facilitate the development of trade and the operation of markets. Thus, Alam (2015) suggests that "the transport system is the key to the movement of goods and people and provides accessibility to the jobs, health, education, and other socio-economic services that are essential to the welfare of the people. Poor transport inhibits growth of cities and makes them dysfunctional. This may have depressing effect on national economic growth." Recent transport infrastructure investments have demonstrated their importance in reducing transport costs of getting both labour and goods to local, national and international markets. For example, the joint KFW/ADB project in Jhenaidah, Kushtia, Meherpur and Chuadanga to develop rural feeder roads and unsurfaced roads leading to markets led to a fall in local transport costs of 10–15 per cent (KfW 2012). The lack of good quality transport links is judged to constrain economic development. At present, the transport sector comprises less than one per cent of gross domestic product (GDP) in Bangladesh, whilst infrastructure performance as a whole places the country in the 130th place in terms of global competitiveness.

12.2.4.1   Future Scenarios: To 2030

Business As Usual The historical baseline rate of transport sector growth has been around six and a half per cent, and the BAU scenario projects a similar rate to 2030. Whilst the majority of growth is in road infrastructure, investment continues in rail infrastructure and the development of urban mass transit systems. Resilience to climate change in design specification is piece-meal.

More Sustainable The rate of transport sector growth averages at seven and a half per cent to 2030. This growth is balanced across both urban and rural areas and includes trans-national initiatives with neighbouring countries. Environmental regulation brings about a mix of transport modes such that rail increases in importance, and road vehicles adopt predominantly low emission models. The importance of climate change resilience is recognised in infrastructure design specification.

Less Sustainable The rate of transport sector growth averages at five and a half per cent to 2030 and is mainly limited to investment in urban roads that, to some extent, ameliorates congestion in these areas. Trans-national transport links are not prioritised. Design specification does not account for climate change impacts resulting in regular transport disruption as a consequence of storm damage.

12.2.5   Factor and Product Market Competition

Competitive factor and product markets ensure that prices in these markets reflect their costs but do not result in higher prices than necessary. For example, labour market prices are often a significant part of product market costs. Thus, competitive labour markets should result in lower product market costs. Similarly, competition in product prices serves to put downward pressure on prices, thereby benefitting the consumer. The delta region is increasingly dependent on national and international factor and product markets. Low-skilled labour supplies the increasingly urbanised manufacturing sector, whilst shrimp production is mainly exported and currently remains competitive in the global market. However, in both cases, competitiveness is won at the expense of ensuring sustainability. At present, competitive drivers—particularly in export-orientated production—dominate at the expense of labour working conditions and environmental sustainability. Export growth has been mainly limited to the ready-made garment sector (Sattar 2015). Economic growth, fuelling growing household incomes, has served to reduce competitive advantage to a small degree.

12.2.5.1   Future Scenarios: To 2030

Business As Usual Economic growth may reduce competitive advantage in some export sectors, relative to neighbouring countries or regions. Market transparency, in terms of price comparison, increases with the spread of IT capabilities and capacity across the country.

More Sustainable Improved regulation of factor and product markets results in more transparency and reliability, whilst maintaining an absence of monopolistic markets. Improved transport infrastructure and resulting connectivity further enhances competition in the delta and nationally.

Less Sustainable Slow progress in transport connectivity continues to limit factor and product market competition, as do trade and non-trade barriers. Aggressive export strategies unsupported by market regulation result in further environmental degradation and consequent vulnerability to climate risks.

12.2.6   Access to, and Cost of, Financial Capital

Financial capital has a critical facilitative role in modern economies. Mansur (2015) states that "the financial sector is a vital part of an economy because of the role it plays in intermediating savings of the private and public sector to productive activities including investment." Investment then results in an increased stock of capital in the economy. In the delta region, as in the rest of the country, access to financial resources will determine whether the poor are able to borrow money for the purpose of investing in equipment (e.g. IT hardware and software, manufacturing machinery, etc.) required to undertake effective business operations. The number of debit and credit accounts per 1,000 population in Bangladesh increased from 242 to 333 and from 51 to 63, respectively, during the period from 2005 to 2010 but reflects a remaining lack of savings or access to finance across the population (Mujeri 2015). Whilst the money market is relatively well developed, capital, bond and insurance markets are insufficiently developed to allow long-term borrowing to be easily secured, thereby limiting the scale of longer-term investments in, for example, energy and transport infrastructure.

12.2.6.1   Future Scenarios: To 2030

Business As Usual Private savings and borrowings continue to grow: debit and credit accounts of 800 and 120 per 1,000 population, respectively, to 2030 give some impetus to the existing growth of small and medium-sized enterprises (SMEs) in the country. Confidence in the capital markets gradually returns following recent scandals relating to insider trading, though regulation remains rather weak and limits the extent of trading activity. Bond and insurance markets develop but are constrained by the weak regulatory regime. Investment into climate resilience continues to be piece-meal.

More Sustainable Regulatory reform across the financial sector encourages the growth of availability in financial capital across both poor and more affluent parts of the population. All individuals have savings accounts by 2030 and the majority have borrowing facilities. Small businesses are routinely able to achieve economies of scale as a consequence of their being able to facilitate expansion. Climate resilience is recognised as being a prerequisite for medium- to longer-term business sustainability and so appropriate investments in adaptation are made.

Less Sustainable The scale of all financial markets remains such that the vast majority of the population still have neither debit nor credit accounts by 2030. Public confidence in the regulatory regime remains low, limiting the scale of financial investments in the capital and bond markets. Investments in climate resilience are regarded as being of secondary importance so that operations remain vulnerable to storms.

12.2.7   Export Market Strategy

Demand for exports frequently represents a significant proportion of aggregate demand and so has implications for employment and household incomes throughout the economy. A potential constraint on economic growth is the balance of trade (exports minus imports). Thus, domestic demand for exports can be accommodated as long as export value is sufficient to offset this in the longer term. Economic development in the delta region is currently significantly orientated towards international markets such as shrimp and garments and therefore somewhat dependent on the income generated. This orientation does, however, mean that environmental degradation—and its negative effects on other farm and non-farm production—is substantial. Between the years 2000 and 2013, the export percentage of GDP has increased from 12 to 20 per cent, whilst the total value of exports has increased fourfold over the period. Export concentration in ready-made clothing garments is very high compared to other countries, though the number of countries exported to has increased from 60 to 111 between 2000 and 2013.

12.2.7.1   Future Scenarios: To 2030

Business As Usual Growth in exports as a percentage of GDP continues but starts to flatten out to 18 per cent by 2030. Gradual improvements in internal connectivity, together with investment in improved IT capacity facilitate this growth. Export concentration falls, though rather slowly, as more high-tech companies grow domestically and supply low-cost outsourcing to international companies.

More Sustainable Trade and non-trade barriers are substantially reduced in order to encourage a diversification in the range of goods and services exported; IT services provide a substantial focus of national export activities to 2030. Substantial investment in infrastructure results in a greater volume of low-cost manufacturing being exported to neighbouring countries. Resulting higher incomes and tax revenues are reflected in increased expenditures on climate resilience investments.

Less Sustainable Regulatory constraints and poor macro-economic management, combined with political instability and low investment in transport infrastructure, result in slow growth in exports and a lingering concentration of export value in clothing garments. The lack of environmental and social regulation in this industry exacerbates this tendency. Exports comprise 15 per cent of GDP by 2030.

12.2.8   Fiscal Strategy and Stability

The collection and use of tax revenues in Bangladesh are direct fiscal instruments with which to pursue redistributive objectives whilst supplying public goods (e.g. education, health, defence, etc.) to the general population. The balance between tax revenues and government expenditure determines the level of debt that the government has to service in future time periods. The levels of investment in developmental priorities such as education, health and transport infrastructure across the delta region are determined by the extent of tax revenue funds available together with the governmental ordering of these priorities. Historical trends in external debt and domestic debt are presented in Fig. 12.2. It shows that in the past 20 years external debt has been substituted for domestic debt as international donor organisations have targeted development priorities elsewhere, and the domestic government has become more confident of servicing shorter-term domestic debt.

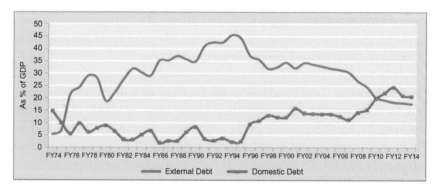

Fig. 12.2 Trends in external and domestic public debt, Bangladesh (1973–2014) (Reprinted with permission from Mansur 2015)

12.2.8.1   Future Scenarios: To 2030

Business As Usual A return to political stability in central government means that it is able to resume raising funds from the sale of government bonds, though resources remain limited. The fiscal deficit averages around five per cent to 2030. Tax revenue rises incrementally over the period as household incomes rise.

More Sustainable As with BAU, bond sales resume. As GDP rises at around seven and a half per cent per annum, resources raised through both bonds and taxes increased quickly; investment in tax collection processes ensures that revenue streams to government are reliable and much less vulnerable to corrupt practices. The fiscal deficit averages below five per cent to 2030.

Less Sustainable Ongoing political instability limits the extent that resources can be raised by bond sales. Growth of tax revenues is also constrained by inefficiencies in the tax collection system slower growth in GDP. The fiscal deficit averages above five per cent to 2030.

12.2.9   Natural Resource Management

Preservation of environmental quality ensures that land stays productive and water resources continue to be available both to consumers and producers. Maintenance of natural resources also ensures that their value is conserved for future users, including following generations. Sufficient control of environmental pollution serves to avoid damage to human health, ecosystems, crops and other productive resources. Ultimately, environmental quality and availability of resources dictate the extent of human activity.

Over-use of nitrogen and phosphorous-based fertilisers in agricultural production risks pollution to water courses with subsequent effects on the availability of clean water resources for human consumption and inputs to industrial production in the region. The external costs of salinity in inland aquaculture are often currently not sufficiently recognised in production decisions. In general, whilst there currently appears to be a well-developed set of environmental regulations over all media, effective inspection and enforcement are often insufficiently implemented to achieve the objectives of the regulations.

12.2.9.1   Future Scenarios: To 2030

Business As Usual The effectiveness of environmental regulation slowly improves to 2030, though population pressure means that environmental degradation continues to reduce the productivity of natural resources. Climate change, particularly sea-level rise, exacerbates this pattern across the delta region.

More Sustainable Implementation of multi-lateral environmental agreements (MEAs) and domestic regulation is increased significantly to 2030. Land and water quality, as well as forest resource conservation, are all improved, with subsequent benefits for human welfare across the delta region and the country more generally.

Less Sustainable The existing range of environmental regulations remain only partially enforced, with adverse effects on a growing population resulting. Natural resources are further depleted and degraded, limiting their productivity and reducing their natural roles in soil quality conservation, water cycling and other regulatory functions.

12.2.10   Population Policy

The demographic transition model (DTM), an established means with which to view demographic change, suggests a well-defined link between such change and economic development (Thompson 1929). In the early stages of this transition, at low levels of economic development, whilst population growth can facilitate economies of scale and specialisation of labour it is now recognised that it can result in over-exploitation of natural resources and lower per capita incomes, with potential implications on longer-term economic growth. High population density in the delta region puts agricultural, aquacultural and forest resources under pressure, limiting

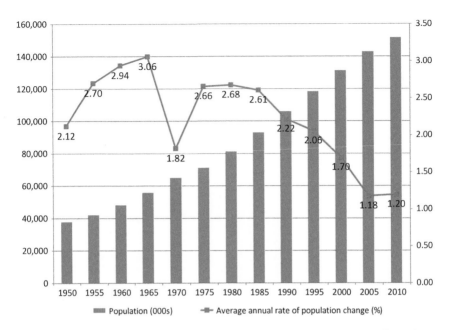

Fig. 12.3 Population numbers and growth in Bangladesh, 1901–2011 (based on data from UN 2015)

potential per capita allocations. Sea-level rise, having adverse impacts on the availability of land to live on, is likely to further exacerbate this pressure. Figure 12.3 shows that population annual growth rates are positive across Bangladesh, though they have slowed from around two per cent to less than one and a half per cent in the 25 years to 2011. Total population has increased from 130 million in 2001 to 150 million in 2011. Population density was 1,015/km² in 2011, which compares with 350/km² in India.

12.2.10.1   Future Scenarios: To 2030

National population projections prepared by Hayes and Jones (2015) for the Government of Bangladesh are presented in Fig. 12.4. Their three scenarios map directly on to the three developed scenarios as follows: High = Less Sustainable; Medium = Business As Usual; Low = More Sustainable.

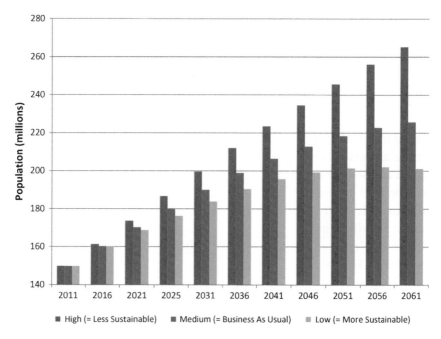

Fig. 12.4 Population projections for Bangladesh (based on data from Hayes and Jones 2015)

Business As Usual Continuing economic growth allows rural-urban migration to somewhat reduce the pressure on rural natural resources. Population growth continues to fall slowly as maternal health-care and education are more widely available across the delta region.

More Sustainable High economic growth results in a fall in fertility rates as government awareness campaigns regarding contraception are effective and as the social acceptability of child marriage declines. Declining rural population growth and further urban development results in a halt to the growth in landlessness.

Less Sustainable Ineffective government policies to reduce fertility rates result in population growth continuing at current rates. Sea-level rise results in reduced land available to inhabit and a consequent increase in landlessness. Failure of farm and non-farm industry to absorb labour results in further rural-urban migration and an expansion of existing urban slum areas.

12.2.11   Governance

As summarised in Fig. 12.5, governance consists of a number of over-riding aspects of government behaviour that affect how citizens, businesses and society more generally are able to function. Poor performance in aspects of governance tends to have a detrimental effect on the ability of economic agents (e.g. consumers and producers) to undertake economic activities, increasing their transaction costs and so reducing economic efficiency and competitiveness. As with Bangladesh as a whole, governance in the delta region is critical in determining the efficiency and effectiveness of economic activity. For example, high levels of corruption may lead to a distortion of the incentives that would otherwise exist to sustainably manage the resources of the Sundarbans. In a comparison of a variety of governance indicators for Bangladesh and two groups of countries, low-income countries[1] and lower middle-income countries (following Hasan et al. 2015), for four of the six indicators, Bangladesh ranks above the average for low-income countries in 2013 (Fig. 12.5). However, none of the indicators for Bangladesh exceeds the average for lower middle-income countries.

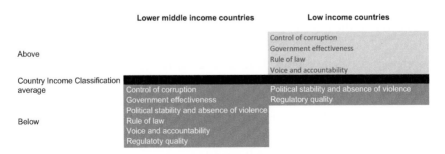

Fig. 12.5   Comparison of World-wide Governance Indicators between Bangladesh and country income groups—2013 (based on data from WGI 2016)

12.2.11.1   Future Scenarios: To 2030

Business As Usual GDP growth rates to 2030 ensure that Bangladesh moves in to the lower middle-income group of countries and governance indicators gradually improve. However, aspects of governance—including regulatory quality—remains below average, negatively impacting on, for example, effective environmental regulation.

More Sustainable The performance of all governance indicators improves to at least average levels for lower middle-income countries. Transaction costs are consequently significantly reduced, resulting in substantial inflow of capital investment from neighbouring countries and high-income countries.

Less Sustainable The performance of governance indicators improves only slowly to 2030 with periodic political instability a feature of the period. Environmental regulation remains of mixed effectiveness, resulting in a low level of pro-active adaptation to climate change. Transaction costs associated with poor governance continue to thwart a range of economically productive initiatives.

12.2.12   Macro-Economic Policy Levers

Table 12.3 indicates macro-economic policy levers that may be used to bring about changes in the macro-economic enabling factors. It is intended that the information in the table is used as an initial stimulus in thinking about pursuing macro-economic objectives that then influence poverty in the delta region. It is, of course, recognised that analysis of poverty alleviation strategies has been undertaken in considerable depth in a number of recent government reports, most notably Sen and Ali (2015); here, a link between that body of work and the scenario-based work undertaken in the current reported research is developed. Thus, it is anticipated that the connection between macro-economic planning, poverty alleviation and bio-physical changes resulting, inter alia from climate change, in the delta region can be directly made.

Table 12.3 Macro-economic enabling factors and policy levers

Macro-economic enabling factor	Policy levers	Application at delta region scale
Patterns of resource and land ownership	Land law reform and implementation	National level reform, perhaps implemented by district authorities
Levels of public health and education provision	Public expenditure	Determined at national level; spend distribution informed by "lagging regions" equity analysis (Khondker and Mahzab 2015)
Level of transport infrastructure investment	Public expenditure	Determined at national level; spend distribution informed by "lagging regions" equity analysis (Khondker and Mahzab 2015)
Factor and product market competition	Consumer market regulation; employment legislation and effective implementation	National level policies
Access to—and cost of—financial capital	Financial market regulation and effective implementation	National level reform, perhaps implemented by district authorities
Export market strategy	Public support to new industries; exchange rate policy	National level policies
Fiscal strategy and stability	Public expenditure and tax policy	National level policies; spend and tax levels informed by "lagging regions" equity analysis (Khondker and Mahzab 2015)
Natural resource management policy	Government regulation and effective implementation	National level policies, perhaps implemented by district authorities
Population policy	Public information and incentives	National level policies, perhaps implemented by district authorities
Governance	Reform of regulatory regimes	National level reform

12.3   Quantified Future Trends for Economic Input Variables to 2030

Quantitative trends for key input variables were derived, constructed on the interpretation of the three scenarios. Thus, the dataset presented in Table 12.4 is based on collective judgement of how the scenario portraits above might affect the variables in question. The values are generated based on knowledge of the dimension of changes that had occurred in these variables in the past decade; this effectively bounded the scale of changes. Consequently, the estimates are considered to be conservative; in any case, they should be seen as indicative only.

12.4   Conclusions

Using the expertise within the project, the main macro-economic factors likely to determine future economic development paths in Bangladesh's delta region and the wider country have been identified. For each of these factors, possible patterns of change are sketched in narrative terms, and the means with which the Government of Bangladesh can influence them are highlighted. The analysis serves to emphasise that local and national administrations in Bangladesh have a range of economic, and other, policy instruments at their disposal to ensure that economic development follows a path that significantly reduces the vulnerability of the population to climate, and other environmental, change.

In a future phase of research, it would be valuable for the relationships between macro-economic factors to be quantified in such a way that advisors to the Government of Bangladesh could undertake simulations that allow testing for the robustness of alternative macro-economic policies. In this way, the inherent trade-offs between these factors could be made explicit and encouraging an informed discussion as to the merits of alternative policy options.

Table 12.4 Percentage change in economic input variables 2015–2030

Economic input variable	Unit	Future scenario		
		Business As Usual	More Sustainable	Less Sustainable
Cost of different fertiliser types	BDT[a]/kg	+10	+20	0
Cost of pesticide for each agriculture crop	BDT/ha	+10	+20	0
Cost of feed for each aquaculture crop	BDT/ha	+10	0	+20
Cost of seed for agriculture crops	BDT/kg	+10	+20	0
Cost of post larvae or fishling for aquaculture products	BDT/individual	+10	0	+20
Daily wage (without food)	BDT/day	+10	+30	0
Cost of diesel	BDT/gallon	+10	+20	0
Employment rate	Per cent population	+10	+30	0
Literacy rate	Per cent population	+4	+8	+2
Children in school	Per cent population	+5	+10	+2
Travel time to major cities	Hours	−30	−50	−10
Remittances	BDT/month	+30	+40	+20
Income from manufacturing, services and livestock/poultry sectors	BDT/month	+110	+165	+65
Household expenses	BDT/month	+10	+30	0
Purchase power parity (PPP) exchange rate for Bangladesh		0	0	0
USD/BDT exchange rate		0	0	0

[a]Bangladeshi taka

Note

1. As defined by the World Bank.

References

Alam, G.M.K. 2015. *Strategy for infrastructure sector*. Background paper for the seventh Five Year Plan. Dhaka: Policy Research Institute of Bangladesh. http://www.plancomm.gov.bd/wp-content/uploads/2015/02/10_Strategy-for-Infrastructure-Development.pdf. Accessed 4 Aug 2016.

GED. 2015. *Seventh five year plan FY2016 – FY2020: Accelerating growth, empowering citizens*. Final draft. Dhaka: General Economics Division (GED), Planning Commission, Government of the People's Republic of Bangladesh. http://plancomm.gov.bd/wp-content/uploads/2015/11/7FYP_after-NEC_11_11_2015.pdf. Accessed 6 Jan 2017.

Hasan, M., J. Rose, and S. Khair. 2015. *Governance and justice. Background paper for the 7th Five Year Plan of the Government of Bangladesh*. Dhaka: Bangladesh Planning Commission, Government of the People's Republic of Bangladesh. http://www.plancomm.gov.bd/wp-content/uploads/2015/02/12_Governance-and-Justice-_Final-Draft.pdf. Accessed 4 Aug 2016

Hayes, G., and G. Jones. 2015. *The impact of the demographic transition on socioeconomic development in Bangladesh: Future prospects and implications for public policy*. Dhaka: The United Nations Population Fund, Bangladesh Country Office. http://www.plancomm.gov.bd/wp-content/uploads/2015/02/22_Impact-of-Demographic-Transition-on-Socioeconomic-Development.pdf. Accessed 4 Aug 2016.

Hossain, M. 2015. *Improving land administration and management in Bangladesh. Background paper for the 7th Five Year Plan of the Government of Bangladesh*. Dhaka: General Economics Division (GED), Planning Commission, Government of the People's Republic of Bangladesh. http://www.plancomm.gov.bd/wp-content/uploads/2015/02/4_Improving-Land-Administration-and-Management_Final.pdf. Accessed 4 Aug 2016.

Joliffe, D., I. Sharif, L. Gimenez, and F. Ahmed. 2013. *Bangladesh – Poverty assessment: Assessing a decade of progress in reducing poverty, 2000–2010*. Bangladesh development series; paper no. 31. Washington, DC: World Bank. https://openknowledge.worldbank.org/handle/10986/16622. Accessed 15 June 2017. Licence Creative Commons Attribution (CC BY 3.0 IGO).

KfW. 2012. *Bangladesh: Joint project – Rural markets and roads, Khulna division. Ex Post-Evaluation Brief*. Frankfurt: KfW Development Bank. https://www.kfw-entwicklungsbank.de/Evaluierung/Ergebnisse-und-Publikationen/PDF-Dokumente-A-D_EN/Bangladesh_M%C3%A4rkte_Stra%C3%9Fen_2012_E.pdf. Accessed 23 May 2017.

Khondker, B.H., and M.M. Mahzab. 2015. *Lagging districts development.* Background paper for the 7th Five Year Plan of the Government of Bangladesh. Dhaka: Bangladesh Planning Commission, Government of the People's Republic of Bangladesh. http://www.plancomm.gov.bd/wp-content/uploads/2015/02/15_Lagging-Regions-Study.pdf. Accessed 4 Aug 2016.

Mansur, A.H. 2015. *Fiscal management and revenue mobilisation.* Background paper for the seventh Five Year Plan. Dhaka: Policy Research Institute of Bangladesh. http://www.plancomm.gov.bd/wp-content/uploads/2015/02/8_Fiscal-Management-and-Revenue-Mobilization.pdf. Accessed 4 Aug 2016.

Mujeri, M.K. 2015. *Improving access of the poor to financial services.* Background paper for the seventh Five Year Plan. Dhaka: General Economics Division, Bangladesh Planning Commission, Government of the People's Republic of Bangladesh. http://www.plancomm.gov.bd/wp-content/uploads/2015/02/1_Improving-Access-of-the-Poor-to-Financial-Services.pdf. Accessed 4 Aug 2016.

Sattar, Z. 2015. *Strategy for export diversification 2015–2020: Breaking into new markets with new products.* Dhaka: Policy Research Institute of Bangladesh. http://www.plancomm.gov.bd/wp-content/uploads/2015/02/7_Strategy-for-Export-Diversification.pdf. Accessed 4 Aug 2016.

Sen, B., and Z. Ali. 2015. *Ending extreme poverty in Bangladesh during the seventh Five Year Plan: Trends, drivers and policies.* Background paper for the seventh Five Year Plan. Dhaka: General Economics Division, Bangladesh Planning Commission, Government of the People's Republic of Bangladesh. http://www.plancomm.gov.bd/wp-content/uploads/2015/02/25_Ending-Extreme-Poverty-in-Bangladesh.pdf. Accessed 4 Aug 2016.

Sen, B., and M. Rahman. 2015. *Earnings inequality, returns to education and demand for schooling: Addressing human capital for accelerated growth in the seventh Five Year plan of Bangladesh.* Background paper for the 7th Five Year Plan of the Government of Bangladesh. Dhaka: Bangladesh Planning Commission, Government of the People's Republic of Bangladesh. http://www.plancomm.gov.bd/wp-content/uploads/2015/02/3_Strategy-for-Education-and-Training_Final-Version.pdf. Accessed 4 Aug 2016.

Thompson, W.S. 1929. Population. *American Journal of Sociology* 34 (6): 959–975. https://doi.org/10.1086/214874.

UN. 2015. *World population prospects: The 2015 revision.* https://esa.un.org/unpd/wpp/. Accessed 5 June 2017.

WGI. 2016. *Worldwide Governance Indicators (WGI).* Washington, DC: World Bank. http://info.worldbank.org/governance/wgi/index.aspx#home. Accessed 16 Sep 2016.

Part 4

Observations and Potential Trends

13

Biophysical Modelling of the Ganges, Brahmaputra, and Meghna Catchment

Paul G. Whitehead

13.1 Introduction

The large river systems of the Ganges, Brahmaputra, and Meghna (GBM) combine to create the GBM delta. Over 670 million people depend on the river basins and the associated delta for their livelihood and well-being. Thus, knowing how such river systems might be impacted by future changes in climate and socio-economics is important for the wide range of stakeholders in the delta region. This chapter summarises the physical aspects of the GBM system and describes the modelling of the catchments for flow and water quality.

The GBM river system extends between the latitude of 22° 30′ N to 31° 30′ N and longitude of 78°0′ E to 92° 0′ E in the countries of India, Nepal, China, Bhutan, and Bangladesh (Fig. 13.1), with a total catchment area of 1,612,000 km². The GBM river system is considered to be one large transboundary river basin, even though the three rivers of this system have

P. G. Whitehead (✉)
School of Geography and the Environment, University of Oxford, Oxford, UK

© The Author(s) 2018
R. J. Nicholls et al. (eds.), *Ecosystem Services for Well-Being in Deltas*,
https://doi.org/10.1007/978-3-319-71093-8_13

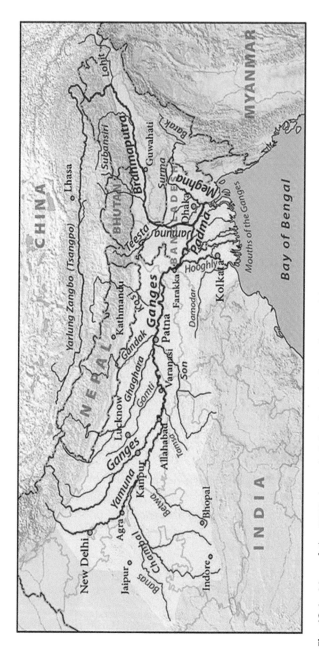

Fig. 13.1 Map of the GBM rivers draining into the Bay of Bengal (Whitehead et al. 2015a—Reproduced with permission of The Royal Society of Chemistry)

distinct characteristics and flow through very different geographical regions for most of their lengths. They join to form the GBM delta before flowing into the Bay of Bengal. The GBM river system is the third largest freshwater outlet to the world's oceans, being exceeded only by the Amazon and the Congo river systems. The headwaters of both the Ganges and the Brahmaputra Rivers originate in the Himalayan mountain range.

13.1.1 Ganges River

The Ganges River originates from the Gangotri glacier in the Himalayas at an elevation of nearly 7,010 m and traverses a length of about 2,550 km (measured along the Bhagirathi and the Hooghly) before it flows southeast into the Bay of Bengal. Along its way, the Ganges is joined by a number of tributaries to form the large fertile alluvial plain in North India. At the Farakka barrage, a major diversion delivers water from the Ganges into the Hooghly River providing water to West Bengal and Kolkata. Approximately 50 per cent of flows are diverted except during high flows ($>70,000$ m^3/s), with the exact diversions varying depending on inflows and season. The Farakka Treaty signed between India and Bangladesh in 1996 was a significant agreement between the two countries and provides an agreed mechanism for sharing the available water (Farakka Treaty 1996). After the Farakka Barrage, the Ganges, Brahmaputra, and Meghna Rivers join and flow into the Bay of Bengal. The main sources of water in the rivers are rainfall, subsurface flow, and snow-melt water from the Himalayas. Average annual rainfall varies between 300 and 2,000 mm, with the western side of the region receiving less rainfall in comparison with the eastern side. Rainfall is concentrated in the monsoon months of June through October, causing low flow conditions in the Ganges River and its tributaries during the dry periods of November to May. Fertile Eutric Cambisols are the main soil type in the lower basin, ideally suited for intensive cultivation. Shallow Luvisols of low fertility dominate the upper basin of the Ganges River. Land use in the India part of the Ganges basin consists of extensive agricultural land with a wide variety of crops, expanding urban areas, and areas of scrub and bare soils. The upper reaches consist of snow and rock, some remaining forest, and urban areas. The major cities located in the river basin are Delhi, Kolkata, Kanpur, Lucknow, Patna, Agra, Meerut, Varanasi, and Allahabad. These cities are expanding at a substantial rate as reflected in the rising population levels and extensive industrial growth. According to census data, the average population density in the Ganges basin is of the order of 520 persons per km². Electronics, leather, textiles, paper, jute, cement, and fertiliser production are some of the industries situated along the course of the river. Disposal of untreated urban wastes and industrial effluents increase pollution loads into the Ganges river system. According to the Central Pollution Control Board Report (CPCB 2003), the total wastewater generation from 222 towns in the Ganges basin is 8,250 ML/d (million litres/day), out of which 2,538 ML/d is directly discharged into the Ganges River, 4491 ML/d is disposed of into tributaries of River Ganges, and 1,220 ML/d is disposed onto land or low-lying areas. The Ganges river system is a very large and complex river system to model and becomes even more complex once it enters the Bangladesh delta region with a complex network of channels and braided river systems. This area is a tidal zone and is included as part of the estuarine modelling component of this research (see Chap. 16).

13.1.2   Brahmaputra River

The Brahmaputra river originates on the northern slope of the Himalayas in China, where it is called Yalung Zangbo. It flows eastwards for about 1,130 km, then turns southwards, and enters Arunachal Pradesh (India) at its northernmost point and flows for about 480 km. Then it turns westwards and flows through Arunachal Pradesh, Assam, and Meghalaya for another 650 km before entering Bangladesh, where it is also called Jamuna, and merging with the Ganges and Meghna rivers. The tributaries of the Meghna river originate in the mountains of eastern India and flow south-west to join the Ganges and Brahmaputra rivers before flowing into the Bay of Bengal forming the greatest deltaic plain in the world. About 80 per cent of Bangladesh is made up of fertile alluvial lowland that becomes part of the Greater Bengal Plain. The country is flat with some hills in the northeast and south-east. About seven per cent of the total area of Bangladesh is covered with rivers and inland water bodies, and the surrounding areas are routinely flooded during the monsoon.

13.2   Precipitation in the GBM Basin

The GBM river basin is unique in the world in terms of diversified climate. For example, the Ganges river basin is characterised by significant snowfall and precipitation in the northwest of its upper region and very high precipitation in the areas downstream in Bangladesh near the delta. High precipitation zones and dry rain shadow areas are located in the Brahmaputra river basin, whereas the world's highest precipitation area is situated in the Meghna river basin. Monsoon precipitation in the Ganges river basin lasts from July to October with only a small amount of rainfall occurring in December and January. The delta region experiences strong cyclonic storms, both before the commencement of the monsoon season, from March to May, and at the end of the monsoon from September to October. Some of these storms result in significant loss of life and the destruction of homes, crops, and livestock, most recently in Cyclone Sidr in 2007.

13.3   GBM River Study

The main objectives of this research with regard to upstream catchments have been to assess the magnitude and variability in the flow of the GBM rivers as a function of future changes in the climate, land use, and socio-economic conditions and to determine the flux of flows, sediments, and nutrient fluxes moving down the rivers into the Bay of Bengal, essential inputs for any analysis of the delta. Flow, sediment, and nutrient fluxes can provide critical information that can assist the Indian and Bangladesh governments to mitigate future impacts. There are several existing modelling studies on the Ganges river, most of which were funded by either government departments or international organisations, such as the World Bank (Sadoff et al. 2013). In order to undertake an assessment of the hydrology and nutrient dynamics in the GBM rivers, the semi-distributed, process-based INCA model (INtegrated Catchment model) for nitrates (N) and phosphates (P) is applied to the whole of the GBM river systems (Whitehead et al. 2015a, b). A set of climate and strategic socio-economic scenarios have then been evaluated to assess the potential impacts on both flows and water quality in the GBM river system.

13.3.1   The INCA N and P Model

Modelling complex river systems such as Ganges, Brahmaputra, and Meghna requires a semi-distributed model that can account for the spatial variability across the catchment. INCA is one such model that has been applied extensively to heterogeneous catchments and has the advantage that it is dynamic, process-based, and integrates hydrology and water quality. The INCA model has been developed over 20 years as part of the UK Research Council (NERC) and EU-funded projects (Whitehead et al. 2015b), and the model simulates hydrology flow pathways in the surface and groundwater systems and tracks fluxes of solutes/pollutants on a daily time step in both terrestrial and aquatic portions of catchments. The model allows the user to specify the spatial nature of a river basin or catchment, to alter reach lengths, rate coefficients, land use, velocity-flow relationships and to vary input pollutant deposition loads from point sources, diffuse land sources, and diffuse atmospheric sources. INCA originally allowed simulation of a single stem of a river in a semi-distributed manner, with tributaries treated as aggregated inputs. The revised version now simulates nutrient dynamics in dendritic stream networks as in the case of the GBM system with many tributaries. The model is based on a series of interconnected differential equations that are solved using numerical integration based on the fourth-order Runge-Kutta technique. The advantage of this technique is that it allows all equations to be solved simultaneously. The INCA model has been set up for the rivers as a multi-reach model with all the sub-catchments and reach boundaries being selected based on a number of factors such as a confluence point with a tributary, a sampling or monitoring point, or an effluent input or an abstraction point associated with a major irrigation scheme or a large city. Figure 13.2 shows the sub-catchments for the Ganges. The land use data have been derived using a 1 km grid resolution DTM with land cover data generated from the

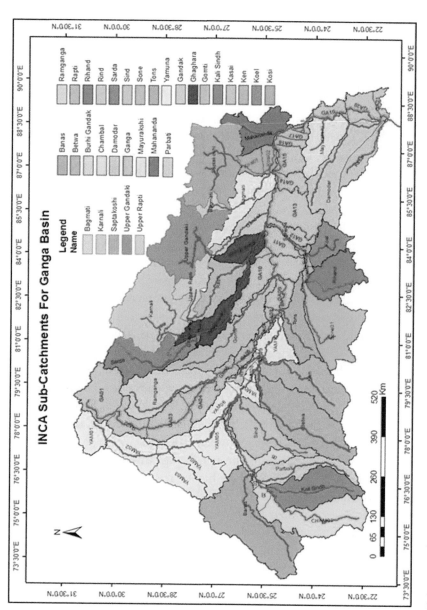

Fig. 13.2 Map showing the multi-branch Ganges river system and sub-catchment (Whitehead et al. 2015a—Reproduced with permission of The Royal Society of Chemistry)

moderate resolution imaging spectroradiometer (MODIS) satellite. Further details of the model setup are given in Futter et al. (2015), Jin et al. (2015), and Whitehead et al. (2015a).

13.3.2   Climate Drivers for the GBM Rivers

In order to run a set of hydrological simulations and climate scenarios, INCA requires a daily time series of climate data, namely, precipitation, hydrologically effective rainfall (HER), temperature, and soil moisture deficit (SMD). The model uses these data to drive the hydrological components of the model which generate the sub-catchment river flows. However, obtaining meteorological data over such a large catchment scale is difficult, especially given the wide spatial differences in topography, altitude, and land use in India, China, Bangladesh, Bhutan, and Nepal. Observational data are available from in-situ weather stations and also from satellite measurements, and these have been integrated into observational datasets which cover the region as part of the Aphrodite online data system.[1] These data have been used to calibrate the climate models both in space and time.

The large-scale general circulation models (GCMs) have been used to simulate climate across the region and to assess the impacts of increasing greenhouse gas concentrations on the global climate system. However, GCMs typically have coarse spatial resolutions with horizontal grid boxes of a few hundred kilometres size and cannot provide the high-resolution climate information that is required for climate impact and adaptation studies. The use of regional climate models (RCMs), which dynamically downscale the GCM simulations through being driven using boundary conditions from GCMs, can provide higher resolution grids (typically 50 km or finer) and are better able to represent features such as local topography and coast lines and their effects on the regional climate, in particular precipitation. There have been relatively few studies focused upon the Ganges river linked to the Bangladesh region which have used RCM output. In this research, an existing set of GCM simulations were used to provide boundary conditions for a RCM for the period 1971–2099 over a South Asia domain, as described in Chap. 11 and Caesar et al. (2015). The GCM is the third climate configuration of the Met Office Unified Model (HadCM3) and is run as a 17-member perturbed physics ensemble driven by the IPCC's Special Report on Emissions Scenarios (SRES) A1B scenario (Nakićenović et al. 2000). SRES A1B was developed for the IPCC and still underpins much recent research into climate impacts. It is a medium-high emissions scenario and is based upon a future assumption of strong economic growth and associated increase in the rate of greenhouse gas emissions. To put this into context with the newer Representative Concentration Pathways (RCPs) used in the IPCC Fifth Assessment Report, SRES A1B lies between the RCP 6.0 and RCP 8.5 in terms of the end of twenty-first-century projected temperature increases and atmospheric carbon dioxide concentrations (see Chap. 11).

The model uses the HER, SMD, and temperature daily time series together with all the reach, land use, and catchment data to simulate flow and water quality at every reach along the whole system for the whole period of 1981–2000. The model outputs are then compared to the observed flow data for the rivers to calibrate and validate the model. The observed flow quality data is sparse on the GBM river systems, although there are several flow gauges on the Ganges and there is a flow gauge on the Brahmaputra at Bahadurabad. In general, the calibration period of 1981–1990 and the validation period of 1991–2000 are both modelled with a good statistical agreement between simulated and observed flow. The Nash-Sutcliffe statistics for the whole period of the observed flow data for the 1981–2000 ranges is 0.55 to 0.75 which, given the complexity of the Ganges and the Brahmaputra, is reasonable. The model captures the main dynamics of the rise to the peaks in monsoon periods and the recession curves towards the dry season, as illustrated for the Brahmaputra in Fig. 13.3.

The fits to the Ganges flow data are of a similar order of magnitude (Whitehead et al. 2015a). In addition to calibrating the flow model, it is necessary to calibrate the water quality model. The water quality data is limited to infrequent observations at several monitoring points along the rivers, and Fig. 13.3 shows the simulated daily concentrations of nitrate-N and ammonium-N from the model at the lower reach of the Brahmaputra River System. The observed data for water quality along the rivers is available from the Indian Central Pollution Control Board.[2] The mean nitrate (as N) in the Brahmaputra at Dhubri is 0.12 mg/l and comparable to 0.11 from the model simulation. A key objective of the river modelling is

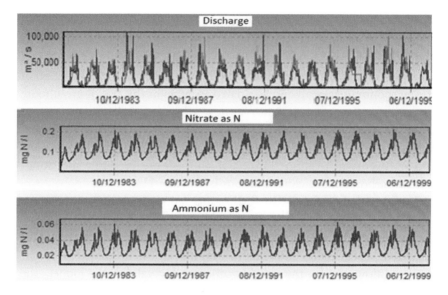

Fig. 13.3 Simulated (blue line) and observed (purple line) daily flows at the flow gauge on the Brahmaputra River System at Bahadurabad for 1981–2000 together with simulated nitrate and ammonium (Whitehead et al. 2015a—Reproduced with permission of The Royal Society of Chemistry)

to estimate the nutrient load flowing down the river into the Bay of Bengal, as the nutrients are crucial for agriculture, fisheries, and ecology (see Chaps. 24, 25, and 26). These are of primary concern within the project, as they relate directly to human well-being and resource availability for people in the delta region. Figure 13.4 shows the simulated and observed nitrogen load in the Ganges which suggested that the model is simulating the nutrient fluxes well.

Finally, Fig. 13.5 shows how the flows of the GBM Rivers combine to generate the total flows entering the Bay of Bengal illustrating the build-up of flows in the monsoon season from May to November. The INCA models have been used to evaluate a set of scenarios for flows and water quality, and these are described in detail in Whitehead et al. (2015a, b) and Jin et al. (2015). These scenarios have been used in the models of the study area in coastal Bangladesh (Chaps. 16 and 17) and the Bay of Bengal (Chap. 14), as well as the integrated assessment described in Chap. 28.

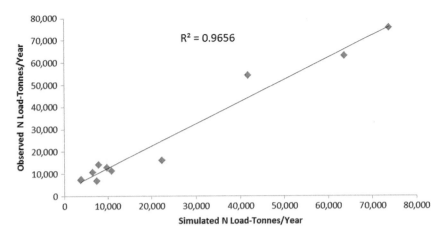

Fig. 13.4 Simulated and observed loads in the Ganges River at Kanpur (Whitehead et al. 2015a—Reproduced with permission of The Royal Society of Chemistry)

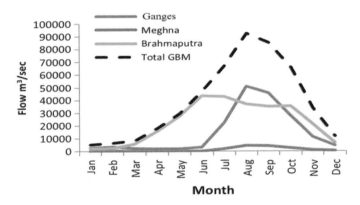

Fig. 13.5 GBM total monthly flows showing contributions from the Ganges, Brahmaputra, and Meghna Rivers for the 1990s (Whitehead et al. 2015a—Reproduced with permission of The Royal Society of Chemistry)

13.4 Summary Results

The INCA model simulated daily flow and water quality for three time slices 1980–2000, 2040–2060, and 2080–2100 using the UK Met Office Hadley Centre model climate simulations. For the 2050s and the 2090s, these results indicate a significant increase in monsoon flows with enhanced flood potential. Dry season flows are predicted to fall with extended drought periods, which could have impacts on water and sediment supply, irrigated agriculture, and saline intrusion. In contrast, the socio-economic changes had relatively little impact on flows, except under the low flow regimes where increased irrigation and dam diversion could further reduce water availability. However, should large-scale water transfers upstream of Bangladesh be constructed, these have the potential to reduce flows and divert water away from the delta region depending on the volume and timing of the transfers. This could have significant implications for the delta in terms of saline intrusion, water supply, agriculture, and maintaining crucial ecosystems such as the Sundarbans mangrove forest, with serious implications for people's livelihoods in the area. The socio-economic scenarios have a significant impact on water quality, altering nutrient fluxes being transported into the delta region. More details of these results can be found in Whitehead et al. (2015a, b) and Jin et al. (2015).

Notes

1. See https://climatedataguide.ucar.edu/climate-data
2. See http://cpcb.nic.in

References

Caesar, J., T. Janes, A. Lindsay, and B. Bhaskaran. 2015. Temperature and precipitation projections over Bangladesh and the upstream Ganges, Brahmaputra and Meghna systems. *Environmental Science-Processes and Impacts* 17 (6): 1047–1056. https://doi.org/10.1039/c4em00650j.
CPCB. 2003. *Status of sewage treatment plants in Ganga Basin.* Delhi: Central Pollution Control Board, Ministry of Environment and Forests, Government of India. www.cpcb.nic.in/newitems/8.pdf. Accessed 24 June 2016.

Farakka Treaty. 1996. Treaty between the government of the People's Republic of Bangladesh and the Government of the Republic of India on sharing of the Ganga/Ganges water at Farakka. 36 I.L.M 523 (1997).

Futter, M.N., P.G. Whitehead, S. Sarkar, H. Rodda, and J. Crossman. 2015. Rainfall runoff modelling of the upper Ganga and Brahmaputra basins using PERSiST. *Environmental Science-Processes and Impacts* 17 (6): 1070–1081. https://doi.org/10.1039/c4em00613e.

Jin, L., P.G. Whitehead, S. Sarkar, R. Sinha, M.N. Futter, D. Butterfield, J. Caesar, and J. Crossman. 2015. Assessing the impacts of climate change and socio-economic changes on flow and phosphorus flux in the Ganga river system. *Environmental Science-Processes and Impacts* 17 (6): 1098–1110. https://doi.org/10.1039/c5em00092k.

Nakićenović, N., J. Alcamo, G. Davis, B. Devries, J. Fenhann, S. Gaffin, K. Gregory, A. Gruebler, T.Y. Jung, T. Kram, E. Lebre Larovere, L. Michaelis, S. Mori, T. Morita, W. Pepper, H. Pitcher, L. Price, K. Riahi, A. Roehrl, H.-H. Rogner, A. Sankovski, M. Schlesinger, P. Shukla, S. Smith, R. Swart, S. Vanrooijen, N. Victor, and Z. Dadi. 2000. *Special report on emissions scenarios, a special report of working group III of the intergovernmental panel on climate change*. Cambridge, UK: Cambridge University Press.

Sadoff, C., N.R. Harshadeep, D. Blackmore, X. Wu, A. O'Donnell, M. Jeuland, S. Lee, and D. Whittington. 2013. Ten fundamental questions for water resources development in the Ganges: Myths and realities. *Water Policy* 15: 147–164. https://doi.org/10.2166/wp.2013.006.

Whitehead, P.G., E. Barbour, M.N. Futter, S. Sarkar, H. Rodda, J. Caesar, D. Butterfield, L. Jin, R. Sinha, R. Nicholls, and M. Salehin. 2015a. Impacts of climate change and socio-economic scenarios on flow and water quality of the Ganges, Brahmaputra and Meghna (GBM) river systems: Low flow and flood statistics. *Environmental Science-Processes and Impacts* 17 (6): 1057–1069. https://doi.org/10.1039/c4em00619d.

Whitehead, P.G., S. Sarkar, L. Jin, M.N. Futter, J. Caesar, E. Barbour, D. Butterfield, R. Sinha, R. Nicholls, C. Hutton, and H.D. Leckie. 2015b. Dynamic modeling of the Ganga river system: Impacts of future climate and socio-economic change on flows and nitrogen fluxes in India and Bangladesh. *Environmental Science-Processes and Impacts* 17 (6): 1082–1097. https://doi.org/10.1039/c4em00616j.

14

Marine Dynamics and Productivity in the Bay of Bengal

Susan Kay, John Caesar, and Tamara Janes

14.1   Introduction

The Bay of Bengal is an important provider of ecosystem services for people in the coastal part of the delta, but tropical cyclones formed in the Bay can lead to highly destructive flooding. This chapter describes the physical conditions and patterns of biological production in the Bay and outlines current trends and changes. The monsoon climate drives seasonal changes in currents and causes large inputs of fresh water from rain and rivers. A fresh water layer on the sea surface suppresses circulation of nutrients for much of the year and so the central Bay has low productivity. The coastal areas are much more productive, driven by inputs of nutrients from rivers and from vertical mixing due to coastal currents. Strong stratification encourages the development of tropical cyclones. Sea-level rise is already occurring in the Bay and is likely to be the most

S. Kay (✉)
Plymouth Marine Laboratory, Plymouth, UK

J. Caesar • T. Janes
Met Office Hadley Centre for Climate Science and Services,
Exeter, Devon, UK

© The Author(s) 2018
R. J. Nicholls et al. (eds.), *Ecosystem Services for Well-Being in Deltas*,
https://doi.org/10.1007/978-3-319-71093-8_14

important effect of climate change for people in coastal Bangladesh. Sea-level rise in the twenty-first century is projected to be in the range of 0.5–1.7 m with associated impacts of increased flooding depth, area and maximum wave height. Other changes are less certain, but could include increased stratification, small changes in biological production and alterations in the frequency and intensity of tropical cyclones.

For the people of coastal Bangladesh, the sea is both a support and a threat. Marine fisheries ecosystem services provide a vital contribution to livelihood and to diet, while flooding associated with storm surge events has regularly caused devastation. This chapter describes the current biophysical situation in the northern Bay of Bengal and the factors influencing change, while Chap. 25 presents the future prospects for fisheries as projected in this research. The research is also linked to Chap. 8 which discusses flooding from both rivers and the sea and Chap. 16 which discusses possible changes in tropical cyclones and coastal flooding.

14.2   Physical Description and Major Influences on Circulation in the Bay

The Bay of Bengal lies in the far north-east of the Indian Ocean, surrounded by land except on the south side, where it is open to the influence of the wider Indian Ocean. The Ganges-Brahmaputra-Meghna (GBM) delta is the northern boundary of the Bay, with delta sediments forming a shallow area that extends about 200 km south from the shoreline (Fig. 14.1). On the western side, along the Indian coast, the sea bed dips quickly to 2,000 m or more after a narrow coastal strip. On the eastern side, there is a wider area of shallower water, the Andaman Sea.

The monsoon atmospheric circulation is a dominating influence on the Bay, with south-west winds and heavy rain in the summer, north-east winds and low rainfall in the winter. Fresh water and nutrient inputs from a number of large rivers also have a strong impact. The GBM system and the Irrawaddy rank 4th and 15th globally in their discharge of water to the sea, an annual average of 1032 km^3 and 393 km^3, respectively (Dai and Trenberth 2002). On the east coast of India, the Godavari discharges another 97 km^3 and the Mahanadi 73 km^3. Monsoon-driven changes in

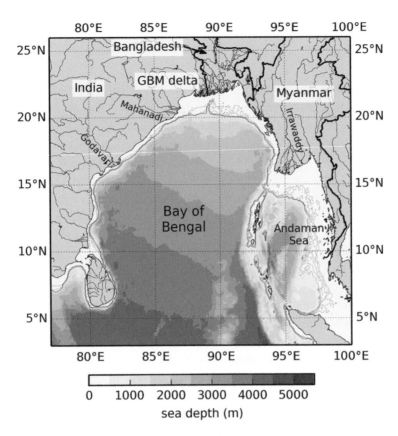

Fig. 14.1 Overview of the Bay of Bengal, showing bathymetry and the location of features mentioned in the text. Contours are shown at 100 m intervals to 500 m; greater depths are shown by colour shading, as in the key. Bathymetry data is taken from the GEBCO 1′ dataset, www.gebco.net

rainfall mean that these discharges are strongly seasonal: the winter discharge rate of the GBM is less than ten per cent of its summer peak (Dai and Trenberth 2002).

The Indian Ocean Dipole (IOD) and El Niño-Southern Oscillation (ENSO) are additional, but weaker, sources of influence on the Bay of Bengal circulation patterns, associated with changes in wind patterns and rainfall at the surface and unusual circulation patterns within the sea (Aparna et al. 2012; Currie et al. 2013).

14.3 Water Structure, Circulation and Productivity

The strongest feature of circulation in the Bay is the East India Coastal Current, which flows north-eastwards along the Indian coast from February to September and south-eastwards from October to January, under the influence of monsoon winds and rain (Chaitanya et al. 2014; Durand et al. 2009; Shankar et al. 1996). The reversal of this current in late summer is partly driven by the influx of river water from the north of the Bay (Diansky et al. 2006); an area of low salinity water moves south along the Indian coast between August and December (Akhil et al. 2014; Chaitanya et al. 2014; Shetye 1993). The current is also influenced by wider circulation in the Indian Ocean (Schott et al. 2009). For example, positive IOD is associated with a weaker southward East India Coastal Current and lower salinity in the northern Bay (Pant et al. 2015).

The Bay of Bengal is a net exporter of fresh water. Salinity is lowest in the north, where monsoon rains and flow from the GBM delta combine to give a great input of fresh water, especially between June and October. Fresh water is dispersed southwards by surface currents along both coasts and by gradual mixing with saltier waters below (Behara and Vinayachandran 2016; Benshila et al. 2014; Shetye 1993). The mixing process is slow and in the post-monsoon season fresh water creates a thin layer of low salinity at the sea surface, overlying the constant-temperature layer below (Felton et al. 2014; Vinayachandran et al. 2002). The warm but saltier water below the surface layer is often referred to as the barrier layer because it inhibits convection and so restricts the vertical mixing of water and heat flow to the deeper water column.

The strong stratification associated with the barrier layer limits the movement of nutrients from deep waters to the sunlit zone near the surface. This lack of nutrients inhibits growth of photosynthesising plankton (phytoplankton) and so biological production is low in the central Bay (Martin and Shaji 2015 and references therein). Nearer the coast, nutrients reach the sunlit zone from river inputs and from vertical mixing due to upwelling caused by coastal currents; this means that productivity is much higher, as shown by patterns of chlorophyll concentration (Fig. 14.2). Observational evidence on chlorophyll and nutrient levels is limited

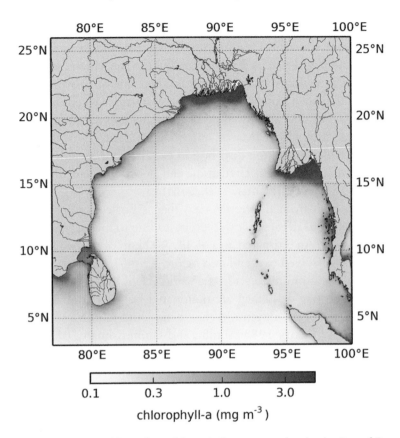

Fig. 14.2 Mean monthly surface chlorophyll concentration in the Bay of Bengal for 2000–2009. The data is taken from the composite satellite dataset compiled by the Ocean Colour Climate Change Initiative, version 2, European Space Agency, available online at http://www.esa-oceancolour-cci.org/

(Narvekar and Kumar 2014; Kumar et al. 2010), but satellite data shows evidence of two chlorophyll peaks, in July–August and in December–February, with the second being higher (Martin and Shaji 2015). The seasonal cycle of phytoplankton growth can be explained by a combination of nutrient and light limitation. Although stratification is relatively weak in the spring, winds are too light to cause much mixing and so productivity is low. Summer brings stronger stratification but also a supply of riverborne nutrients that trigger a bloom, which is then suppressed as sediment in the water limits the supply of light. The waters clear during the autumn and the north-east winds of winter lead to mixing of the water column and a new supply of nutrients, leading to the second and stronger bloom (Narvekar and Kumar 2014; Kumar et al. 2010). The winter bloom starts later and persists longer in the north-east than the north-west; this has been attributed to a combination of the eastward advection of river-sourced nutrients and an upwelling of nutrients due to subsurface currents in the east in winter (Martin and Shaji 2015). IOD/ENSO events can influence phytoplankton distribution patterns in the southern Bay of Bengal through changes in circulation and hence nutrient distribution (Martin and Shaji 2015; Currie et al. 2013).

Phytoplankton from the fertile areas that is not consumed by other organisms falls and decays at lower levels, leading to low oxygen levels in the deeper water. An estimated area of 389,000 km^2 of sea bed in the Bay of Bengal has dissolved oxygen levels below 0.5 ml/l, forming one of the largest oxygen minimum zones in the world (Helly and Levin 2004).

Tides in the Bay of Bengal are dominated by a semidiurnal (twice-daily) signal, with tidal amplitudes increasing south to north through the Bay (Murty and Henry 1983; Sindhu and Unnikrishnan 2013), as discussed in Chap. 8.

14.4 Climate Change Trends

Climate change will bring multiple changes to the Bay of Bengal. The most significant for coastal Bangladesh is sea-level rise, which is already being experienced at rates of up to 5 mm/year in the GBM delta (Antony et al. 2016; Unnikrishnan and Shankar 2007). The Fifth Assessment Report of the Intergovernmental Panel on Climate Change (Church et al. 2013) includes projections for sea-level rise at Haldia in the northern Bay (22.0°N, 88.1°E), for different Representative Concentration Pathways (RCPs) of greenhouse gases. The mean estimates for sea-level rise over the twenty-first century are 0.38 m, 0.48 m and 0.63 m under RCPs 2.6, 4.5 and 8.5 respectively, with a range of uncertainty approximately ±0.2 m (+0.3 m for RCP 8.5). These projections do not include the possible collapse of parts of the Antarctic ice sheet, which could add around 0.5 m more (Church et al. 2013; Levermann et al. 2014). Subsidence will also add to the sea-level rise experienced locally; this has been estimated for the GBM delta as about 3 mm/year (Brown and Nicholls 2015). The worst case rise from these figures, using the highest RCP 8.5 estimate and including the ice sheet melt, is 1.72 m by 2100. A more conservative estimate, using the mean RCP 4.5 rise but including Antarctic melting, gives a local rise of 1.27 m; the best case, lowest RCP 2.6 and no Antarctic contribution, would be 0.47 m. Any of these scenarios would have a considerable effect on the coastal zone, but the details of its impact will depend on the actions taken to adapt to the changing sea levels, such as raising of embankments and capture of river sediment (Brammer 2014). The effect of sea-level rise on coastal flooding is discussed in Chap. 16.

Rising sea surface temperatures have already been observed in the southern part of the Bay of Bengal, below 15°N, with an increase of about 0.2 °C between the late 1980s and the first decade of the twentieth century (Balaguru et al. 2014). Temperatures in the northern part of the Bay did not increase significantly in this period; however, a temperature rise is projected across the whole Bay in the twenty-first century, in the range 1–2 °C by mid-century and 2–3 °C by the end of the century under a medium greenhouse gas emission scenario (Fernandes et al. 2016 and see Chap. 25, Fig. 25.2). Increased temperatures are expected to lead to stronger stratification, which could reduce nutrient flows to the surface and hence give lower productivity. This has been reported for the western Indian Ocean (Roxy et al. 2016), but projections for the Bay of Bengal suggest that changes there will be small (Bopp et al. 2013; Fernandes et al. 2016, see also Chap. 25). There have been few modelling studies for this area and, as noted above, understanding of production in the Bay is based on limited observational evidence, so any conclusion must be tentative. In addition, productivity in the northern Bay will be sensitive to changes in riverborne nutrients, which will be affected by changes in precipitation and by other anthropogenic pressures. Changes in population, land use and sewage treatment can all affect nutrient loadings at the river mouth, and upstream dams can affect the amount and seasonal pattern of discharge to the sea (see Chap. 13 and Whitehead et al. 2015a, b). Any changes in primary production could have an effect on potential fish catch, but research carried out in the study area suggests that the effects of fisheries management policies are likely to be more important (Chap. 25).

14.5   Tropical Cyclones in the Bay of Bengal

The GBM region experiences relatively few tropical cyclones compared with other terrestrial basins: the number of named storms in the North Indian Ocean for 1980–2009 was only seven per cent of the global total (Knapp et al. 2010; Diamond and Trewin 2011). However, the large coastal population around the Bay of Bengal means that the impact of cyclone-induced storm surge can be very large. The main seasons for tropical cyclones in the northern Indian Ocean are April–May (pre-monsoon) and October–November (post-monsoon), with the strongest cyclones tending to occur post-monsoon when the available heat energy in the upper levels of the ocean is the greatest (Balaguru et al. 2014; Karim and Mimura 2008). The strong stratification of the sea surface following the monsoon rain and river input has been linked to the development of intense cyclones, as it limits the loss of heat from the cyclone system to the sea (Neetu et al. 2012).

Observation-based studies have generally shown no significant change in the overall number of tropical cyclones and depressions, but an increase in the frequency of intense tropical cyclone events in the Bay of Bengal during the month of November (Unnikrishnan et al. 2006; Karim and Mimura 2008). The observed increase in intensity since the 1970s has been related to increasing sea surface temperatures in the Indian Ocean, which provides more energy to fuel cyclones (Webster et al. 2005; Balaguru et al. 2014). Changes in tropical cyclone frequency, track and intensity could have significant implications for coastal Bangladesh, but changes in storminess are among the most uncertain of IPCC projections (IPCC 2013), and tropical cyclones are particularly difficult to forecast. Projected change in the frequency and intensity of tropical cyclones is discussed further in Chap. 16.

14.6   Summary of Issues for the Bay of Bengal

Climate change will have multiple effects in the Bay of Bengal. Current work suggests that changes in primary production will be small, but this is based on a limited understanding of present-day production and should be considered tentative. Changes in tropical cyclones are to be expected but little is known with confidence. Relative sea-level rise is expected to be in the range of 0.5–1.7 m by 2100, and this will undoubtedly have a significant effect on the lives of people in the coastal zone of the delta.

References

Akhil, V.P., F. Durand, M. Lengaigne, J. Vialard, M.G. Keerthi, V.V. Gopalakrishna, C. Deltel, F. Papa, and C.D.B. Montegut. 2014. A modeling study of the processes of surface salinity seasonal cycle in the Bay of Bengal. *Journal of Geophysical Research-Oceans* 119 (6): 3926–3947. https://doi.org/10.1002/2013jc009632.

Antony, C., A.S. Unnikrishnan, and P.L. Woodworth. 2016. Evolution of extreme high waters along the east coast of India and at the head of the Bay of Bengal. *Global and Planetary Change* 140: 59–67. https://doi.org/10.1016/j.gloplacha.2016.03.008.

Aparna, S.G., J.P. McCreary, D. Shankar, and P.N. Vinayachandran. 2012. Signatures of Indian Ocean Dipole and El Nino-Southern Oscillation events in sea level variations in the Bay of Bengal. *Journal of Geophysical Research-Oceans* 117. https://doi.org/10.1029/2012jc008055.

Balaguru, K., S. Taraphdar, L.R. Leung, and G.R. Foltz. 2014. Increase in the intensity of postmonsoon Bay of Bengal tropical cyclones. *Geophysical Research Letters* 41 (10): 3594–3601. https://doi.org/10.1002/2014gl060197.

Behara, A., and P.N. Vinayachandran. 2016. An OGCM study of the impact of rain and river water forcing on the Bay of Bengal. *Journal of Geophysical Research: Oceans* 121 (4): 2425–2446. https://doi.org/10.1002/2015JC011325.

Benshila, R., F. Durand, S. Masson, R. Bourdalle-Badie, C.D.B. Montegut, F. Papa, and G. Madec. 2014. The upper Bay of Bengal salinity structure in a high-resolution model. *Ocean Modelling* 74: 36–52. https://doi.org/10.1016/j.ocemod.2013.12.001.

Bopp, L., L. Resplandy, J.C. Orr, S.C. Doney, J.P. Dunne, M. Gehlen, P. Halloran, C. Heinze, T. Ilyina, R. Seferian, J. Tjiputra, and M. Vichi. 2013. Multiple stressors of ocean ecosystems in the 21st century: Projections with CMIP5 models. *Biogeosciences* 10 (10): 6225–6245. https://doi.org/10.5194/bg-10-6225-2013.

Brammer, H. 2014. Bangladesh's dynamic coastal regions and sea-level rise. *Climate Risk Management* 1: 51–62. https://doi.org/10.1016/j.crm.2013.10.001.

Brown, S., and R.J. Nicholls. 2015. Subsidence and human influences in mega deltas: The case of the Ganges-Brahmaputra-Meghna. *Science of the Total Environment* 527: 362–374. https://doi.org/10.1016/j.scitotenv.2015.04.124.

Chaitanya, A.V.S., M. Lengaigne, J. Vialard, V.V. Gopalakrishna, F. Durand, C. Kranthikumar, S. Amritash, V. Suneel, F. Papa, and M. Ravichandran. 2014. Salinity measurements collected by fishermen reveal a "river on the sea" flowing along the eastern coast of India. *Bulletin of the American Meteorological Society* 95 (12): 1897–1908. https://doi.org/10.1175/bams-d-12-00243.1.

Church, J.A., P.U. Clark, A. Cazenave, J.M. Gregory, S. Jevrejeva, A. Levermann, M.A. Merrifield, G.A. Milne, R.S. Nerem, P.D. Nunn, A.J. Payne, W.T. Pfeffer, D. Stammer, and A.S. Unnikrishnan. 2013. Sea level change. In *Climate change 2013: The physical science basis. Contribution of working group I to the fifth assessment report of the intergovernmental panel on climate change*, ed. T.F. Stocker, D. Qin, G.-K. Plattner, M. Tignor, S.K. Allen, J. Boschung, A. Nauels, Y. Xia, V. Bex, and P.M. Midgley. Cambridge, UK/New York: Cambridge University Press.

Currie, J.C., M. Lengaigne, J. Vialard, D.M. Kaplan, O. Aumont, S.W.A. Naqvi, and O. Maury. 2013. Indian Ocean Dipole and El Nino/Southern Oscillation impacts on regional chlorophyll anomalies in the Indian Ocean. *Biogeosciences* 10 (10): 6677–6698. https://doi.org/10.5194/bg-10-6677-2013.

Dai, A., and K.E. Trenberth. 2002. Estimates of freshwater discharge from continents: Latitudinal and seasonal variations. *Journal of Hydrometeorology* 3 (6): 660–687. https://doi.org/10.1175/1525-7541(2002)003<0660:eofdfc>2.0.co;2.

Diamond, H.J., and B.C. Trewin. 2011. Tropical cyclones [in "state of the climate in 2010"]. *Bulletin of the American Meteorological Society* 92 (6): S114–S131. https://doi.org/10.1175/1520-0477-92.6.S1.

Diansky, N.A., V.B. Zalesny, S.N. Moshonkin, and A.S. Rusakov. 2006. High resolution modeling of the monsoon circulation in the Indian Ocean. *Oceanology* 46 (5): 608–628. https://doi.org/10.1134/s000143700605002x.

Durand, F., D. Shankar, F. Birol, and S.S.C. Shenoi. 2009. Spatiotemporal structure of the East India coastal current from satellite altimetry. *Journal of Geophysical Research-Oceans* 114. https://doi.org/10.1029/2008jc004807.

Felton, C.S., B. Subrahmanyam, V.S.N. Murty, and J.F. Shriver. 2014. Estimation of the barrier layer thickness in the Indian Ocean using Aquarius salinity. *Journal of Geophysical Research-Oceans* 119 (7): 4200–4213. https://doi.org/10.1002/2013jc009759.

Fernandes, J.A., S. Kay, M.A.R. Hossain, M. Ahmed, W.W.L. Cheung, A.N. Lázár, and M. Barange. 2016. Projecting marine fish production and catch potential in Bangladesh in the 21st century under long-term environmental change and management scenarios. *ICES Journal of Marine Science* 73 (5): 1357–1369. https://doi.org/10.1093/icesjms/fsv217.

Helly, J.J., and L.A. Levin. 2004. Global distribution of naturally occurring marine hypoxia on continental margins. *Deep-Sea Research Part I-Oceanographic Research Papers* 51 (9): 1159–1168. https://doi.org/10.1016/j.dsr.2004.03.009.

IPCC. 2013. *Climate change 2013: The physical science basis.* Contribution of working group I to the fifth assessment report of the intergovernmental panel on climate change. Cambridge, UK/New York: Cambridge University Press.

Karim, M.F., and N. Mimura. 2008. Impacts of climate change and sea-level rise on cyclonic storm surge floods in Bangladesh. *Global Environmental Change-Human and Policy Dimensions* 18 (3): 490–500. https://doi.org/10.1016/j.gloenvcha.2008.05.002.

Knapp, K.R., M.C. Kruk, D.H. Levinson, H.J. Diamond, and C.J. Neumann. 2010. The international best track archive for climate stewardship (IBTrACS). *Bulletin of the American Meteorological Society* 91 (3): 363–376. https://doi.org/10.1175/2009BAMS2755.1.

Kumar, S.P., M. Nuncio, J. Narvekar, N. Ramaiah, S. Sardesai, M. Gauns, V. Fernandes, J.T. Paul, R. Jyothibabu, and K.A. Jayaraj. 2010. Seasonal cycle of physical forcing and biological response in the Bay of Bengal. *Indian Journal of Marine Sciences* 39 (3): 388–405.

Levermann, A., R. Winkelmann, S. Nowicki, J.L. Fastook, K. Frieler, R. Greve, H.H. Hellmer, M.A. Martin, M. Meinshausen, M. Mengel, A.J. Payne, D. Pollard, T. Sato, R. Timmermann, W.L. Wang, and R.A. Bindschadler. 2014. Projecting Antarctic ice discharge using response functions from SeaRISE ice-sheet models. *Earth System Dynamics* 5 (2): 271–293. https://doi.org/10.5194/esd-5-271-2014.

Martin, M.V., and C. Shaji. 2015. On the eastward shift of winter surface chlorophyll-a bloom peak in the Bay of Bengal. *Journal of Geophysical Research-Oceans* 120 (3): 2193–2211. https://doi.org/10.1002/2014jc010162.

Murty, T.S., and R.F. Henry. 1983. Tides in the Bay of Bengal. *Journal of Geophysical Research-Oceans and Atmospheres* 88 (NC10): 6069–6076. https://doi.org/10.1029/JC088iC10p06069.

Narvekar, J., and S.P. Kumar. 2014. Mixed layer variability and chlorophyll a biomass in the Bay of Bengal. *Biogeosciences* 11 (14): 3819–3843. https://doi.org/10.5194/bg-11-3819-2014.

Neetu, S., M. Lengaigne, E.M. Vincent, J. Vialard, G. Madec, G. Samson, M.R.R. Kumar, and F. Durand. 2012. Influence of upper-ocean stratification on tropical cyclone-induced surface cooling in the Bay of Bengal. *Journal of Geophysical Research-Oceans* 117. https://doi.org/10.1029/2012jc008433.

Pant, V., M.S. Girishkumar, T.V.S.U. Bhaskar, M. Ravichandran, F. Papa, and V.P. Thangaprakash. 2015. Observed interannual variability of near-surface salinity in the Bay of Bengal. *Journal of Geophysical Research-Oceans* 120 (5): 3315–3329. https://doi.org/10.1002/2014jc010340.

Roxy, M.K., A. Modi, R. Murtugudde, V. Valsala, S. Panickal, S.P. Kumar, M. Ravichandran, M. Vichi, and M. Levy. 2016. A reduction in marine primary productivity driven by rapid warming over the tropical Indian Ocean. *Geophysical Research Letters* 43 (2): 826–833. https://doi.org/10.1002/2015gl066979.

Schott, F.A., S.-P. Xie, and J.P. McCreary Jr. 2009. Indian Ocean circulation and climate variability. *Reviews of Geophysics* 47. https://doi.org/10.1029/2007rg000245.

Shankar, D., J.P. McCreary, W. Han, and S.R. Shetye. 1996. Dynamics of the east India coastal current: Analytic solutions forced by interior Ekman pumping and local alongshore winds. *Journal of Geophysical Research-Oceans* 101 (C6): 13975–13991.

Shetye, S.R. 1993. The movement and implications of the Ganges-Brahmaputra runoff on entering the Bay of Bengal. *Current Science* 64: 32–38.

Sindhu, B., and A.S. Unnikrishnan. 2013. Characteristics of tides in the Bay of Bengal. *Marine Geodesy* 36 (4): 377–407. https://doi.org/10.1080/01490419.2013.781088.

Unnikrishnan, A.S., and D. Shankar. 2007. Are sea-level-rise trends along the coasts of the north Indian Ocean consistent with global estimates? *Global and Planetary Change* 57 (3–4): 301–307. https://doi.org/10.1016/j.gloplacha.2006.11.029.

Unnikrishnan, A.S., K. Rupa Kumar, S.R. Fernandes, G.S. Michael, and S.K. Patwardhan. 2006. Sea level changes along the Indian coast: Observations and projections. *Current Science* 90: 362–368.

Vinayachandran, P.N., V.S.N. Murty, and V.R. Babu. 2002. Observations of barrier layer formation in the Bay of Bengal during summer monsoon. *Journal of Geophysical Research-Oceans* 107 (C12). https://doi.org/10.1029/2001jc000831.

Webster, P.J., G.J. Holland, J.A. Curry, and H.R. Chang. 2005. Changes in tropical cyclone number, duration, and intensity in a warming environment. *Science* 309 (5742): 1844–1846. https://doi.org/10.1126/science.1116448.

Whitehead, P.G., E. Barbour, M.N. Futter, S. Sarkar, H. Rodda, J. Caesar, D. Butterfield, L. Jin, R. Sinha, R. Nicholls, and M. Salehin. 2015a. Impacts of climate change and socio-economic scenarios on flow and water quality of the Ganges, Brahmaputra and Meghna (GBM) river systems: Low flow and flood statistics. *Environmental Science-Processes and Impacts* 17 (6): 1057–1069. https://doi.org/10.1039/c4em00619d.

Whitehead, P.G., S. Sarkar, L. Jin, M.N. Futter, J. Caesar, E. Barbour, D. Butterfield, R. Sinha, R. Nicholls, C. Hutton, and H.D. Leckie. 2015b. Dynamic modeling of the Ganges river system: Impacts of future climate and socio-economic change on flows and nitrogen fluxes in India and Bangladesh. *Environmental Science-Processes and Impacts* 17 (6): 1082–1097. https://doi.org/10.1039/c4em00616j.

15

A Sustainable Future Supply of Fluvial Sediment for the Ganges-Brahmaputra Delta

Stephen E. Darby, Robert J. Nicholls,
Md. Munsur Rahman, Sally Brown, and Rezaul Karim

15.1 Introduction

The world's deltas are facing a major sustainability challenge. Specifically, most of the world's large deltas (24 of 33) are threatened by a combination of rising sea levels and ground surface subsidence, increasing the possibility of submergence for around two and a half million square kilometres

S. E. Darby (✉)
Geography and Environment, Faculty of Social, Human and Mathematical Sciences, University of Southampton, Southampton, UK

R. J. Nicholls • S. Brown
Faculty of Engineering and the Environment and Tyndall Centre for Climate Change Research, University of Southampton, Southampton, UK

Md. Munsur Rahman • R. Karim
Institute of Water and Flood Management, Bangladesh University of Engineering and Technology, Dhaka, Bangladesh

© The Author(s) 2018
R. J. Nicholls et al. (eds.), *Ecosystem Services for Well-Being in Deltas*,
https://doi.org/10.1007/978-3-319-71093-8_15

of land (Syvitski et al. 2009), thus presenting a major issue for approximately 500 million inhabitants. Recent work has warned that the current scale of delta submergence (loss of elevation) is unprecedented in the last 7,000 years (Giosan et al. 2014). It has also been predicted that the extent of flood prone areas in deltas will further increase, perhaps by as much as 50 per cent, as sea levels continue to rise due to anthropogenic climate change (Syvitski et al. 2009).

At one level, understanding of the factors driving relative sea-level rise is well developed. Changes in delta surface elevation are controlled by a balance between rates of (i) eustatic sea-level rise and losses in surface elevation (i.e., natural sediment compaction, as well as accelerated subsidence driven by activities such as groundwater extraction) and (ii) gains in surface elevation as a result of the deposition of (largely) fluvial sediment. What is clear is that, as the only factor that could potentially offset losses in delta surface elevation, a sustainable supply of fluvial sediment is critical in generating the deposition needed to prevent 'drowning' (Ericson et al. 2006; Syvitski et al. 2009). River sediments therefore have considerable economic value, not only as a natural agent of flood mitigation but because deposited river sediments also carry nutrients (carbon, nitrogen, phosphorus) that help maintain agricultural productivity (Jin et al. 2015; Whitehead et al. 2015). As such, these river sediments can appropriately be described as 'brown gold'.

The combined sediment loads (more than 1 gigatonne per year; Islam et al. 1999) from the Ganges and Brahmaputra rivers (note that the Meghna is excluded from consideration as its sediment flux of around 13 million tonnes (Mt) per year is negligible in comparison to either the Ganges or Brahmaputra) have built one of the world's largest and most populous river delta systems (Woodroffe et al. 2006). Under natural conditions, these massive sediment loads would drive sediment deposition on the delta surface at sufficiently high rates (~3.5 mm per year) to compensate for slow sea-level rise and natural compaction-driven subsidence (Goodbred and Kuehl 1999; Wilson and Goodbred 2015). For these reasons, the Ganges-Brahmaputra-Meghna (GBM) delta presents an ideal system to investigate whether climate-driven changes in future fluvial sediment flux could compensate for (or compound) the adverse impacts of accelerated global sea-level rise and anthropogenic subsidence, particularly as the lives and livelihoods of so many people are at stake.

Within this context, this chapter reviews the policy implications arising from new insights, developed through state-of-the-art modelling into the prediction of future sediment loads supplied by the Ganges and Brahmaputra Rivers to the GBM delta. In particular, the implications of the new predictions of future sediment supply are outlined in terms of managing the sediment to maximise its potential for offsetting relative sea-level rise and to ensure that agriculture is able to maximise the value of the natural fertilisation afforded by the nutrients bound to the finer fraction of that sediment load (Jin et al. 2015; Whitehead et al. 2015).

15.2   Prognosis: New Insights

As part of the wider research, a climate-driven hydrological water balance and sediment transport model (HydroTrend; Kettner and Syvitski 2008) was employed to simulate future climate-driven water discharges and sediment loads flowing from the catchments upstream into the GBM delta (see Darby et al. 2015 for a detailed overview). Specifically, HydroTrend was parameterised using high-quality topographic data and forced with daily temperature and precipitation data obtained from downscaled Regional Climate Model (RCM) simulations for the period 1971–2100 (see Chap. 11 and Caesar et al. 2015). Note that the RCM (Jones et al. 2004) has a relatively high spatial resolution (0.22° × 0.22°, approximately 25 km) and covers a large South Asian domain (with rotated pole coordinates of 260° longitude and 70° latitude). This is important because it allows for the development of full mesoscale circulations and thereby captures important regional atmospheric dynamics relevant to the GBM catchments.

The model simulations were run for the period 1971–2100 using observed greenhouse gas forcing for the historical period and the SRES A1B emissions scenario (Nakićenović et al. 2000) for the future period. As discussed in Chap. 11, the SRES A1B scenario represents a medium-high emissions scenario that is consistent with observed carbon emissions over the past two decades and other existing climate modelling. Furthermore, the HadCM3 simulations used to drive the RCM use a perturbed physics ensemble (PPE) approach, whereby key climate model parameters, which have an associated uncertainty, are perturbed within an ensemble of simulations to produce a range of projections which reflect the uncertainty in the parameters.

The Met Office perturbed versions of HadCM3 with associated HadRM3P simulations for the 130-year period from 1971 to 2100 to create 17 ensemble results. Three members from this ensemble were selected, referred to as the Q0, Q8 and Q16 runs, respectively (see Table 15.1). The Q0 run represents exhibits a mid-range climate sensitivity to the A1B emissions forcing; Q16 has the highest climate sensitivity (i.e., it is the ensemble member that exhibits the highest global temperature response to the A1B emissions forcing); and finally, the Q8 run, although it has similar sensitivity to Q0, exhibits a different precipitation response. Specifically, unlike the other ensemble members, the Q8 run shows a mid-century decrease in precipitation (Table 15.1). The inclusion of the Q8 run therefore enables the impacts on sediment transfer processes of this possible climate response to be considered, even if the likelihood of this response can be considered to be relatively low.

It was found that fluvial sediment delivery rates to the GBM delta associated with these climate data sets were all projected to increase under the influence of anthropogenic climate change, albeit with the magnitude of the increase varying across the Ganges and Brahmaputra catchments (Fig. 15.1). Of the two study basins, the Brahmaputra's fluvial sediment load is predicted to be more sensitive to future climate change. By the middle part of the twenty-first century, model results suggest that sediment loads will increase (relative to the 1981–2000 baseline period) over a range of between 16 and 18 per cent (depending on climate model run) for the Ganges, but by between 25 and 28 per cent for the Brahmaputra. The simulated increase in river sediment supply from the two catchments

Table 15.1 Overview of the change in temperature and precipitation with respect to the annual mean for the 1981–2000 baseline period under the Q0, Q8 and Q16 Met Office RCM runs for the South Asian domain used in this study (for the SRES A1B emissions scenario) (Darby et al. 2015—Published by the Royal Society of Chemistry)

Climate model run	Mid-century (2041–2060)		End of century (2080–2099)	
	Temperature increase (K)	Precipitation increase (%)	Temperature increase (K)	Precipitation increase (%)
Q0	2.3	11.1	4.1	5.1
Q8	2.6	−9.9	4.1	12.7
Q16	2.6	12.2	4.6	29.5

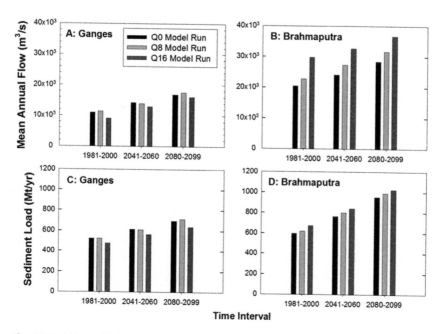

Fig. 15.1 Water discharges and river sediment loads as simulated by HydroTrend for a baseline (1981–2000) and two future (2041–2060 and 2080–2099) 20-year periods for the Ganges and Brahmaputra basins, with HydroTrend forced using climate data for a range of climate model runs (Q0, Q8 and Q16) under the SRES A1B emissions scenario. Subplots (**a**) and (**b**) show the mean annual flow discharges for the Ganges and Brahmaputra basins, respectively, while subplots (**c**) and (**d**) show the mean annual sediment fluxes for the Ganges and Brahmaputra basins, respectively (Darby et al. 2015—Published by the Royal Society of Chemistry)

further increases towards the end of the twenty-first century, reaching between 34 and 37 per cent for the Ganges and between 52 and 60 per cent for the Brahmaputra by the 2090s. The variability in these changes across the three climate change simulations is small compared to the temporal changes (Fig. 15.1).

This research has, therefore, shown that substantial increases in sediment loads are predicted to occur under the medium to high climate change scenarios that were explored. Specifically, the increases in end of century sediment loads that are projected from the Ganges (which range from an additional 161 Mt per year under the Q16 run to 191 Mt per year for Q8) and Brahmaputra (352 Mt per year under the Q16 run to 373 Mt per year under the Q8 run) amount to a combined increase of between 513 Mt per year and 564 Mt per year. This represents a potential increase of around 50 per cent over and above contemporary sediment loads, raising the question of the best way to manage the additional sediment supply to help alleviate problems in the delta, both today and in the future.

15.3   Policy Implications

Both the present and the additional river sediments that are projected to be supplied to the GBM delta from the Ganges and Brahmaputra catchments represent a resource of significant value through the potential they afford to (i) promote delta building and hence offset relative sea-level rise and (ii) act as vectors of nutrient deposition, offering a 'free' source of natural fertilisation for productive agricultural soils.

However, in some senses, establishing the precise supply of river sediments to the delta in the future is irrelevant as local communities can only derive benefits from those sediments if they are deposited and retained on the delta surface. Large areas of coastal Bangladesh have been protected by polders since the 1960s, excluding sedimentation processes from the land surface therefore not counteracting subsidence due to compaction. In the following sections, the scale of subsidence, the current trends of erosion and accretion, and the potential adaptations that could be implemented to help ensure that the value of these natural sediment services is fully realised are briefly reviewed.

15.3.1   Delta Building

As noted previously, sediment deposition in practice is the only means by which rising sea levels (driven by global warming and accelerated subsidence of the delta surface) can be offset to help slow or prevent relative sea levels continuing to rise in the future. In Table 15.2, reliable subsidence data for the study area (Brown and Nicholls 2015) shows that the median value of the subsidence is 2.6 mm per year, a value that needs to be added to the rate of climatic sea-level rise (SLR) in order to estimate the net relative sea-level rise (RSLR).

Table 15.2 Subsidence rates (mm/year) by geological era within the study area (extracted from Brown and Nicholls 2015)

Geological era	No of data	Min	Max	Mean	Median	Standard deviation
Recent (<1000 y)	16	1.3	10	4	2.6	2.9
Holocene (excluding recent)	6	1.3	4	2.6	2.4	1.1
Pleistocene	0	–	–	–	–	–
Pliocene	0	–	–	–	–	–
All records	22	1.3	10	3.6	2.6	2.6

Despite a number of basin scale and local level anthropogenic interventions in the GBM delta, delta accretion has been found to be the dominant process over the last 200 years, particularly in the Central Estuarine System (CES) of Bangladesh coast, although estimates indicate that the average annual area of land accreted can vary according to the time period and area under analysis (between 3 km²/year and 24 km²/year), as shown in Table 15.3. Consequently, the incoming sediment flux is sufficient to enable RSLR to be offset to some extent.

While both the fine (silts and clays) and coarse (sand) sediment load contributes to delta building, it is generally recognised that it is the sand fraction that is most important (Paola et al. 2011; Giosan et al. 2014; Nittrouer and Viparelli 2014). However, the quantity of coarse sediment which reaches the delta plain (estimated at between 10 and 25 per cent of the total sediment load by Okada et al. 2016) and which potentially contributes to delta building is not simply a function of the overall rate at which sediment is supplied by the rivers from upstream. Instead, in many of the world's deltas, including the GBM, engineering structures and management practices are key factors in controlling local flow dynamics and the exchange of sediment between rivers and the delta plains (Hung et al. 2014; Auerbach et al. 2015). In this respect the way in which the delta's flood defence, irrigation and drainage infrastructure (i.e., the GBM delta's canal and dyke networks) is located, built and operated will play a critical role in determining the potential for sediment deposition and hence delta building.

Although such water engineering infrastructure is essential to protect communities from extreme flooding and to enable productive agriculture,

Table 15.3 Estimates of net accretion within the study area over the last 200 years

Study period	Length of the study (year)	Net accretion (km²)	Accretion rate (km²/year)	Study area	Source
1973–2000	27	510	21	CES	MES II (2001)
1977–2010	34	139	4.08	WES-CES Island	Alam and Uddin (2013)
1940–1963	23	279	12.1	CES	Eysink (1983)
1776–1996	220	2197	9.9	CES	EGIS (1997)
1792–1984	192	1346	7	–	Allison (1998)
1840–1984	144	638	4.4	–	Allison (1998)
2007–2013	6	120	20	Char Island	Hussain et al.(2014)
1776–1943	167	760	4.6	CES	Sarker et al. (2011)
1943–1973	30	1100	42	CES	Sarker et al. (2011)
1973–2008	35	595	17	CES	Sarker et al. (2011)
1750–2000	250	2146	8.58	–	Rashid et al. (2011)
1990–1995	5	16.6	3.1	Bhola Island	Krantz (1999)
1973–2010	37	870	23.5	CES	Sarkar et al. (2013)
2007–2011	4	16.2	3.4	Urir Char	Taguchi et al. (2013)

dyke and canal networks also disconnect rivers from their delta plains and limit the amount of sediment reaching the delta surface. In policy terms, a trade-off must be made between achieving maximum sediment deposition to promote delta building versus the imperative to prevent flooding of agricultural areas and, in so doing, limit sediment deposition. In other deltas around the world, including the Mississippi (Paola et al. 2011; Giosan et al. 2014; Nittrouer and Viparelli 2014) and the Mekong (Manh et al. 2014; Chapman and Darby 2016; Chapman et al. 2016), the balance of that trade-off has been switching to a greater recognition of the importance of promoting natural sediment deposition for land building as a key adaptation strategy in the face of rising sea levels. This is because policy makers are recognising that it is not in the long-term economic interest to trade-off present-day requirements against the future sustainability of the delta. There are more than ten years of experience in Bangladesh with relatively small-scale tidal river management, or controlled flooding, and sedimentation in polders (Nowreen et al. 2014; Auerbach et al. 2015; Amir et al. 2013). Developing this at a much larger scale is recommended with full consideration of the technical and social challenges. This includes the land-building potential of present and future sediment supply and hence what might be sustainable in the long term.

15.3.2 Natural Fertilisation

Although infrastructure such as dykes and polders can be effective in protecting communities and farmland from flooding, the exclusion of flood water from polder compartments means that sediments are also excluded as well. The exclusion of coarse sediments (see Sect. 15.3.1) is an important consideration in terms of long-term delta building, but the nutrients that are bound to the fine-grained sediments have made deltaic soils and ecosystems some of the most productive on the planet, underpinning the provisioning ecosystem services in deltas (Chap. 1). This link between the sediment and nutrient transport and deposition (Jin et al. 2015; Whitehead et al. 2015) and agricultural productivity (Lázár et al. 2015) has been highlighted specifically in this research. Moreover, recent work has also been undertaken in the similar context of the Mekong delta, where the economic value of the role that fine-grained sediment deposition plays in underpinning agricultural production has been established. In that research, Chapman et al. (2016) and Chapman and Darby (2016) estimated that the nutrients contained within natural sediment deposits provide about half of the fertilisation required to sustain the annual rice crops, amounting to an economic value of $USD 26 million per year in one single province of the Mekong delta. Importantly, they found that poorer farmers were more reliant on natural sediment deposition as they are less able to afford the purchase of artificial fertilisers that can sustain yields when natural sediment deposition is prevented. Thus, promoting a more natural reconnection of rivers to their delta plains not only has long-term benefits in terms of delta building but also has an immediate benefit for livelihoods, particularly for those who are the poorest and least resilient to fluctuations in fertiliser prices.

15.4   Conclusion

This research shows that, over the course of the remainder of the present century, the supply of fluvial sediment being delivered to the apex of the GBM delta is likely to increase substantially (by around 50 per cent by the 2090s) as a consequence of medium-high anthropogenic climate change. An increase in the climate-driven supply of fluvial sediment to the GBM delta has the potential, through accelerated aggradation on the delta surface, to buffer some of the adverse impacts of climate change that are associated with rising sea levels in the Bay of Bengal and which threaten the vulnerable GBM delta. The projected increase in sediment flux emanating from the GBM delta's sub-continental scale catchments therefore represents a potentially beneficial impact of climate change (for the delta and its inhabitants). However, these potential beneficial impacts of climate change remain subject to uncertainty and can only be realised if more sediment actually reaches the delta. This may not be the case if anthropogenic disturbances within the feeder catchments, notably due to existing and proposed future construction of major dams, result in the delta becoming increasingly disconnected from the sediment supply that sustains it. Disconnection also occurs within the delta due to flood defences and polders built since the 1960s.

In terms of specific policy interventions, the key 'no regrets' adaptation (i.e., an adaptation that is viable irrespective of the actual future trajectory of river sediment loads) that is required is to ensure that the supply of river sediment from the GBM catchments upstream is actually retained on delta surface. This means promoting controlled flooding onto the delta surface (a policy that is consistent with the idea of 'working with the river' and which is increasingly being adopted by other major global delta plans, such as the Mississippi and Mekong), including in poldered areas. Bangladesh has begun to explore this approach with tidal river management, but this needs to be greatly enhanced to realise. This is necessary not only to help build the delta land surface up in an attempt to offset rising sea levels and subsidence but to enable farmers to benefit from the free nutrients transported by those sediments.

References

Alam, M.S., and K. Uddin. 2013. A study of morphological changes in the coastal areas and offshore islands of Bangladesh using remote sensing. *American Journal of Geographic Information System* 2 (1): 15–18. https://doi.org/10.5923/j.ajgis.20130201.03.

Allison, M.A. 1998. Historical changes in the Ganges-Brahmaputra delta front. *Journal of Coastal Research* 14 (4): 1269–1275.

Amir, M.S.I.I., M.S.A. Khan, M.M.K. Khan, M.G. Rasul, and F. Akram. 2013. Tidal river sediment management-A case study in southwestern Bangladesh. *International Journal of Environmental, Chemical, Ecological, Geological and Geophysical Engineering* 7 (3): 176–185.

Auerbach, L.W., S.L. Goodbred, D.R. Mondal, C.A. Wilson, K.R. Ahmed, K. Roy, M.S. Steckler, C. Small, J.M. Gilligan, and B.A. Ackerly. 2015. Flood risk of natural and embanked landscapes on the Ganges-Brahmaputra tidal delta plain. *Nature Climate Change* 5 (2): 153–157. https://doi.org/10.1038/nclimate2472.

Brown, S., and R.J. Nicholls. 2015. Subsidence and human influences in mega deltas: The case of the Ganges-Brahmaputra-Meghna. *Science of the Total Environment* 527: 362–374. https://doi.org/10.1016/j.scitotenv.2015.04.124.

Caesar, J., T. Janes, A. Lindsay, and B. Bhaskaran. 2015. Temperature and precipitation projections over Bangladesh and the upstream Ganges, Brahmaputra and Meghna systems. *Environmental Science-Processes and Impacts* 17 (6): 1047–1056. https://doi.org/10.1039/c4em00650j.

Chapman, A., and S. Darby. 2016. Evaluating sustainable adaptation strategies for vulnerable mega-deltas using system dynamics modelling: Rice agriculture in the Mekong Delta's An Giang Province, Vietnam. *Science of the Total Environment* 559: 326–338. https://doi.org/10.1016/j.scitotenv.2016.02.162.

Chapman, A.D., S.E. Darby, H.M. Hồng, E.L. Tompkins, and T.P.D. Van. 2016. Adaptation and development trade-offs: Fluvial sediment deposition and the sustainability of rice-cropping in An Giang Province, Mekong Delta. *Climatic Change* 137 (3): 593–608. https://doi.org/10.1007/s10584-016-1684-3.

Darby, S.E., F.E. Dunn, R.J. Nicholls, M. Rahman, and L. Riddy. 2015. A first look at the influence of anthropogenic climate change on the future delivery of fluvial sediment to the Ganges-Brahmaputra-Meghna delta. *Environmental Science-Processes and Impacts* 17 (9): 1587–1600. https://doi.org/10.1039/c5em00252d.

EGIS. 1997. *Morphological dynamics of the Brahmaputra-Jamuna river. Prepared for Water Resources Planning Organization (WARPO), Dhaka.* Report prepared by environment and GIS support project water sector planning, Delft Hyudraulics. Dhaka: Water Resources Planning Organization (WARPO), Ministry of Water Resources, Government of the People's Republic of Bangladesh.

Ericson, J.P., C.J. Vorosmarty, S.L. Dingman, L.G. Ward, and M. Meybeck. 2006. Effective sea-level rise and deltas: Causes of change and human dimension implications. *Global and Planetary Change* 50 (1–2): 63–82. https://doi.org/10.1016/j.gloplacha.2005.07.004.

Eysink, W.D. 1983. *Basic considerations on the morphology and land accretion potentials in the estuary of the Lower Meghna River.* Land reclamation project technical report 15. Dhaka: Bangladesh Water Development Board (BWDB).

Giosan, L., J. Syvitski, S. Constantinescu, and J. Day. 2014. Climate change: Protect the world's deltas. *Nature* 516: 31–33. https://doi.org/10.1038/516031a.

Goodbred, S.L., and S.A. Kuehl. 1999. Holocene and modern sediment budgets for the Ganges-Brahmaputra river system: Evidence for highstand dispersal to flood-plain, shelf, and deep-sea depocenters. *Geology* 27 (6): 559–562. https://doi.org/10.1130/0091-7613(1999)027<0559:hamsbf>2.3.co;2.

Hung, N.N., J.M. Delgado, A. Guntner, B. Merz, A. Bardossy, and H. Apel. 2014. Sedimentation in the floodplains of the Mekong Delta, Vietnam part II: Deposition and erosion. *Hydrological Processes* 28 (7): 3145–3160. https://doi.org/10.1002/hyp.9855.

Hussain, M.A., Y. Tajima, K. Gunasekara, S. Rana, and R. Hasan. 2014. Recent coastline changes at the eastern part of the Meghna Estuary using PALSAR and Landsat images. *IOP Conference Series: Earth and Environmental Science* 20 (1): 012047. https://doi.org/10.1088/1755-1315/20/1/012047.

Islam, M.R., S.F. Begum, Y. Yamaguchi, and K. Ogawa. 1999. The Ganges and Brahmaputra rivers in Bangladesh: Basin denudation and sedimentation. *Hydrological Processes* 13 (17): 2907–2923. https://doi.org/10.1002/(sici)1099-1085(19991215)13:17<2907::aid-hyp906>3.0.co;2-e.

Jin, L., P.G. Whitehead, S. Sarkar, R. Sinha, M.N. Futter, D. Butterfield, J. Caesar, and J. Crossman. 2015. Assessing the impacts of climate change and socio-economic changes on flow and phosphorus flux in the Ganga river system.

Environmental Science-Processes and Impacts 17 (6): 1098–1110. https://doi.
org/10.1039/c5em00092k.

Jones, R.G., M. Noguer, D.C. Hassell, D. Hudson, S.S. Wilson, G.J. Jenkins, and J.F.B. Mitchell. 2004. *Generating high resolution climate change scenarios using PRECIS*. Exeter: Met Office Hadley Centre. http://www.metoffice.gov. uk/media/pdf/6/5/PRECIS_Handbook.pdf. Accessed 19 July 2016.

Kettner, A.J., and J.P.M. Syvitski. 2008. HydroTrend v.3.0: A climate-driven hydrological transport model that simulates discharge and sediment load leaving a river system. *Computers and Geosciences* 34 (10): 1170–1183. https://doi.org/10.1016/j.cageo.2008.02.008.

Krantz, M. 1999. *Coastal erosion on the island of Bhola, Bangladesh*. B178. Goterborg: Earth Science Centre, Goterborg University. www.gu.se/ digitalAssets/1347/1347852_b178.pdf. Accessed 19 July 2016.

Lázár, A.N., D. Clarke, H. Adams, A.R. Akanda, S. Szabo, R.J. Nicholls, Z. Matthews, D. Begum, A.F.M. Saleh, M.A. Abedin, A. Payo, P.K. Streatfield, C. Hutton, M.S. Mondal, and A.Z.M. Moslehuddin. 2015. Agricultural livelihoods in coastal Bangladesh under climate and environmental change – A model framework. *Environmental Science-Processes and Impacts* 17 (6): 1018–1031. https://doi.org/10.1039/c4em00600c.

Manh, N.V., N.V. Dung, N.N. Hung, B. Merz, and H. Apel. 2014. Large-scale suspended sediment transport and sediment deposition in the Mekong Delta. *Hydrology and Earth System Sciences* 18 (8): 3033–3053. https://doi. org/10.5194/hess-18-3033-2014.

MES II. 2001. *Hydro-morphological dynamics of the Meghna Estuary, Meghna Estuary Study*. Meghna Estuary Study (MESII) project report. Dhaka: Bangladesh Water Development Board.

Nakićenović, N., J. Alcamo, G. Davis, B. Devries, J. Fenhann, S. Gaffin, K. Gregory, A. Gruebler, T.Y. Jung, T. Kram, E. Lebre Larovere, L. Michaelis, S. Mori, T. Morita, W. Pepper, H. Pitcher, L. Price, K. Riahi, A. Roehrl, H.-H. Rogner, A. Sankovski, M. Schlesinger, P. Shukla, S. Smith, R. Swart, S. Vanrooijen, N. Victor, and Z. Dadi. 2000. *Special report on emissions scenarios, a special report of working group III of the Intergovernmental Panel on Climate Change*. Cambridge: Cambridge University Press.

Nittrouer, J.A., and E. Viparelli. 2014. Sand as a stable and sustainable resource for nourishing the Mississippi River Delta. *Nature Geoscience* 7 (5): 350–354. https://doi.org/10.1038/ngeo2142.

Nowreen, S., M.R. Jalal, and M.S.A. Khan. 2014. Historical analysis of rationalizing South West coastal polders of Bangladesh. *Water Policy* 16 (2): 264–279. https://doi.org/10.2166/wp.2013.172.

Okada, S., A. Yorozuya, H. Koseki, S. Kudo, and K. Muraoka. 2016. Comprehensive measurement techniques of water flow, bedload and suspended sediment in large river using Acoustic Doppler Current Profiler. In *River sedimentation*, ed. S. Wieprecht, S. Haun, K. Weber, M. Noack, and K. Terheiden, 258–258. Boca Raton: CRC Press.

Paola, C., R.R. Twilley, D.A. Edmonds, W. Kim, D. Mohrig, G. Parker, E. Viparelli, and V.R. Voller. 2011. Natural processes in delta restoration: Application to the Mississippi Delta. *Annual Review of Marine Science* 3: 67–91. https://doi.org/10.1146/annurev-marine-120709-142856.

Rashid, S., M.S. Alam, and S.D. Shamsuddin. 2011. Development of islands in the Meghna Estuary over the past 250 years. In *Climate change: Issues and perspectives for Bangladesh*, ed. R. Ahmed and S.D. Shamsuddin. Dhaka: Sahitya Prkash.

Sarkar, M.H., M.R. Akhand, S.M.M. Rahman, and F. Molla. 2013. Mapping of coastal morphological changes of Bangladesh using RS, GIS and GNSS technology. *Journal of Remote Sensing and GIS* 1 (2): 27–34.

Sarker, M.H., J. Akter, M.R. Ferdous, and F. Noor. 2011. Sediment dispersal processes and management in coping with climate change in the Meghna Estuary, Bangladesh. In *Sediment problems and sediment management in Asian river basins*, IAHS Publication, ed. D.E. Walling, vol. 349, 203–217. Wallingford: International Association of Hydrological Sciences.

Syvitski, J.P.M., A.J. Kettner, I. Overeem, E.W.H. Hutton, M.T. Hannon, G.R. Brakenridge, J. Day, C. Vorosmarty, Y. Saito, L. Giosan, and R.J. Nicholls. 2009. Sinking deltas due to human activities. *Nature Geoscience* 2 (10): 681–686. https://doi.org/10.1038/ngeo629.

Taguchi, Y., M.A. Hussain, Y. Tajima, M.A. Hossain, S. Rana, A.K.M.S. Islam, and M.A. Habib. 2013. *Detecting recent coastline changes around the Urir Char Island at the eastern part of Meghna Estuary using PALSAR images*. Proceedings of the 4th International Conference on Water & Flood Management (ICWFM-2013), 9–11 Mar 2013, Dhaka.

Whitehead, P.G., S. Sarkar, L. Jin, M.N. Futter, J. Caesar, E. Barbour, D. Butterfield, R. Sinha, R. Nicholls, C. Hutton, and H.D. Leckie. 2015. Dynamic modeling of the Ganga river system: Impacts of future climate and socio-economic change on flows and nitrogen fluxes in India and Bangladesh. *Environmental Science-Processes and Impacts* 17 (6): 1082–1097. https://doi.org/10.1039/c4em00616j.

Wilson, C.A., and S.L. Goodbred Jr. 2015. Construction and maintenance of the Ganges-Brahmaputra Meghna Delta: Linking process, morphology, and stratigraphy. *Annual Review of Marine Science* 7: 67–88, ed. C.A. Carlson and S.J. Giovannoni.

Woodroffe, C.N., R.J. Nicholls, Y. Saito, Z. Chen, and S.L. Goodbred. 2006. Landscape variability and the response of Asian megadeltas to environmental change. In *Global change and integrated coastal management: The Asia-Pacific region*, ed. N. Harvey, 277–314. Dordrecht: Springer.

16

Present and Future Fluvial, Tidal and Storm Surge Flooding in Coastal Bangladesh

Anisul Haque, Susan Kay, and Robert J. Nicholls

16.1 Introduction

Inundation in coastal Bangladesh can be caused by a number of flood drivers as explained in Chap. 8: (i) fluvial floods, (ii) tidal floods, (iii) fluvio-tidal floods and (iv) storm surge floods. The dominance of a specific flood driver depends on location. The first three types of flooding can occur at the same time under monsoon conditions and during spring tides. Fluvial and fluvio-tidal floods are linked to the monsoon and can persist for weeks or longer, whereas a tidal flood is a short-term (daily)

A. Haque (✉)
Institute of Water and Flood Management, Bangladesh University of Engineering and Technology, Dhaka, Bangladesh

S. Kay
Plymouth Marine Laboratory, Plymouth, UK

R. J. Nicholls
Faculty of Engineering and the Environment and Tyndall Centre for Climate Change Research, University of Southampton, Southampton, UK

© The Author(s) 2018
R. J. Nicholls et al. (eds.), *Ecosystem Services for Well-Being in Deltas*,
https://doi.org/10.1007/978-3-319-71093-8_16

293

flood linked to tidal fluctuations. Storm surge flooding is caused by the landfall of tropical cyclones in coastal Bangladesh (Chap. 14).

In this research, quantitative information on the magnitude and extent of all these flood types both today and through the twenty-first century is developed using a hydrodynamic numerical model together with a series of scenarios. This chapter describes the methods, the main findings and characteristics of the dominant present and future flooding patterns in the coastal region. It analyses the fluvial, tidal and fluvio-tidal flooding using one model framework, while flooding due to cyclones is considered independently.

16.2   Analysis of Fluvial, Tidal and Fluvio-tidal Flooding

In this section, the occurrence of fluvial, tidal and fluvio-tidal floods in coastal Bangladesh are analysed. These floods occur during the monsoon season, when the three types of flooding can occur simultaneously, although the relative severity depends on location. Generally tidal flooding occurs close to the coast, whereas fluvial flooding occurs further inland along the rivers and estuaries. Where fluvial flooding is influenced by the tide, the flood is characterised as a compound fluvio-tidal flood (Chap. 8).

To analyse the extent of flooding in the region, a well-known opensource numerical model (Delft 3D Flow) was applied. In the model, the study region is represented by all the rivers and estuaries of the coastal zone which have a width greater than 100 m. Fluvial flows enter the model domain through three major rivers—Ganges, Brahmaputra and Upper Meghna (see Chap. 8, Fig. 8.1). For all present-day conditions simulated by the model, measured discharges from the Bangladesh Water Development Board (BWDB) are specified in these three locations, whereas, for all future scenarios, discharges in these locations are provided from the simulation results of the hydrological model, INCA (Whitehead et al. 2015a, b; see Chap. 13). As measured data for the sea level is not available, sea level variation is provided from the simulation results of an ocean model GCOMS (see Chap. 14) for both present-day conditions and future scenarios. The region is discretised into 896,603 grid points, where the grid size varies from 243 to 1,164 m in the longitudinal direction and 186 to 1,704 m in the lateral direction. The coarser grid size is used in the ocean, and the finer grid size is used for the land areas to capture the details of the river/estuarine systems and topographic variation.

Topography in the model is provided by the digital elevation model (DEM) available in the database of Water Resources Planning Organisation (WARPO). The DEM in its current form has a 50 × 50 m resolution. For the river bathymetry, combinations of secondary and primary data are used. Secondary data are collected from BWDB, and primary data for 294 locations along the rivers/estuaries of the coastal zone are measured as part of this research. Ocean bathymetry is provided from the open access General Bathymetric Chart of the Oceans (GEBCO). The coastal zone in Bangladesh contains 139 polders, of which 103 are in the study area. In the model, polder locations are specified from the polder map available in the WARPO database, and design polder heights are specified from the data collected from BWDB. Subsidence and morphological changes (bank line shifting, bed level changes and floodplain sedimentation) are not considered in any of the flood simulations. However, wetting and drying phenomena are considered during the rise and recession phases of fluvial floods, during rising and falling tides of tidal floods and during the landfall and decay stages of cyclone generated storm surge floods. The model is calibrated for a base year condition (year 2000) and validated for present-day extreme condition (year 1998). With a simulation time step of ten minutes, total run time for one-year flood model simulation is 48 hours using a PC with core i7 processor and 32 MB RAM. A discussion on model calibration and validation can be found in Haque et al. (2016).

The present-day conditions and future scenarios are shown in Table 16.1. The present-day conditions (base year for average condition and present-day extreme for a high fluvial flood year) are classified following the BWDB definition of flooding (see Chap. 8 and BWDB 2012). Future flooding scenarios are based on the scenario descriptions

Table 16.1 Scenario descriptions and the inundation characteristics for fluvial and related flooding

Scenario name	Description	Date of flood initiation	Date of flood peak	Date of flood recession	Maximum land area inundated (flood peak) (km²)	Excess/deficit relative to base year (%)
Base year	Year: 2000 Inundation: Average	First week July	Second week September	Fourth week October	1,204	–
Present extreme	Year: 1998 Inundation: Extreme	Second week July	Third week September	Fourth week October	2,425	+101
Mid-century	Year: 2052 Temp. rise: 2.64 °C Sea-level rise: 26 cm Scenario: Q16	Second week July	Fourth week August	Fourth week October	690	–42
End-century	Year: 2088 Temp. rise: 4.63 °C Sea-level rise: 54 cm Scenario: Q16	First week June	First week August	First week November	1,657	+38
End-century extreme	Year: 2088 Temp. rise: 4.63 °C Sea-level rise: 1.5 m Scenario: Q16	First week June	First week August	First week November	3,990	+231

and quantification undertaken during the project (Chaps. 9, 11 and 14). In total there are five simulations for fluvial and related flooding and five simulations for storm surge flooding due to cyclonic events.

Sea-level rise projections for these scenarios were based on the Fifth Assessment Report of the Intergovernmental Panel on Climate Change (AR5, Church et al. 2013, see also Chap. 14). Mid-century rise was taken as 0.24 m, compared to a baseline in 2000, and end-century rise as 0.54 m. The 0.54 m value is quite conservative and in particular omits any potential contribution from melting of the West Antarctic ice sheet. The extreme end-century value used in some simulations in this chapter, 1.5 m, is based on the high-end estimates from AR5 with an additional 0.5 m for ice sheet melting (Kay et al. 2015). Subsidence is added to local sea-level rise but has not been included in the research presented due to the uncertainty in predicting subsidence in protected and non-protected land. However, subsidence rates are estimated to be of the order 2.5 mm/year (Brown and Nicholls 2015) so could add an additional 0.25 m by the end of the century (see also Chap. 15).

The areal extent of inundation due to these present and future conditions is given in Table 16.1. Compared to the base year condition, the present-day extreme flood is doubled in the inundated area. The present-day extreme flood condition corresponds roughly to a flood with 100 year return period (Islam and Chowdhury 2002). In the future there is a tendency for a large increase in flooded area, but not in all cases. By the end of the century (end-century extreme scenario), the flooded area increases by 231 per cent compared to the base year and 65 per cent compared to the present extreme. This scenario might cause overtopping of the coastal embankments (Nihal et al. 2016), and needs further evaluation. It should be noted that while sea-level rise dominates the change, increases in upstream discharges are also significant (see Whitehead et al. 2015a, b).

Maps of the flooded areas are shown in Fig. 16.1. Unprotected areas in the north of the study area are inundated due to changes in the discharge of the upstream rivers during the monsoon season (May to October). At the onset of monsoon, river water inundates any unprotected land and rises to its peak, generally in August or September (see Table 16.1). The influence of tides is minimal in these areas, except when monsoon winds and resulting elevated sea levels during the spring tide can delay drainage

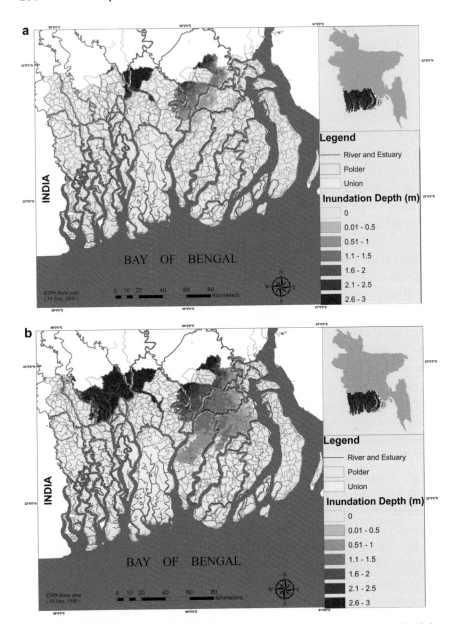

Fig. 16.1 Inundation maps for different flooding scenarios as described in Table 16.1 (**a**) base year, (**b**) present-day extreme, (**c**) mid-century, (**d**) end-century, (**e**) end-century extreme

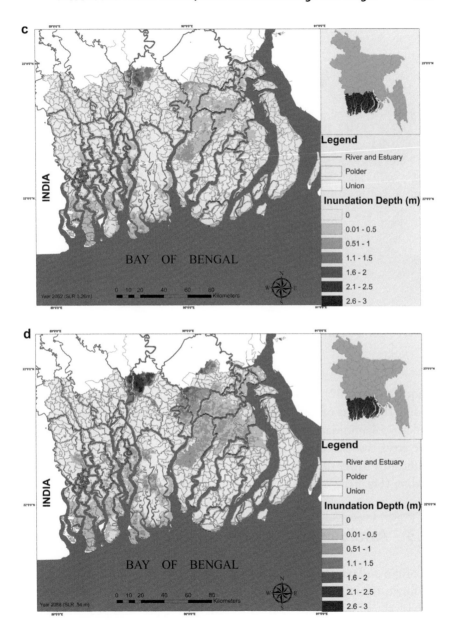

Fig. 16.1 (continued)

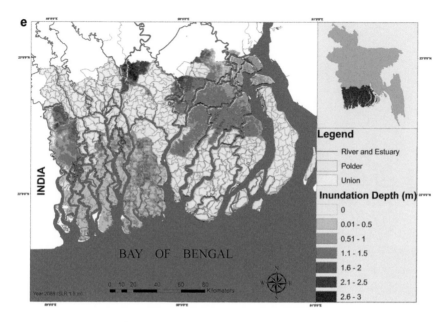

Fig. 16.1 (continued)

(Haque et al. 2002). Tidal flooding impacts unprotected areas close to the sea. The banks of the lower reaches of rivers (estuaries) are affected by regular tidal inundation. There is negligible influence of fluvial flow in these areas, and seasonal variations are not visible. Flood depths are also significantly lower than those that occur with the monsoon-related fluvial flooding (Fig. 16.2).

16.3   Analysis of Storm Surge Flooding

The origin and nature of storm surge flooding is distinct from fluvial, tidal and fluvio-tidal flooding. Storm surge flooding is generated by the landfall of tropical cyclones which normally occur during pre- and post-monsoon seasons (April–May and October–November), as discussed in Chap. 8. Both extreme and more frequently occurring cyclones are considered.

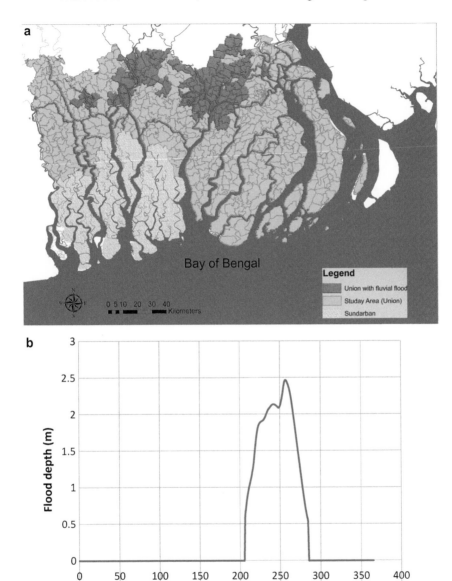

Fig. 16.2 Comparison of fluvial and tidal flooding characteristics (**a**) and (**c**) location and (**b**) and (**d**) typical hydrographs

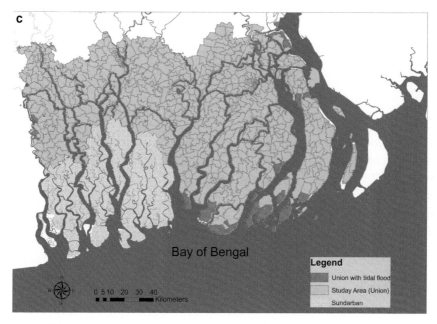

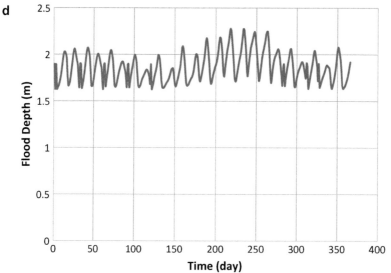

Fig. 16.2 (continued)

Storm surges are simulated by applying a numerical model (Delft Dashboard and Delft 3D Flow) as discussed in Nihal et al. (2015). Simulating a full population of extreme events for flood analysis represent a major activity which was beyond the resources available for this research. Hence, a combination of an historical analogue approach to capture extreme events and simulation of more common cyclones is adopted. Cyclone Sidr made landfall on the Bangladesh coast on 15 November 2007 as a Category 4 Cyclone, killing more than 3,000 people in Bangladesh. It is considered to be the most severe cyclone to make landfall in Bangladesh since the 1991 surge which killed over 100,000 people (Chap. 8). Cyclone Sidr was selected as a representation of a typical high strength cyclone. Two future scenarios were also developed based on Cyclone Sidr as defined in Table 16.2 (*SIDR-mid-century* and *SIDR-end-century*).

In addition, two examples of less severe cyclone scenarios were extracted from the model analysis of cyclones and resulting extreme sea levels by Kay et al. (2015). These are generated using a shelf-sea model (POLCOMS) driven with the regional climate scenarios described in Chap. 11. They are termed *GENERATED* scenarios and comprise two scenarios: *GENERATED-mid-century* and *GENERATED-end-century* (Table 16.2). These modelled cyclones were characterised by low atmospheric pressure (<965mb) coinciding with abnormally high sea level (>0.7 m above annual normal); two typical events were selected for analysis (Fig. 16.3). Flood extents were then modelled using the Delft3D model, using the shelf-sea model to provide boundary conditions at 86.414°E, 19.967°N and 92.9433°E, 19.9751°N. The process of surge propagation inland through the estuaries and the land are simulated by Delft3D. Due to the structured grid and grid size (explained in Sect. 16.2), the complex geometrical variation of the coast is largely linearised. This means that localised flood effects due to local physical settings (e.g. road and canal network inside poldered area) are not always captured by the model, but it provides a useful national case study of how extreme sea-level events may develop due to twenty-first century climate change for coastal Bangladesh.

The modelled *GENERATED* cyclones have lower wind speeds than the modelled Cyclone *SIDR* (Fig. 16.4) and landfall locations to the west

Table 16.2 Scenario descriptions, inundated area and the maximum surge depth due to selected present and potential cyclonic events

Cyclone	Description	Total area inundated (km^2)	Excess/ deficit compared to *SIDR* (%)	Maximum flood depth (m)
SIDR (base case)	Cyclone *SIDR* made landfall east of the Sundarbans on 15 November 2007	914	–	5.6
SIDR-mid-century	A *SIDR*-like cyclone (similar strength and landfall location) with hydrodynamic setting as the mid-century scenario Q16 with 26 cm SLR (Table 16.1)	2,063	+125	5.9
SIDR-end-century	A *SIDR*-like cyclone (similar strength and landfall location) with hydrodynamic setting as the end-century scenario Q16 with 54 cm SLR (Table 16.1)	3,491	+282	6.0
GENERATED-mid-century	Cyclone scenario extracted from the Bay of Bengal POLCOMS Model. A specific cyclone from year 2062 (low temperature rise and 26-cm sea-level rise which is a Q16 scenario) is selected	1,305	+42	4.2
GENERATED-end-century	Cyclone scenario extracted from the Bay of Bengal POLCOMS Model. A specific cyclone from year 2088 (high temperature rise and 54 cm sea-level rise which is a Q16 scenario) is selected	690	–24	4.6

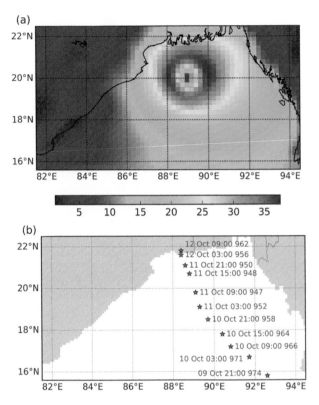

Fig. 16.3 Example of a simulated cyclone event, October 2028. (**a**) Wind speed (m/s) at 09:00 on 11 October. (**b**) Minimum pressure at 6 hourly intervals from 9 October 15:00 to 12 October 03:00 2028. Data is taken from the simulations described in Kay et al. (2015)

of *SIDR* (mid-century cyclone at the western boundary of the Sundarbans and end-century cyclone in the Indian section of the Sundarbans). These differences influence the total inundated area and flood depth—the *GENERATED-mid-century* inundates an area 42 per cent greater than Cyclone *SIDR* (compared to 125 per cent greater area for a *SIDR-mid-century* cyclone), whereas the *GENERATED-end-century* cyclone floods an area 24 per cent smaller (compared to 282 per cent greater area for a *SIDR-end-century* cyclone). The maximum flood depth of the *GENERATED* cyclones produces a smaller flood depth (around 4.5 m)

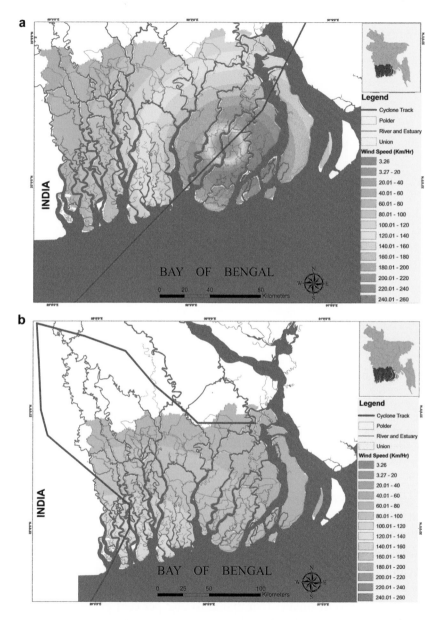

Fig. 16.4 Wind speed during cyclone events for (**a**) *SIDR*, (**b**) *GENERATED-mid-century*, (**c**) *GENERATED-end-century* scenarios

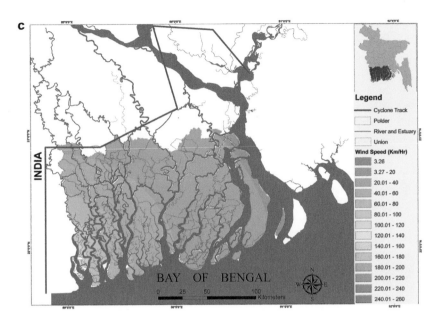

Fig. 16.4 (continued)

compared to *SIDR*-strength cyclones (around 6 m). So the modelling suggests that, with a moderate estimate of sea-level rise, weaker cyclones at end-century may still produce lower flood depths than extreme events in the present day.

The inundation associated with extreme storm surges is shown in Fig. 16.5 and reported in Table 16.2. Cyclones *SIDR*, *SIDR-mid-century* and *SIDR-end-century* have similar cyclone wind speed, pressure drop and landfall locations. A cyclone with *SIDR* strength and in the same landfall location in mid-century floods an extra 125 per cent of land, mainly due to sea-level rise (see Table 16.2). The corresponding increase of maximum flood depth is 30 cm. Similarly, by end-century, the land that is inundated is nearly quadrupled combined with an increase of 33 cm to maximum surge height under 54 cm sea-level rise.

Other similar studies show the importance of landfall location of cyclones on the extent of land inundation, in addition to sea-level rise

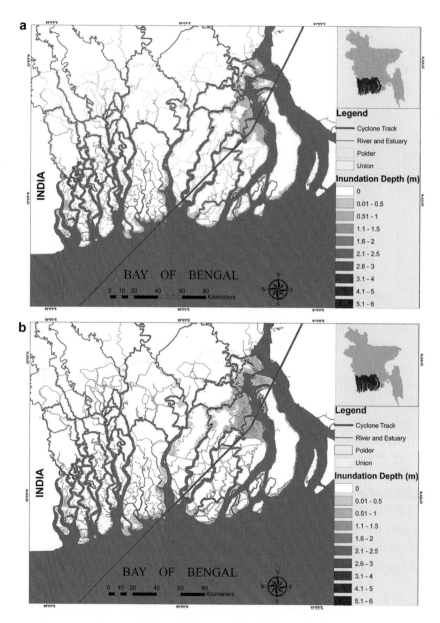

Fig. 16.5 Flood (or inundation) depth and inundated areas of cyclones (**a**) *SIDR*, (**b**) *SIDR-mid-century*, (**c**) *SIDR-end-century*, (**d**) *GENERATED-mid-century* and (**e**) *GENERATED-end-century* scenarios

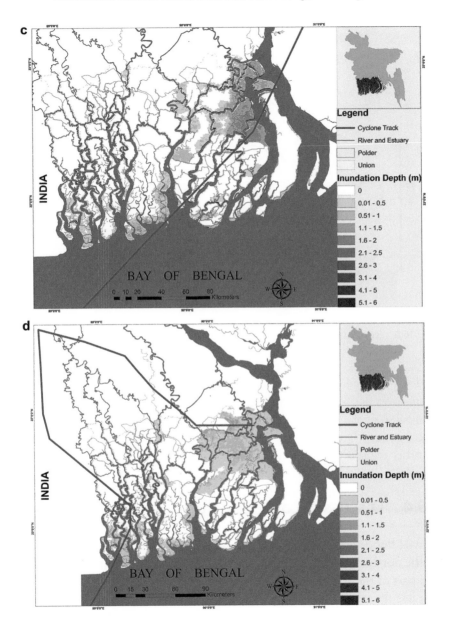

Fig. 16.5 (continued)

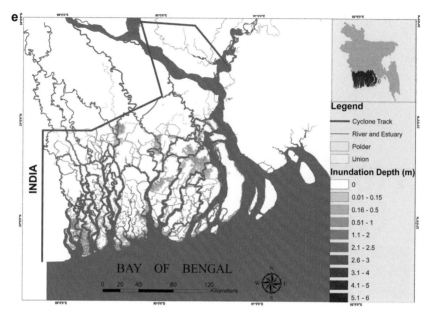

Fig. 16.5 (continued)

and wind speed. When a cyclone makes landfall in the Sundarbans, the mangroves act as a buffer which reduces the extent of land inundation (see Sakib et al. 2015). Another study shows that due to net land subsidence, cyclone landfall in the central region is more vulnerable compared to other regions (Nihal et al. 2015).

16.4   Discussion and Conclusions

Fluvial floods occur from July to October, and the flood area is confined to the northern part of the coastal region. Tidal floods occur in the southern part of the coastal region along the banks of the estuaries and tend to last less than a day. The magnitude of the land inundation due to fluvial and tidal flood is aggravated over the twenty-first century by sea-level rise and larger upstream river flows. In an extreme present-day flood situation (e.g. for a 100-year flood), total inundated area is approximately 101 per cent more than an average situation. Due to sea-level rise and change in upstream river flows, this value may rise up to 231 per cent (for an end-century scenario with a 1.5 m sea-level rise). Some specific future flood events cause overtopping of coastal embankments.

Storm surge floods in the region are caused by the landfall of tropical cyclones, which normally occur April to May (pre-monsoon) and October to November (post-monsoon). The areal extent of inundation and flood depth due to storm surge depends on the cyclone wind speed, landfall location and landfall time (Chap. 8). Sea-level rise generally increases the area of land inundation. A *SIDR*-strength cyclone repeated in mid-century with a sea-level rise of 26 cm increases the area of land inundation by 125 per cent. A similar cyclone at the end-century, with a sea-level rise of 54 cm, increases the area of land inundation by 282 per cent. Corresponding increase of maximum flood depth is 30 and 33 cm in mid- and end-century, respectively. Hence, changes in cyclone-induced flooding are of greater magnitude than fluvial flooding.

The analysis shows that the effects of sea-level rise are not simple inundation as shown in the first inundation assessments of Bangladesh as published by Milliman et al. (1989) and Huq et al. (1995). This reflects the extensive polders and embankment systems that exist in coastal Bangladesh (Chap. 8). Further, the areas outside the polders, such as the Sundarbans, can accrete with the sea level (see Chap. 26). However, given the large areas potentially flooded, adaptation and planning for increased flooding are essential. More consideration needs to be given to understanding and predicting embankment breaching as this will become more likely over time if no planning is undertaken. Upgrading the embankments, as is already proposed in the Bangladesh Delta Plan 2100 is one approach, combined with other improvements such as improved warning systems (Dasgupta et al. 2014). More fundamentally there is a need to develop more sustainable long-term management techniques such as 'tidal river management'. This is controlled or engineered sedimentation within polders to build land elevation (Amir et al. 2013; Auerbach et al. 2015). It has been employed locally to date and given the large amount of natural sediment available could be employed on a much larger scale (see Chap. 15).

References

Amir, M.S.I.I., M.S.A. Khan, M.M.K. Khan, M.G. Rasul, and F. Akram. 2013. Tidal river sediment management-A case study in southwestern Bangladesh. *International Journal of Environmental, Chemical, Ecological, Geological and Geophysical Engineering* 7 (3): 176–185.

Auerbach, L.W., S.L. Goodbred, D.R. Mondal, C.A. Wilson, K.R. Ahmed, K. Roy, M.S. Steckler, C. Small, J.M. Gilligan, and B.A. Ackerly. 2015. Flood risk of natural and embanked landscapes on the Ganges-Brahmaputra tidal delta plain. *Nature Climate Change* 5 (2): 153–157. https://doi.org/10.1038/nclimate2472.

Brown, S., and R.J. Nicholls. 2015. Subsidence and human influences in mega deltas: The case of the Ganges-Brahmaputra-Meghna. *Science of the Total Environment* 527: 362–374. https://doi.org/10.1016/j.scitotenv.2015.04.124.

BWDB. 2012. *Annual flood report 2012*. Dhaka: Bangladesh Water Development Board (BWDB). http://www.ffwc.gov.bd/index.php/reports/annual-flood-reports. Accessed 9 May 2016.

Church, J.A., P.U. Clark, A. Cazenave, J.M. Gregory, S. Jevrejeva, A. Levermann, M.A. Merrifield, G.A. Milne, R.S. Nerem, P.D. Nunn, A.J. Payne, W.T. Pfeffer, D. Stammer, and A.S. Unnikrishnan. 2013. Sea level change. In *Climate change 2013: The physical science basis. Contribution of working group I to the fifth assessment report of the Intergovernmental Panel on Climate Change*, ed. T.F. Stocker, D. Qin, G.-K. Plattner, M. Tignor, S.K. Allen, J. Boschung, A. Nauels, Y. Xia, V. Bex, and P.M. Midgley. Cambridge, UK/New York: Cambridge University Press.

Dasgupta, S., M. Huq, Z.H. Khan, M.M.Z. Ahmed, N. Mukherjee, M.F. Khan, and K. Pandey. 2014. Cyclones in a changing climate: The case of Bangladesh. *Climate and Development* 6 (2): 96–110. https://doi.org/10.1080/17565529.2013.868335.

Haque, A., M. Salehin, and J.U. Chowdhury. 2002. Effects of coastal phenomena on the 1998 flood. In *Engineering concerns of flood: A 1998 perspective*, ed. M.A. Ali, S.M. Seraj, and S. Ahmad. Dhaka: Directorate of Advisory, Extension and Research Services, Bangladesh University of Engineering and Technology.

Haque, A., S. Sumaiya, and M. Rahman. 2016. Flow distribution and sediment transport mechanism in the estuarine systems of Ganges-Brahmaputra-Meghna delta. *International Journal of Environmental Science and Development* 7 (1): 22–30. https://doi.org/10.7763/IJESD.2016.V7.735.

Huq, S., S.I. Ali, and A.A. Rahman. 1995. Sea-level rise and Bangladesh: A preliminary analysis. *Journal of Coastal Research SI* 14: 44–53.

Islam, A.K.M.S., and J.U. Chowdhury. 2002. Hydrologic characteristics of the 1998 flood in major rivers. In *Engineering concerns of flood: A 1998 perspective*, ed. M.A. Ali, S.M. Seraj, and S. Ahmad. Dhaka: Directorate of Advisory, Extension and Research Services, Bangladesh University of Engineering and Technology.

Kay, S., J. Caesar, J. Wolf, L. Bricheno, R.J. Nicholls, A.K.M.S. Islam, A. Haque, A. Pardaens, and J.A. Lowe. 2015. Modelling the increased frequency of extreme sea levels in the Ganges-Brahmaputra-Meghna delta due to sea level rise and other effects of climate change. *Environmental Science-Processes and Impacts* 17 (7): 1311–1322. https://doi.org/10.1039/c4em00683f.

Milliman, J.D., J.M. Broadus, and G. Frank. 1989. Environmental and economic implications of rising sea level and subsiding deltas: The Nile and Bengal examples. *Ambio* 18 (6): 340–345.

Nihal, F., M. Sakib, W.E. Elahi, A. Haque, M. Rahman, and R.A. Rimi. 2015. *Sidr like cyclones in Bangladesh coast.* 2nd International Conference on Environment, Technology and Energy, November 22–23, Colombo.

Nihal, F., M. Sakib, S. Noor, A. Haque, M. Rahman, W.E. Elahi, and U. Halder. 2016. *Climatic impacts on the fluvial and tidal inundation patterns in the Ganges-Brahmaputra-Meghna delta.* Proceedings of the 2016 International Conference on Disaster Management and Civil Engineering (ICDMCE'2016), April 12–13, Kyoto.

Sakib, M., F. Nihal, A. Haque, M. Rahman, and M. Ali. 2015. Sundarban as a buffer against storm surge flooding. *World Journal of Engineering and Technology* 3: 59–64. https://doi.org/10.4236/wjet.2015.33C009.

Whitehead, P.G., E. Barbour, M.N. Futter, S. Sarkar, H. Rodda, J. Caesar, D. Butterfield, L. Jin, R. Sinha, R. Nicholls, and M. Salehin. 2015a. Impacts of climate change and socio-economic scenarios on flow and water quality of the Ganges, Brahmaputra and Meghna (GBM) river systems: Low flow and flood statistics. *Environmental Science-Processes and Impacts* 17 (6): 1057–1069. https://doi.org/10.1039/c4em00619d.

Whitehead, P.G., S. Sarkar, L. Jin, M.N. Futter, J. Caesar, E. Barbour, D. Butterfield, R. Sinha, R. Nicholls, C. Hutton, and H.D. Leckie. 2015b. Dynamic modeling of the Ganga river system: Impacts of future climate and socio-economic change on flows and nitrogen fluxes in India and Bangladesh. *Environmental Science-Processes and Impacts* 17 (6): 1082–1097. https://doi.org/10.1039/c4em00616j.

17

Modelling Tidal River Salinity in Coastal Bangladesh

Lucy Bricheno and Judith Wolf

17.1 Introduction

In the Ganges-Brahmaputra-Meghna (GBM) delta, river water is widely used to irrigate crops. In recent years there has been an increase in river salinity (Dasgupta et al. 2014). This can affect soil salinity by inundation (if no defences or polders are breached) or by seepage through embankments which has led to an increase in soil salinities, particularly in the west, close to the Sundarbans forest region. This change has potential consequences for land use practices and livelihoods for the delta population. For example, some landowners and farmers have already chosen to convert freshwater rice paddies to brackish water shrimp farms, which may take decades to reverse. Understanding how changing climate (changes in the monsoon rains, leading to changing freshwater flow) may affect future salinity levels is therefore an important topic. In this research, the Finite Volume Coastal Ocean Model (FVCOM)

L. Bricheno (✉) • J. Wolf
National Oceanographic Centre, Liverpool, UK

model has been used to make future projections of river salinity which are used in the assessment of future soil salinity and agricultural productivity (see Chaps. 18 and 24).

17.2   The Ganges-Brahmaputra-Meghna (GBM) Delta

With an average freshwater discharge of around 40,000 m³/s/year, the GBM river system has the third largest discharge worldwide. The GBM river delta is a low-lying fertile area covering over 100,000 km² in India and Bangladesh and is thus classified as a megadelta (Woodroffe et al. 2006). This large discharge forms a large offshore freshwater plume, and the freshwater signal can be observed far offshore, for example, consider World Ocean Atlas/observations from the Soil Moisture and Ocean Salinity (SMOS) satellite in Fig. 17.1. The GBM delta is generally migrating eastwards, and the majority of the freshwater is channelled through the Padma (or Lower Meghna) River (Brammer 2014). There is some limited discharge in the western estuarine section, with the only source of freshwater, the Gorai River. There are some defunct river channels behaving as tidal creeks, but not as sources of freshwater.

Northern Bay of Bengal is mesotidal, with tidal ranges at the coast, varying from two metres to over four metres. Large tides at the coast can penetrate far inland, with tidal variability being observed as far as 200 km

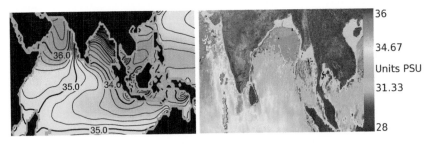

Fig. 17.1 Annual average sea surface salinity from World Ocean Atlas (left) and SMOS (right) (Reproduced with permission from Antonov et al. 2010, and data from BEC 2010, respectively)

from the coast. Water levels in the delta are also significantly affected by the large volume of river discharge, which can raise mean water depths by as much as four metres.

17.3   River Salinity Within the Delta

Salinity is defined as the salt concentration (e.g. sodium and chloride) in water. It is measured in unit of PSU (practical salinity unit[1]). The averaged salinity in the global ocean is 35.5 PSU, while freshwater (e.g. inland lakes or rivers) has salinity close to 0 PSU. Observations of river salinity in Bangladesh and the study area are limited. There are a few point samples of electrical conductivity available, often only during the dry season when salinities are high. Even with this limited data, a clear spatial pattern emerges; Dasgupta et al. (2014) present a contour map, showing a strong longitudinal gradient, with the highest salinities (30+ PSU) in the west of the GBM delta and fresher river waters (<5 PSU) in the east. There are also higher salinities around the coast of around 10 PSU in places. However, this broad pattern does not capture the complex pattern of river salinity in the delta system.

To help address this and provide calibration for the high resolution model FVCOM (Chen et al. 2003), a field survey of salinity levels and river soundings was carried out (see Jahan et al. 2015, data available on request). Using this data the spatial variability of salinity of and changing river salinity over a tidal and seasonal cycle for the delta can be simulated. River cross-section soundings are also primary input to any river/estuary hydraulic/hydrodynamic model. As for most developing countries, there is a lack of sufficient bathymetric data in Bangladesh. The bathymetry of the whole Sundarbans area has never been measured fully, so a bathymetric survey was commissioned. The location of historic (prior to this research) and new survey cross-sections is shown in Fig. 17.2. Besides taking sections, the surveyors also collected soil and water samples and point soundings along the length of the Ganges River.

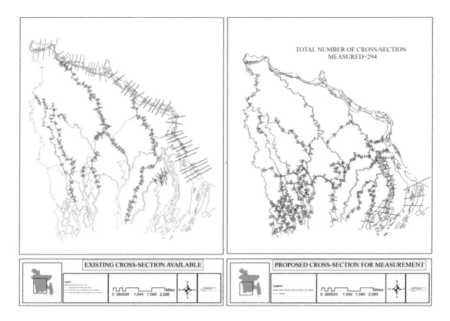

Fig. 17.2 Mean high water salinity observed during March to May 2014 (left). Modelled mean surface salinity during March to May 2000 (right). Both colour scales show salinity between 0 and 35 PSU

17.4   FVCOM Model

The FVCOM model is an unstructured finite volume, 3D baroclinic model which was developed for use in estuarine systems and has since been applied to wider areas (Chen et al. 2006). It can model the effects of tides, freshwater flow, saline water intrusion and sea-level rise. At the Bay of Bengal open boundary, the model is forced by hourly ocean tidal data and daily temperature and salinity from the Bay of Bengal Global Coastal Ocean Modelling System (GCOMS) model (see Chap. 14 and Kay et al. 2015). Figure 17.3 shows the observed river salinity data compared with a FVCOM simulation for 2000. The observations are single snapshots in time, while the modelled results represent the modelled mean for three months. The spatial distribution of the observed salinity is wall captured in the model. The modelled salinity is slightly high in the far western section (around the Sundarbans) while salinities are under-predicted (i.e. too fresh

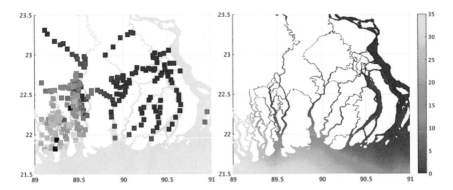

Fig. 17.3 Locations of river cross-sections: existing sections (left), newly surveyed sections (right) (Reproduced under Creative Commons Attribution 4.0 International License from Bricheno et al. 2016)

in the model) in the Balaswar river (centred at 89.5 East). This disparity between the model and observations may be due to problems in channel connectivity—where freshwaters are unable to penetrate far enough to the west due to small river channels missing in the model. Overall the root mean square error with respect to observations is 5.5 PSU. Even though a direct comparison is not possible, the two results are correlated with an R-squared value of 0.69 (where 0 is uncorrelated, and 1 is perfectly correlated).

Figure 17.4 shows the final model bathymetry developed for this research. It extends inland to the tidal limit and the confluence of the Ganges, Meghna and Brahmaputra, avoiding the need for a separate river model to introduce the freshwater flow. The model is run with 11 vertical levels and variable horizontal resolution between 10 km (at the open boundary) and 50 m, within narrow river channels. The challenge in constructing this model grid was to generate a good-quality mesh and bathymetry which allows the model to produce accurate flows through the delta region. The observations of tidal level within the delta are used to validate the basic hydrodynamics (Bricheno et al. 2016). The modelled sea surface elevations have been compared against tidal data and water level observations inland. At the open coast, where tides are dominant, the model performs well, giving confidence in its abilities. FVCOM was able to accurately model the tidal ranges in the central delta; however

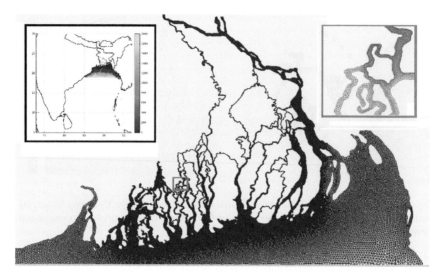

Fig. 17.4 Unstructured model grid location within the Bay of Bengal (left inset map) and close-up of Sundarbans channels (right inset)

there was some under-prediction of the tidal range in the Sundarbans forest region, and within the Lower Meghna.

River discharge is applied as an upstream boundary; a volume of freshwater is introduced to the model at a single grid point at the northern boundary of the model (indicated in Fig. 17.5). The daily discharge volume rate comes from the hydrological INCA model (see Chap. 13 and Whitehead et al. 2015).

Following the model validation, FVCOM is then run for discrete future time slices (described in Table 17.1) to examine projected changes in tidal range and river salinity. Meteorological forcing is not applied locally in the delta model, but the input to INCA comes from the Regional Climate Model (Chap. 11 and Caesar et al. 2015). The model runs are therefore consistent with the scenarios of changing sea level and freshwater flow used elsewhere in the research.

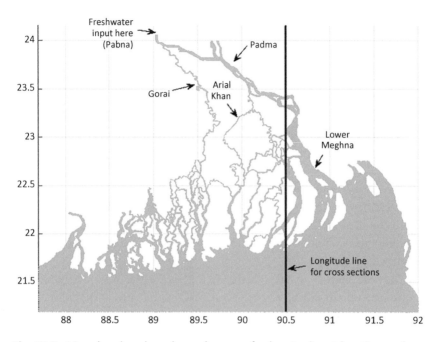

Fig. 17.5 Map showing river channel names, freshwater input location and section line discussed in the text

Table 17.1 Summary of scenario runs with annual river discharge over time discussed in this chapter

Scenario	Description	Climate and management	MSLR (cm)	Year	Annual river discharge (m³)	'Wet' or 'dry'
1	Baseline	Q0+business as usual	0.0	2000–2001	9,928,407	
2	Mid-century	Q0+business as usual	31.96	2047–2048	13,979,424	Wet
3	Mid-century	Q8+less sustainable	27.06	2050–2051	10,011,085	Dry
4	End century	Q8+more sustainable	58.77	2082–2083	16,517,208	Wet
5	End century	Q0+business as usual	59.01	2097–2098	10,978,254	Dry

17.5   Baseline River Salinity Modelling

Water levels in the delta are controlled by a balance between river and tidal flow, acting on different timescales. Throughout the year the situation can change, from tides controlling the water levels in the dry season to dominance by river flow during the monsoon. Some sensitivity to model bottom roughness is seen in the simulations, though the magnitude of these changes is small in the context of tidal range and seasonal water level changes associated with freshwater.

Figure 17.6 shows modelled surface salinity from the baseline simulation described in Table 17.1. As there is a continuous data set from the model, the mean and maximum are calculated from the full annual data set, while Fig. 17.2 compares point observations with a modelled map for the March to May period.

It may be seen that the salinity is higher in the west of the study region around the Sundarbans forest. This is because the majority of freshwater discharges occur through the eastern distributaries, especially the Meghna Estuary, and the western delta is composed mainly of tidal creeks with little or no freshwater input. In the FVCOM delta model, there is only one point of freshwater input (marked on Fig. 17.5). The network of river distributaries then carries this to the Bay of Bengal. The model may be missing additional freshwater inputs from smaller distributaries and Indian rivers that enter the

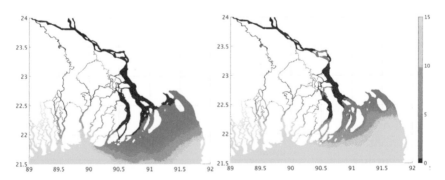

Fig. 17.6 Mean high water salinity during 2000 (left) and maximum high water salinity during 2000 (PSU) (right)

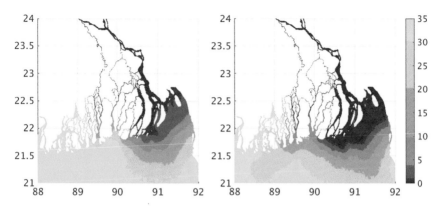

Fig. 17.7 Spring (March–April–May) mean salinity (left) and autumn (September–October–November) mean salinity (right) during 2000 (PSU)

west of the delta. Freshwater input to the model is also limited, as surface precipitation and evaporation are not included.

Figure 17.7 shows how seasonal discharge associated with monsoon rains alters river surface salinity in the delta. The spring mean (an average salinity for March, April, May) represents the end of the dry season, when sea water has been able to penetrate further inland, unopposed by river outflow. The second panel shows average salinity for the autumn (September, October, November) after a period of heavy rainfall during the monsoon. At this point in the year, a large volume of river water has been discharged through the delta, creating a freshwater plume offshore. Though the large channels of the lower Meghna remain relatively fresh throughout the year, the salinity in the narrow channels in the central and western sectors of the delta are more strongly affected by these seasonal fluctuations.

Another controlling factor on river salinity is the effect of tidal mixing. Figure 17.8 shows how the freshwater plume moves in and out over the course of a 12.4 hour cycle. The largest differences are seen along the sharp front, where fresh river waters of <3 PSU meet ocean waters of >30 PSU. Further inland the impact of the tide is also observed: leading to differences in salinity of 3 PSU and above.

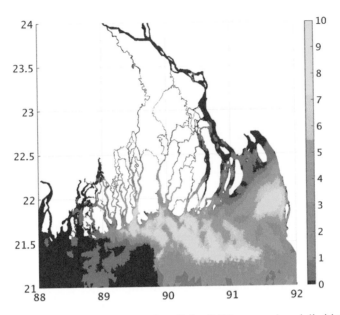

Fig. 17.8 Map showing the change in salinity (PSU) comparing daily high water and low water salinity for a 25 hour period on a spring tide during the dry season

These processes operate on very different timescales: day-to-day, the tides dominate while the annual river discharge cycle controls local salinity on a longer annual scale.

17.5.1 The Freshwater Plume

Further offshore, away from the narrow river channels, freshwater enters the Bay of Bengal and spreads out into a coastal plume. Figure 17.9 shows the extent of the freshwater plume offshore and its structure in the vertical during the dry (upper) and wet season (lower). The modelled plume is quite well mixed in the vertical, though some structure of fresher river waters overlying saltier sea water is observed in the July snapshot. When the water column stratifies following the monsoon, the freshwater layer can be between 5 and 10 m thick in the vertical. The extent and position of this freshwater plume may have implications for coastal water quality, nutrient fluxes and fisheries.

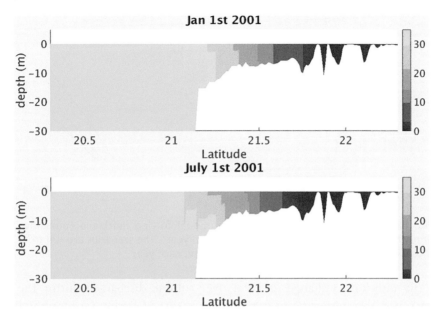

Fig. 17.9 Vertical profiles of salinity along 90.5E comparing 1 January 2001 (upper) and 1 July 2001 (lower)

17.5.2 Salinity Projections

In order to investigate potential future river salinities, five scenarios were run: one as a historic baseline and four futures (Table 17.1). The future scenarios were designed based on the discharge rates provided by the INCA model to encapsulate the full range of possible climate states and wet/dry years. Cumulative river discharge was therefore used to guide the selection of scenario years; for a mid-century period (2030–2059) and a late-century period (2070–2099), all river discharge years were examined, and those with maximum/minimum cumulative annual discharge were selected to be simulated in FVCOM. The annual cumulative discharge volume for each scenario year is shown in Fig. 17.10. A rate of changing sea-level rise consistent with the discharge year was applied at the open ocean boundary. Mean sea-level rise (MSLR), as described in Chap. 14 and Kay et al. (2015), was applied. The total MSLR for each year is shown in Table 17.1

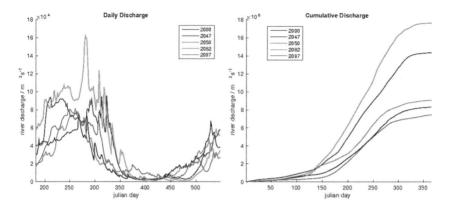

Fig. 17.10 Daily river discharge used as model forcing (left) and cumulative annual discharge (right) in m³ for five scenario years. The scenarios are run from monsoon to monsoon period, starting on 1 June each year

In addition to changes in total river volume discharged during the year, it is also important to consider the nature of freshwater input and tidal influence. The timing and intensity of freshwater input is closely related to the intensity of the monsoon as illustrated by the left hand panel of Fig. 17.10 which shows the daily river discharge entering the model at a site close to the city of Pabna, situated on the Padma (Ganges) River. In 'wet' years (blue and green curves) there is an early and sharp monsoon onset around the middle of July (day 200). A more gradual and later onset (around day 220—early August) occurs during the two 'dry years' (pink and red curves). The future tidal range is projected to become larger, although not in a spatially homogenous fashion. Combined with an increase in mean sea level, this leads to an increase of salt intrusion into the delta (e.g. see Fig. 17.11).

The spatial structure of the front is complex and 'pulses' in and out on a tidal time scale of 12.4 hours. There is also considerable vertical structure in the profile of salinity, with a fresher layer overlying salty waters. Tides advect salt water in and out, but also contribute to mixing in the vertical.

Figure 17.12 shows the distribution of mean and maximum annual salinity in the study area under the baseline scenario. A strong east-west profile is observed, with freshwaters in the eastern section, dominated by

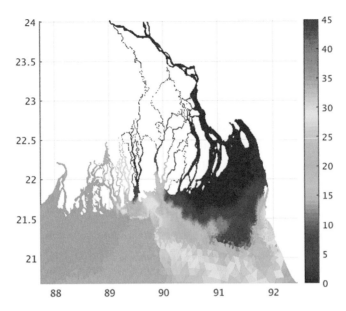

Fig. 17.11 Snapshot of surface salinity (PSU) during the wet season under scenario 5 (July 2098)

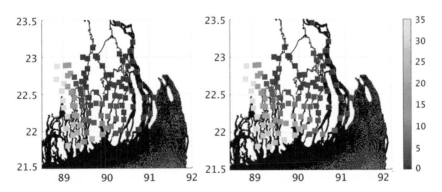

Fig. 17.12 Mean (left) and maximum (right) salinity during the baseline simulation 2000–2001 (Scenario 1 in Table 17.1)

outfall from the Meghna. The drier western section is saltier overall, with maximum values approaching those of seawater. There may be some over-prediction of salinity in the very west of the model, due to the limited input of freshwater in this area.

Table 17.2 Ocean salinity forcing levels used for modelled scenarios

Scenario years	2000–2001	2047–2048	2050–2051	2082–2083	2097–2089
Mean 'ocean' salinity (PSU)	34.39	34.10	34.38	33.99	34.22

Table 17.1 shows that the annual discharge in scenarios 1 and 5 are similar. However, surface salinity is relatively high across much of the delta. This may be due to a change in ocean conditions which are represented by salinity evolution at the open boundary of the model. Table 17.2 shows the mean salinity in the forcing model GCOMS at 19°N, which is used to drive the FVCOM model.

The numbers in Table 17.2 indicate that mean ocean salinities do not vary greatly over time. This suggests that the rise observed in river salinity may therefore be largely related to hydrodynamics, higher mean sea levels and changes in tidal conditions allowing ocean water to penetrate further inland. This is shown most clearly in the west of the study area, where the largest increases in salinity levels along the rivers are projected. In the east of the study area (e.g. at the mouth of the Meghna River), this potential increase is offset due to additional freshwater discharge meaning salinity levels remain relatively unchanged over time.

The maximum annual salinities for four future scenarios within the delta are shown in Fig. 17.13. The top row (mid-century) simulations see higher maximum values in the Western Estuarine System when compared with the baseline year, while the eastern estuary remains largely unchanged.

In the far future simulations (bottom row), the effect of salt intrusion is seen more strongly, particularly in the 'dry' year scenario (5—low freshwater input and no change in management). Under this scenario, high values exceeding 15 PSU are observed throughout the central estuarine section, and salinities exceeding 5 PSU are seen at the mouth of Lower Meghna.

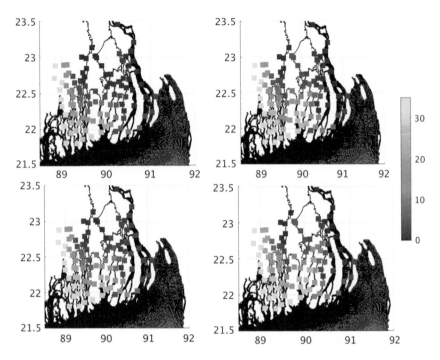

Fig. 17.13 Annual maximum salinities for 103 selected points under the four future scenarios described in Table 17.1. Clockwise from top left: S1, S2, S4, S3

17.6 Conclusion

River salinity in the GBM delta can be simulated with a variable resolution hydrodynamic model. Salt intrusion has a pronounced spatial pattern, with the saltiest river waters observed in the western estuarine section and the Sundarbans forest. Waters become progressively fresher moving towards the east and the mouth of the Lower Meghna. There is a strong seasonal signal in the freshwater distribution controlled by variable river discharge from the monsoon to the dry season including within the freshwater plume which alters size, position and vertical structure throughout the year.

Tidal excursion of the freshwater front can alter local river salinity by between two and five PSU over the course of a day. River salinity is largely controlled by a combination of annual monsoon discharge and tidal processes. In the western and central estuarine sections salt intrusion at the coast is the dominant factor. In the eastern section, the increased river flow and increased sea level are more balanced, so there is little future change in salinity in the mouth of the Meghna.

Under all future projections of freshwater input and ocean changes, salinity is predicted to increase in the river channels. This increase is more pronounced in the central and western estuarine section with implications for agriculture, shrimp farming and local well-being (see Chaps. 21, 24 and 27).

Note

1. A unit based on the properties of seawater conductivity. It is equivalent to parts per thousand or grams per kilogramme. The highest observed salinities (more than 40 PSU) are found in the Dead Sea.

References

Antonov, J.I., D. Seidov, T.P. Boyer, R.A. Locarnini, A.V. Mishonov, H.E. Garcia, O.K. Baranova, M.M. Zweng, and D.R. Johnson. 2010. Volume 2: Salinity. In *World Ocean Atlas 2009; NOAA Atlas NESDIS 69*, ed. S. Levitus, vol. 184. Washington, DC: U.S. Government Printing Office.

BEC. 2010. *Sea surface salinity data.* http://bec.icm.csic.es/. Accessed 14 June 2017.

Brammer, H. 2014. Bangladesh's dynamic coastal regions and sea-level rise. *Climate Risk Management* 1: 51–62. https://doi.org/10.1016/j.crm.2013.10.001.

Bricheno, L.M., J. Wolf, and S. Islam. 2016. Tidal intrusion within a mega delta: An unstructured grid modelling approach. *Estuarine Coastal and Shelf Science* 182: 12–26. https://doi.org/10.1016/j.ecss.2016.09.014.

Caesar, J., T. Janes, A. Lindsay, and B. Bhaskaran. 2015. Temperature and precipitation projections over Bangladesh and the upstream Ganges, Brahmaputra and Meghna systems. *Environmental Science-Processes and Impacts* 17 (6): 1047–1056. https://doi.org/10.1039/c4em00650j.

Chen, C.S., H.D. Liu, and R.C. Beardsley. 2003. An unstructured grid, finite-volume, three-dimensional, primitive equations ocean model: Application to coastal ocean and estuaries. *Journal of Atmospheric and Oceanic Technology* 20 (1): 159–186. https://doi.org/10.1175/1520-0426(2003)020<0159:augfvt> 2.0.co;2.

Chen, C., R.C. Beardsley, and G. Cowles. 2006. *An unstructured grid, finite-volume coastal ocean model. FVCOM user manual.* 2nd ed. Cambridge, MA: School of Marine Science and Technology at the University of Massachusetts-Dartmouth (SMAST/UMASSD).

Dasgupta, S., F.A. Kamal, Z.H. Khan, S. Choudhury, and A. Nishat. 2014. *River salinity and climate change: Evidence from coastal Bangladesh.* Policy working paper series 6817. Washington, DC: The World Bank. http://documents. worldbank.org/curated/en/522091468209055387/River-salinity-and-climate-change-evidence-from-coastal-Bangladesh. Accessed 11 Apr 2014.

Jahan, M., M.M.A. Chowdhury, S. Shampa, M.M. Rahman, and M.A. Hossain. 2015. *Spatial variation of sediment and some nutrient elements in GBM delta estuaries: A preliminary assessment.* Proceedings of the International Conference on Recent Innovation in Civil Engineering for Sustainable Development (IICSD-2015), December 11–13, Dhaka University of Engineering and Technology, Gazipur.

Kay, S., J. Caesar, J. Wolf, L. Bricheno, R.J. Nicholls, A.K.M.S. Islam, A. Haque, A. Pardaens, and J.A. Lowe. 2015. Modelling the increased frequency of extreme sea levels in the Ganges-Brahmaputra-Meghna delta due to sea level rise and other effects of climate change. *Environmental Science-Processes and Impacts* 17 (7): 1311–1322. https://doi.org/10.1039/c4em00683f.

Whitehead, P.G., E. Barbour, M.N. Futter, S. Sarkar, H. Rodda, J. Caesar, D. Butterfield, L. Jin, R. Sinha, R. Nicholls, and M. Salehin. 2015. Impacts of climate change and socio-economic scenarios on flow and water quality of the Ganges, Brahmaputra and Meghna (GBM) river systems: Low flow and flood statistics. *Environmental Science-Processes and Impacts* 17 (6): 1057–1069. https://doi.org/10.1039/c4em00619d.

Woodroffe, C.N., R.J. Nicholls, Y. Saito, Z. Chen, and S.L. Goodbred. 2006. Landscape variability and the response of Asian megadeltas to environmental change. In *Global change and integrated coastal management: The Asia-Pacific region*, ed. N. Harvey, 277–314. Dordrecht: Springer.

18

Mechanisms and Drivers of Soil Salinity in Coastal Bangladesh

Mashfiqus Salehin, Md. Mahabub Arefin Chowdhury, Derek Clarke, Shahjahan Mondal, Sara Nowreen, Mohammad Jahiruddin, and Asadul Haque

18.1 Introduction

The coastal area of Bangladesh covers about one-fifth of the country and represents more than 30 percent of the country's cultivable lands (Rasel et al. 2013). Livelihoods in this region are thus largely dependent on agricultural practices; however, dry season agricultural productivity in this region is low compared to the national average (see Chap. 24), which is considered to be one of the major reasons for high incidence of poverty (Lázár et al. 2015). Soil salinity is the dominant factor behind the low

---

The original version of this chapter was revised. The second author's name was incorrect. The chapter has been updated with the correct name and affiliation of the author. An erratum to this chapter can be found at https://doi.org/10.1007/978-3-319-71093-8_30

M. Salehin (✉) • Md. Mahabub Arefin Chowdhury • S. Mondal • S. Nowreen
Institute of Water and Flood Management, Bangladesh University of Engineering and Technology, Dhaka, Bangladesh

D. Clarke
Faculty of Engineering and the Environment and Tyndall Centre for Climate Change Research, University of Southampton, Southampton, UK

© The Author(s) 2018
R. J. Nicholls et al. (eds.), *Ecosystem Services for Well-Being in Deltas*,
https://doi.org/10.1007/978-3-319-71093-8_18

crop productivity, which is further compounded by inappropriate and/or faulty water control structures in some areas.

A total of 1.05 million hectares of land out of 2.88 million hectares in the Khulna and Barisal divisions are affected by different degrees of soil salinity within the coastal and offshore lands (SRDI 2010) (Fig. 18.1). The spatial pattern of soil salinity is similar to that of river water salinity (Fig. 18.2) and groundwater salinity (Fig. 18.3). Large areas of land remain fallow in the dry season (January–May) because of soil salinity, lack of good quality irrigation water, and problems with water control (mostly drainage) (Mondal et al. 2006; SRDI 2010).

The coverage of dry season irrigated *Boro* crop within the study area (29.3 percent) is much lower than the national average of around 63

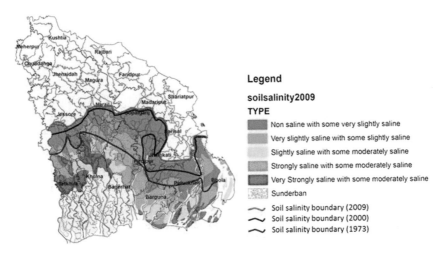

Legend

soilsalinity2009
TYPE

▪	Non saline with some very slightly saline
▪	Very slightly saline with some slightly saline
▪	Slightly saline with some moderately saline
▪	Strongly saline with some moderately saline
▪	Very Strongly saline with some moderately saline
▪	Sunderban
∿	Soil salinity boundary (2009)
∿	Soil salinity boundary (2000)
∿	Soil salinity boundary (1973)

Fig. 18.1 Soil salinity in the south-west coastal region in 2009. Figure shows different degrees of salinity in different places and the increase in soil salinity from 1973 to 2009 (Based on data from the Bangladesh Soil Resources Development Institute)

M. Jahiruddin
Department of Soil Science, Bangladesh Agricultural University, Mymensingh, Bangladesh

A. Haque
Department of Soil Science, Patuakhali Science and Technology University, Patuakhali, Bangladesh

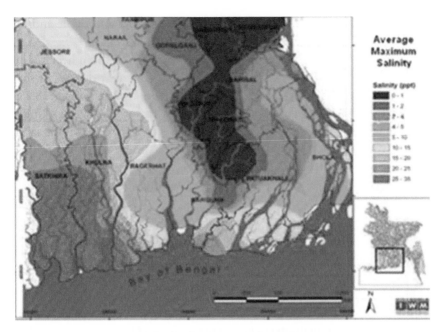

Fig. 18.2 Maximum average river salinity (ppt) in the south-west coastal region (Based on data from the Bangladesh Water Development Board)

percent. With poor dry season surface water resources, groundwater is the primary source of irrigation water in many areas. However, shallow aquifers in most of the south-west coastal region are affected by different levels of salinity due to seawater intrusion and interaction with saline surface water (Fig. 18.3). This has resulted in the extraction of groundwater from deeper aquifers using tube wells.

18.2 Factors Influencing Soil Salinity in Coastal Bangladesh

There has been a progressive increase in soil salinity in terms of intensity, and affected area over the last four decades with saline areas increasing from 0.83 million hectares in 1973 to about 1.05 million hectares in 2009 (SRDI 2010). There was an increase in affected area by 0.19 million

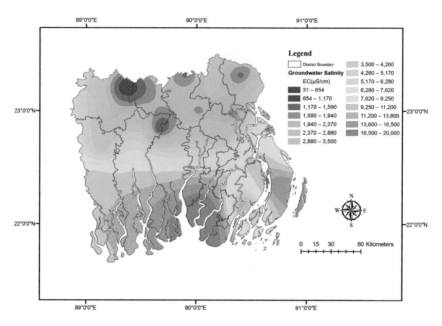

Fig. 18.3 Shallow groundwater salinity in the study area (Based on data from the Bangladesh Water Development Board)

hectares from 1973 to 2000 and a further increase by 0.04 million hectares from 2000 to 2009 (SRDI 2010).

There are numerous interacting drivers that influence soil salinity in Bangladesh (Table 18.1) including irregular rainfall, evaporation, saline river water inundation in both unprotected and protected areas (polders), depth to and salinity of groundwater, irrigation using saline water, storm surge inundation, and brackish shrimp farming (SRDI 2010; Rashid and Islam 2007; Haque et al. 2008; Rasel et al. 2013). As these factors vary spatially in the south-west coastal region, their relative effects on soil salinity are also spatially variable and complex.

Seasonal variation in soil salinity is quite distinct; top soil salinity gradually increases from January, reaches a maximum usually in April or May, and then starts to gradually decrease with the onset of monsoon and reaches the seasonal low usually in September or October when salts are sufficiently flushed out by monsoon rainfall. However, at some places greater accumulation of salts occurs due to a combination of the factors listed in Table 18.1.

Table 18.1 Salinity mechanisms and processes in coastal Bangladesh

Mechanism/ drivers	Processes	Long-term change factors
Climate variability	Irregular rainfall (less rainfall not being able to flush out salts in monsoon; less rainfall forcing more irrigation in dry season)	Increased irrigation
Groundwater salinity	Accumulation of salts from capillary rise of saline groundwater table	Increased sea levels Increased river salinity Increased pumping in upstream freshwater zones
Depth to water table	Capillary rise (water table is < 2 m from surface)	Increased pumping Sea-level rise
Cyclonic storm surges	Overtopping of polders Trapped saline water in low-lying floodplain areas with silted up drainage canals Persistent inundation of tidal plains through embankment breaches	Higher sea levels Increased frequency of surges
River water salinity	Direct tidal inundation in unprotected areas Tidal inundation in polders through embankment breach and/or due to faulty water control structures and poor management Lateral seepage of saline river water through soil/embankment	Reduced river flows due to climate variability Reduced river flows due to upstream diversion Reduced river flows due to upstream dams Sea-level rise
Salinity of irrigation water	Irrigation with saline river or groundwater	Increased irrigation for leaching Saline intrusion into groundwater aquifers
Brackish shrimp cultivation	Deliberate introduction of saline water inside polders Lateral seepage from shrimp ghers to adjacent land Contamination of shallow groundwater	National policies Market forces/prices Power relations

Modified from Salehin et al. (2014) and Clarke et al. (2015)

18.2.1 Groundwater Salinity and Depth to Groundwater Table

Capillary rise causes salt accumulation from shallow groundwater which commonly occurs in irrigated areas in the dry *Rabi* season (Brammer 2014). Once the water table reaches a critical depth below the ground surface, evaporation of this water can occur via capillary rise, transporting soluble salts with it to the active root zone and top soil surface (Beltrán 1999; Ayers and Westcot 1994; Gupta and Khosla 1996). The critical depth may vary from 1 m in coarse textured soil to a few meters in fine textured soils (Gupta and Khosla 1996).

Two important considerations thus are the following: at what depth is the groundwater and how saline is the groundwater? Depth of groundwater can be influenced by increased sea level in the future thus contributing to increased salt accumulation in the soil. Groundwater salinity is anticipated to be negatively impacted by lateral seawater intrusion due to accelerated sea-level rise, increased river salinity resulting from reduction of river flows and sea-level rise, and prolonged inundation induced by cyclonic storm surges. Figure 18.4 shows that in dry season (January–May), the depth to groundwater is within 0–1.5 m in many places in the eastern part of the region. In the monsoon season (July–November), the water table becomes shallow in most regions, but the key difference is that the shallow groundwater is primarily fresh rainwater.

Murshid (2012) indicated that groundwater pumping from shallow wells less than 200 m from a polder boundary can induce an inflow of brackish water under the polder and into the irrigation well. Additionally, groundwater salinity is anticipated to be negatively impacted by rising sea level and subsidence which would push the seawater interface further inland. This would be exacerbated by an increase in river salinity due to reduction in river flows.

18.2.2 River Water Salinity

There are major river channels along with numerous small rivers and estuarine/tidal creeks that carry saline water from the sea to interior lands in the dry season due to tidal exchange (see Chap. 17). The coastal land

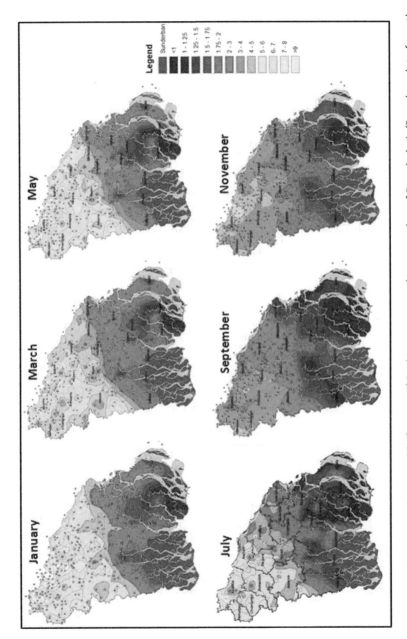

Fig. 18.4 Depth to groundwater table (in meters) in the greater south-west region of Bangladesh (Based on data from the Bangladesh Water Development Board)

elevation is typically between 0.9 and 2.1 m above mean sea level (Iftekhar and Islam 2004; Haque 2006). Tide or surges can rise up to 1.3 m above the general ground level in the dry season and can inundate wide areas (Rasel et al. 2013). Tides and surges of brackish sea water (19–28 parts per thousand (ppt)) can propagate up to 200 km inland in times of reduced river discharge (Mondal et al. 2006; Dasgupta et al. 2014). Inundation can also take place inside the polders through embankment breaches or due to poor management of the outlet gates and via seepage through soils or embankments.

This mechanism is exacerbated by reduced dry reason river flows. Fresh water diversions from the river Ganges has meant river flows into the western part of the south-west coastal zone is substantially reduced in the dry season. This results in high river water salinity in and around Satkhira, Khulna, and Bagerhat districts. In contrast, in the eastern area, fresh water delivery by the Padma and Lower Meghna rivers means that salinity is concentrated nearer the coast (Fig. 18.2).

There is a strong association between river water salinity and soil salinity in the eastern part (Satkhira, Khulna, and Bagerhat districts) and southernmost regions of Barguna, Patuakhali, and Bhola districts (Fig. 18.5). However, a more detailed analysis shows that the association is weaker further inland from the sea where river salinity is lower, yet soil salinity remains high. This implies that factors other than river salinity also play important roles in determining soil salinity. A multiple regression analysis shows that soil salinity can be strongly related to depth to the groundwater table and rainfall in the dry season as well as river water salinity.

The fact that river water salinity plays a dominant role in controlling soil salinity has significant implications. A probable future scenario for this region comprises reduced dry season river flows (caused by climatic change and increased damming and diversion upstream) and accelerated relative sea-level rise (caused by global sea-level rise, subsidence and reduced sediment delivery due to river damming), and the soil salinity problem is expected to further intensify and extend over wider areas (see Chap. 17).

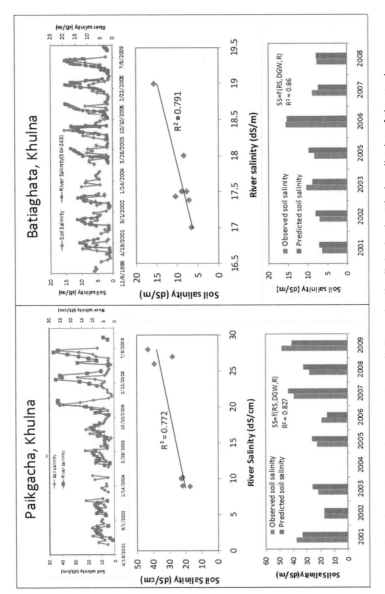

Fig. 18.5 Statistical associations between river salinity and soil salinity in the Khulna district of the study area

18.2.3 Cyclonic Storm Surges

Cyclonic storm surges have been another cause of increase in soil salinity in the surge affected areas in Bangladesh (see Chap. 8). The impacts are frequently prolonged for several years. For example, Cyclone Aila (2007) inundated large areas in Dacope *Upazila* of Khulna district via overtopping of polders and breaches in the embankments. As rehabilitation of the polder dykes was slow, the tidal floodplain areas were repeatedly inundated by saline tidal water for one to two years, with saline water trapped in low-lying areas with silted up drainage canals. It took two to three monsoons to flush out the accumulated salts before agriculture could be partially restored, which caused prolonged suffering to the local people whose livelihoods are intertwined with agriculture (Kabir et al. 2015).

18.2.4 Irrigation Water Quality and Monsoon Rainfall

Clarke et al. (2015) showed that accumulation of salts on the agricultural land in the dry season is controlled by the amount and quality of irrigation water applied. An increase in the salinity of sources of dry season irrigation water will lead to increased salt accumulation in soils. Other key factors were the effectiveness of the monsoon rainfall in removing water by leaching/disposal through effective and well-maintained drainage systems. Their analysis showed that irrigating with water at up to four parts per thousand (ppt) can be sustainable, but if the dry season irrigation water quality goes above 5 ppt, the monsoon rainfall is unable to flush out the salt deposits. It was found that agricultural productivity in the Barisal, Patuakhali, and Bhola districts is likely to fall by 25 percent by 2050, with some regions expected to experience dry season crop yield reductions of 50 percent. Regions which are already experiencing severe salinity in Barisal and Khulna divisions are expected to see salinities of greater than 20 ppt by the end of the century, effectively curtailing dry season agriculture (Fig. 18.6).

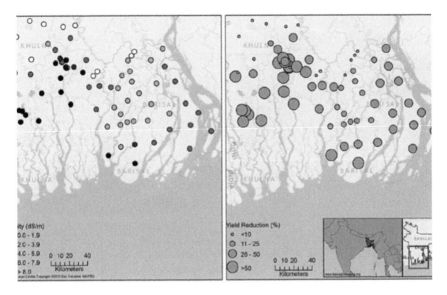

Fig. 18.6 Projections of future 2050 river water salinities (left) and simulated crop yield reduction due to increased irrigation water salinity (right) (Clarke et al. 2015—Published by The Royal Society of Chemistry)

18.2.5   Brackish Shrimp Cultivation

Shrimp farming is the third largest earner of foreign currency, with production primarily based in southern regions around Satkhira, Khulna, and Bagerhat (Begum and Alam 2002). Shrimp culture is one of the most attractive land use practices as the profits generated exceed those of traditional agriculture (Hossain et al. 2004). However, shrimp culture has not been free from its negative effects, especially with the brackish shrimp (locally known as *Bagda*) culture. There has been a rapid growth of *Bagda* over more than two decades in the southern part of the coastal region by taking advantage of high river water salinity (Fig. 18.7) (see Chap. 20). Extensive shrimp ponds (*ghers*) were constructed with saline water brought from surrounding rivers via canals. Unfortunately, this kind of shrimp culture has led to the deterioration of coastal soils and ground/surface water sources through increased salinization (Rahman et al. 2013; Kabir and Eva 2014; Karim et al. 1990). Brackish shrimp culture concentrates in the saline areas where soil salinity is principally impacted by river water salinity in the dry season.

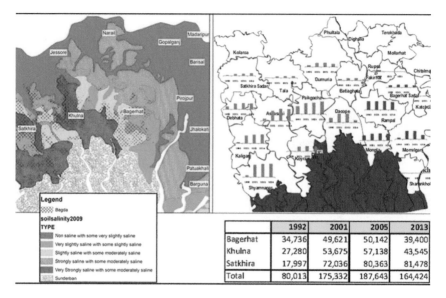

	1992	2001	2005	2013
Bagerhat	34,736	49,621	50,142	39,400
Khulna	27,280	53,675	57,138	43,545
Satkhira	17,997	72,036	80,363	81,478
Total	80,013	175,332	187,643	164,424

Fig. 18.7 Growth of brackish water shrimp culture in Satkhira, Khulna, and Bagerhat districts in Khulna Division. The left figure shows brackish shrimp culture concentrated in the saline areas where soil salinity is principally impacted by river water salinity. The right figure shows a continual increase in brackish shrimp culture in Satkhira district, while Khulna and Bagerhat districts show a recent decline

18.3   Conclusions

Soil salinity is a major constraint of agricultural production and human livelihoods in the coastal region of south-west Bangladesh, but there are multiple relationships between groundwater salinity, surface water salinity, human management (or mismanagement) of water resources, and the resulting accumulation of salts in the soils of the farmlands. When considering future trajectories of environmental change, it will be vital to assess the interactions between the drivers of soil salinity and to be aware of the interactions between surface water and groundwater. Increases in river and groundwater salinity will have significant future implications given their important roles in affecting salt accumulation in soil. River salinity may increase due to accelerated sea-level rise and reduction of dry season river flows. Groundwater is closely linked to the surface hydrology, and it cannot be assumed to be independent of external factors and environmental change. Aquifer water quality can be affected by lateral seawater intrusion due to accelerated sea-level rise, increased river salinity being exchanged with shallow groundwater, cyclonic storm surges, human activities such as groundwater pumping, and deliberate inundation of farmland for brackish aquaculture. Strategic water plans must take note that poor management of surface water and groundwater resource use may result in a long-term degradation of soil quality which will be difficult to reverse.

References

Ayers, R.S., and D.W. Westcot. 1994. *Water quality for agriculture*. FAO irrigation and drainage paper 29. Rome: Food and Agriculture Organisation (FAO). http://www.fao.org/docrep/003/T0234E/T0234E00.htm. Accessed 22 May 2017.

Begum, A., and S.M.N. Alam. 2002. *Social and economic impacts of shrimp disease among small-scale, coastal farmers and communities in Bangladesh*. Primary aquatic animal health care in rural, small-scale, aquaculture development. Rome: Food and Agriculture Organisation (FAO). http://www.fao.org/docrep/005/Y3610E/y3610E18.htm. Accessed 27 Apr 2017.

Beltrán, J.M. 1999. Irrigation with saline water: Benefits and environmental impact. *Agricultural Water Management* 40 (2-3): 183–194. https://doi.org/10.1016/s0378-3774(98)00120-6.

Brammer, H. 2014. Bangladesh's dynamic coastal regions and sea-level rise. *Climate Risk Management* 1: 51–62. https://doi.org/10.1016/j.crm.2013.10.001.

Clarke, D., S. Williams, M. Jahiruddin, K. Parks, and M. Salehin. 2015. Projections of on-farm salinity in coastal Bangladesh. *Environmental Science-Processes and Impacts* 17 (6): 1127–1136. https://doi.org/10.1039/c4em00682h.

Dasgupta, S., F.A. Kamal, Z.H. Khan, S. Choudhury, and A. Nishat. 2014. *River salinity and climate change: Evidence from coastal Bangladesh*. Policy working paper series 6817. Washington, DC: The World Bank. http://documents.worldbank.org/curated/en/522091468209055387/River-salinity-and-climate-change-evidence-from-coastal-Bangladesh. Accessed 11 Apr 2014.

Gupta, S.K., and B.K. Khosla. 1996. *Salinity control in the root zone of irrigated agriculture*. Proceedings of the workshop on water logging and soil salinity in irrigated agriculture, March 12–15, New Delhi.

Haque, S.A. 2006. Salinity problems and crop production in coastal regions of Bangladesh. *Pakistan Journal of Botany* 38 (5): 1359–1365.

Haque, M.A., D.E. Jharna, M.N. Uddin, and M.A. Saleque. 2008. Soil solution electrical conductivity and basic cations composition in the rhizosphere of lowland rice in coastal soils. *Bangladesh Journal of Agricultural Research* 33 (2): 243–250.

Hossain, S., S.M.N. Alam, C.K. Lin, H. Demaine, Y. Sharif, A. Khan, N.G. Das, and M.A. Rouf. 2004. Integrated management approach for shrimp culture development in the coastal environment of Bangladesh. *World Aquaculture*, World Aquaculture Society, Louisiana State University, Baton Rouge. 35–44. https://www.was.org/Magazine/Contents.aspx?Id=8. Accessed 22 May 2017.

Iftekhar, M.S., and M.R. Islam. 2004. Managing mangroves in Bangladesh: A strategy analysis. *Journal of Coastal Conservation* 10: 139. https://doi.org/10.1652/1400-0350(2004)010[0139:MMIBAS]2.0.CO;2.

Kabir, H., and I.J. Eva. 2014. Environmental impacts on shrimp aquaculture: The case of Chandipur village at Debhata upazilla of Satkhira district, Bangladesh. *Journal of the Asiatic Society of Bangladesh, Science* 40 (1): 107–119.

Kabir, T., M. Salehin, and G. Kibria. 2015. *Delineation of physical factors of cyclone aila and their implications for different vulnerable groups.* Proceedings of the 5th International Conference on Water & Flood Management (ICWFM-2015), organized by IWFM, BUET, Dhaka.

Karim, Z., S.G. Hussain, and M. Ahmed. 1990. *Salinity problems and crop intensification in the coastal regions of Bangladesh.* Soils publication no. 33. Dhaka: Bangladesh Agricultural Research Council (BARC).

Lázár, A.N., D. Clarke, H. Adams, A.R. Akanda, S. Szabo, R.J. Nicholls, Z. Matthews, D. Begum, A.F.M. Saleh, M.A. Abedin, A. Payo, P.K. Streatfield, C. Hutton, M.S. Mondal, and A.Z.M. Moslehuddin. 2015. Agricultural livelihoods in coastal Bangladesh under climate and environmental change – A model framework. *Environmental Science-Processes and Impacts* 17 (6): 1018–1031. https://doi.org/10.1039/c4em00600c.

Mondal, M.K., T.P. Tuong, S.P. Ritu, M.H.K. Choudhury, A.M. Chasi, P.K. Majumder, M.M. Islam, and S.K. Adhikary. 2006. Coastal water resource use for higher productivity: Participatory research for increasing cropping intensity in Bangladesh. In *Environment and livelihoods in tropical coastal zones: Managing agriculture-fishery-aquaculture conflicts*, ed. C.T. Hoanh, T.P. Tuong, J.W. Gowing, and B. Hardy. Wallingford: CAB International.

Murshid, S.M. 2012. *Impact of sea level rise on agriculture using groundwater in Bangladesh.* MSc thesis CoMEM programme, University of Southampton. https://repository.tudelft.nl/islandora/object/uuid:e484b9b8-e1d1-40b1-99ee-5c7a274ef500/?collection=research. Accessed 22 May 2017.

Rahman, M.M., V.R. Giedraitis, L.S. Lieberman, T. Akhtar, and V. Taminskiene. 2013. Shrimp cultivation with water salinity in Bangladesh: The implications of an ecological model. *Universal Journal of Public Health* 1 (3): 131–142. http://dx.doi.org/10.13189/ujph.2013.010313.

Rasel, H.M., M.R. Hasan, B. Ahmed, and M.S.U. Miah. 2013. Investigation of soil and water salinity, its effect on crop production and adaptation strategy. *International Journal of Water Resources and Environmental Engineering* 5 (8): 475–481. https://doi.org/10.5897/IJWREE2013.0400.

Rashid, M., and M.S. Islam. 2007. *Adaptation to climate change for sustainable development of Bangladesh agriculture.* Bangladesh country paper for the 3rd session of the Technical Committee of Asian and Pacific Center for Agricultural Engineering and Machinery (APCAEM), November 20–21, Beijing.

Salehin, M., M.S. Mondal, D. Clarke, A. Lazar, M. Chowdhury, and S. Nowreen. 2014. *Spatial variation in soil salinity in relation to hydro-climatic factors in southwest coastal Bangladesh.* Deltas in Times of Climate Change II, September 24–26, Rotterdam.

SRDI. 2010. *Saline soils of Bangladesh.* Dhaka: Soil Resources Development Institute (SRDI), Ministry of Agriculture, Government of the People's Republic of Bangladesh.

19

Population Dynamics in the South-West of Bangladesh

Sylvia Szabo, Sate Ahmad, and W. Neil Adger

19.1   Introduction

During the last decades, Bangladesh has undergone a demographic transition, with declining fertility accompanied by lower mortality and increased life expectancy. During this time, total fertility rate (TFR) dropped from 6.36 in 1950–1955 to 2.23 in 2010–2015 and is projected to reach 1.67 in 2050–2055 (UN 2013). While life expectancy has shown important improvements, maternal and child mortality remain high by international

---

S. Szabo (✉)
Department of Development and Sustainability, Asian Institute of Technology, Bangkok, Thailand

S. Ahmad
Faculty of Agricultural and Environmental Sciences, University of Rostock, Rostock, Germany

W. Neil Adger
Geography, College of Life and Environmental Sciences, University of Exeter, Exeter, UK

© The Author(s) 2018   **349**
R. J. Nicholls et al. (eds.), *Ecosystem Services for Well-Being in Deltas*,
https://doi.org/10.1007/978-3-319-71093-8_19

standards. Malnutrition, which is a key cause of child mortality and poor health, continues to constitute a key public health challenge. Population trends have been uneven across different geographic areas, and this uneven progress towards demographic transitions is likely to be further challenged by the consequences of environmental and climate change (Szabo et al. 2015a). This chapter examines key population dynamics in the south-west coast of Bangladesh with a specific focus on the components of population change, that is, fertility, mortality and migration, and discusses future population prospects and resulting policy implications.

Existing literature highlights the complex interlinkages between population dynamics, human well-being and sustainability (de Sherbinin et al. 2007; Hummel et al. 2013; Szabo et al. 2016b). For example, population density has been found to be both positively and negatively associated with economic growth and individual well-being (Tiffen 1995; Ahlburg et al. 2013). On one hand, population growth and population density may have a positive effect on people's quality of life because of greater provision of and accessibility to services, such as food, water and health-care. On the other hand, however, rapid population growth, including growth of populations in urban areas, caused by natural increase and rural to urban migration, is likely to pose important challenges to the sustainability of urban settlements (Szabo et al. 2016b). Urbanisation is also linked to high mobility and migration rates, which are likely to contribute to increased household income, and thus poverty reduction through provision of remittances (Ratha 2013).

This chapter focuses specifically on the population dynamics and prospects in the study area (see Chap. 4, Fig. 4.2). In this region, the interlinkages between population dynamics and sustainability have been and are likely to continue to be of critical importance. The impacts of environmental and climate change, combined with a relatively rapid pace of urbanisation, and falling fertility and mortality rates will mean that economic growth of the region and Bangladesh, as a whole, will be affected (Szabo et al. 2015a). Past evidence suggests that population growth in the area has been accompanied by improved human well-being (Hossain et al. 2016). This has been possible, at least partially, because of the contribution of local ecosystem services. For example, Hossain and Szabo (2017) estimated that in the Narail district north of the study area, the total yearly market output of provisioning services was approximately USD 173 million. Examining population dynamics and prospects in the region will allow better local and national planning, thus contributing to further development of the country.

19.2   Population Growth and Structure

The total population of the south-west coast of Bangladesh is around 14 million which accounts for ten per cent of the national population. The most populous districts are Khulna and Barisal which includes the two cities of the same names, while the least populous districts are Jhalokati and Barguna. However, Jhalokati has the highest population density, approximately 966 people per square kilometre, while Bagerhat district is the least dense at around 373 people per square kilometre (Szabo et al. 2015a). Both Barisal and Khulna Divisions have experienced substantial population decline, while Dhaka and Chittagong (outside of the study area) had a net increase in terms of the share of total national population (Szabo et al. 2015a) (also see Fig. 19.1). The coastal districts

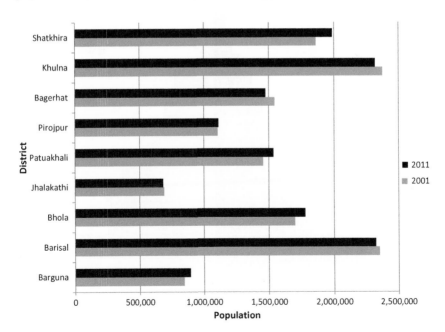

Fig. 19.1   Population change in the study area between 2001 and 2011

appear to have relatively low rates of population growth; within all the districts the annual average growth rate has decreased between 2001 and 2011 compared to the intercensal period 1991–2001 (BBS 2012). For the entire south-west, the average annual population growth rate decreased from 1.03 to 0.15 per cent (Szabo et al. 2015a). Moreover, four districts—Barisal, Jhalokati, Bagerhat and Khulna—have seen a negative average annual growth rate between 2001 and 2011. Thus, in order to understand the demographics of this area, it is important to examine the interplay of the different components of population change, namely, fertility, mortality and migration.

In terms of population structure (Fig. 19.2), while the current population of the region is relatively young, the declining fertility rates have already had an effect on the population pyramid. More specifically, it can be seen that the bottom of the pyramid (ages 0–4) is significantly smaller compared to the two adjacent age groups. The female and male populations show approximately the same distribution, suggesting no

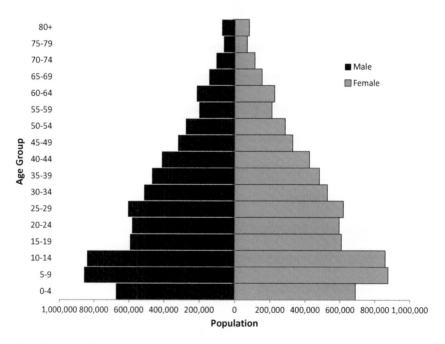

Fig. 19.2 Population structure in the study area in 2011

disproportionate out-migration from one gender group. It is however important to highlight that this population structure is expected to change as Bangladesh, and the study area, continue undergoing the demographic transition. In particular, given both low fertility rates and rising environmental vulnerability, it may be expected that the population pyramid will dramatically change.

19.3   Components of Population Change

19.3.1   Fertility

Recent changes in fertility patterns have arguably been the most significant factor affecting population change in Bangladesh, including the south-west coast of the country. The coastal delta region is currently experiencing rapid demographic transition, and because of the continuous decline in TFR and consequently natural growth rate, the region is likely to see a less rapid population increase or even, in the long term, a population decline. As can be seen in Fig. 19.3, fertility rates have been decreasing since the early 1990s in both Khulna and Barisal Divisions. In Khulna, TFR dropped from 3.1 in 1993/1994 to 1.9 in 2014, while in Barisal it declined from 3.5 in 1993/1994 to 2.2 in 2014

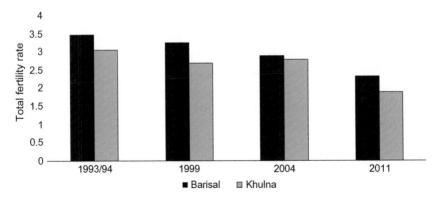

Fig. 19.3 Recent trends in total fertility rate in Barisal and Khulna Divisions (Reproduced with permission from Szabo et al. 2015a)

(Mitra et al. 1994; NIPORT et al. 2016). There exist interesting inter-district differentials in age-specific fertility with Satkhira showing the highest level of adolescent fertility and Pirojpur the highest level of fertility in older ages (40–49).

Factors affecting fertility levels in Bangladesh as a whole have been examined extensively given the relatively rapid pace of demographic transition since the 1970s. Less evidence exists regarding the specific study area; however, it is sensible to assume that similar factors apply at least at a strategic level. In his comparison of socio-economic development across South Asia, Asadullah et al. (2014) highlighted that the decline in fertility was mainly possible because of the increasingly easy access to contraceptives and reproductive health social awareness campaigns. Similarly, Chowdhury et al. (2013) point out that contraceptive use amongst couples in Bangladesh increased from less than 10 per cent in 1970 to 61 per cent in 2010, which led not only to a lower TFR but also contributed to reductions in mortality rates and gains in the overall socio-economic development. The recent trends in TFR indicate that the delta population is likely to experience smaller households and changing household-level dependency ratios. A recent paper by Szabo et al. (2016b) examining demographic patterns in a larger delta region covering almost five divisions highlighted a declining trend in average household size, from 5.1 in 2000 to 4.8 in 2010.

19.3.2   Mortality

Infant mortality and under-five mortality rates are overall lower when compared to the national average, with districts such as Barguna and Satkhira having the lowest infant mortality (Table 19.1). This might be explained by the fact that in some very important socio-demographic indicators, Barisal and Khulna Divisions show better scores than other divisions. For example, Barisal has the lowest percentage of ever-married women aged 15–49 with no education (15.1 per cent), followed by Khulna Division (21.6 per cent) (NIPORT et al. 2016). Moreover, if Bangladesh is divided into west and east, then the west (including the study area) has already achieved replacement fertility, while the divisions in the east such as Sylhet and Chittagong are lagging behind (Rabbi and Kabir 2015).

Table 19.1 Selected maternal and child mortality indicators (based on NIPORT et al. 2012)

Division	District	Indicator			Married female adolescents aged 15–19	Current use of contraception (any method)
		Infant mortality	Neonatal mortality	Under-five mortality		
		Deaths per 1,000 live births			Percentage	
Barisal	Barguna	30	20	41	52	68
	Barisal	45	29	58	35	62
	Bhola	35	24	54	43	66
	Jhalokati	39	32	56	36.4	60
	Patuakhali	41	33	64	46	65
	Pirojpur	46	37	55	38	63
Khulna	Bagerhat	49	37	60	41	67
	Khulna	46	31	53	43	68
	Satkhira	31	21	32	50	70
Study area (average)		40	29	53	43	66
National		45	32	56	41	63

The two districts of Barguna and Satkhira have done relatively well in some of the development indicators as well, which may at least in part, help to explain the lower infant mortality within the study area (Table 19.1). For example, within the Barisal Division, Barguna has the second highest percentage of girls (79.4 per cent) aged 12–15 who completed primary education, the highest percentage of mothers receiving check-ups within two months of delivery (40.6 per cent), as well as within the first two days by any healthcare provider (NIPORT et al. 2012). In addition, the current use of contraception (any method) is relatively higher in the study area (approximately 66 per cent compared to the national average of 63 per cent) (NIPORT et al. 2011; also see Table 19.1). It is reasonable to expect such factors to play an important role in decreasing infant and under-five mortality.

Life expectancy figures for the study are scarce. According to World Health Organization (WHO) data, in Bangladesh the current life expectancy for females is 73.1 and 70.6 for males (compared to 57.7 for females and 56.7 for males in 1985–1990) (WHO 2014; UN 2013). Based on mortality data from the most recent population census, current life expectancy in the study area was estimated. According to these estimates, life expectancy for males varies from 70.9 for Khulna Division to 68.2 in Barisal Division; for females, life expectancy at birth was 73.1 in Khulna and 69.8 in Barisal. These considerable improvements in life expectancy can be explained by a number of factors, as highlighted by Chowdhury et al. (2013).

Environmental factors can affect mortality and population health in different ways. In terms of negative effects, direct factors such as cyclones and flooding can cause death and disease. There were between 225,000 and 500,000 deaths as a result of the 1970 Bhola cyclone; 3,300 and 190 deaths are associated with Cyclones Sidr (2007) and Aila (2009), respectively. The largest flood in Bangladesh in 1998 led to 138,000 deaths and affected 52 per cent of the entire population of Bangladesh at that time (Cash et al. 2013). Other than such rapid onset hazards, slow onset hazards also have the potential to affect population health indirectly. For example, salinisation of water and soil affects water quality for both agriculture (ultimately leading to food insecurity) and domestic consumption. Salinity in drinking water has been found to be associated with increased risk of pre-eclampsia and gestational hypertension in the coastal population of Bangladesh (Khan et al. 2014), while Dasgupta et al. (2016) suggest that salinity in drinking water is a significant determinant of infant mortality in coastal Bangladesh.

However, it should be acknowledged that through socio-economic development and adaptation, coastal Bangladesh is better prepared to cope with the aftermath of natural hazards and environmental disasters (Cash et al. 2013; Haque et al. 2012). In particular, over the last 50 years Bangladesh has learned how to adapt to cyclones, contributing to significant declines in cyclone-related deaths; for example, setting up cyclone shelters and low tech but reliable early warning systems (Haque et al. 2012). These advances in tackling mortality rates can be attributed to enhanced disaster management, greater focus on poverty reduction, a key vulnerability factor, interventions which draw on social capital of the coastal communities and identifying and targeting main drivers of mortality (Cash et al. 2013).

19.3.3 Migration

Migration is arguably the most complex and challenging component of population change as it is influenced by a diverse set of drivers. Black et al. (2011) identify five families of drivers of migration in the form of economic, political, social, demographic and environmental. Such "micro-scale" and "meso-scale" factors result in macro-scale effects in terms of population growth and distribution across a country. However, according to the classic economic model, migration decision-making process depends on individual choice (Harris and Todaro 1970). Nevertheless, research in the last four decades in this area has found that the process of decision making often depends on the family or household situations well as on the presence of social networks (Jampaklay et al. 2007; Winkels 2008). There is evidence that the migration decision-making process is not undertaken by individuals alone, but it occurs at a group level. This suggests that the classic push and pull factors may operate within a larger social context (Seto 2011).

In Bangladesh migration is a historical phenomenon. Migration patterns are not homogenous within the coastal districts as local push factors may influence people's decision to migrate. The census of 2011(BBS 2012) reports that 9.7 per cent of the total population of Bangladesh are lifetime internal migrants. Using a residual method, a crude net migration rate of around minus 4.5 persons (per 1,000 population) for Barisal Division and around plus 1 person (per 1,000 population) for Khulna Division for the year 2001 was estimated (Szabo et al. 2015a). However, calculations for 2010/2011 revealed an increase in out-migration rates, with a crude net migration of minus 12.5 persons (per 1,000 population) for the Barisal Division and minus 5.5 persons (per 1,000 population) for the Khulna Division, meaning that more people are leaving these divisions than are coming in as migrants.

Environmental stresses, economic vulnerability and prospects of remittance income are thought to result in high out-migration rates in coastal areas (Szabo et al. 2015b; Szabo et al. 2016a). As the Ganges-Brahmaputra-Meghna (GBM) delta constitutes a vulnerable social-ecological system (Sebesvari et al. 2016), there is a relatively high potential for migration. A large portion of internal migration in Bangladesh is not permanent, being either seasonal or temporary. A study in Bangladesh has shown that migrant characteristics, environmental change-related factors, conflict and adaptation strategies, as well as social networks —all have significant impacts on temporary migration (Joarder and Miller 2013). Migrants who had been previously working in agriculture or fishing are more likely to migrate permanently. Households who had reported asset loss due to environmental hazards have a higher probability of becoming permanent migrants, while events such as loss of livestock and crop failure were associated with higher likelihood of becoming temporary migrants.

Another study found that permanent migration from hazard-prone areas are quite low despite the major threats with the exception of erosion of land on which people live or where salinity intrusion makes agriculture impossible (Penning-Rowsell et al. 2013). Certain pull factors encouraging families and people to stay behind are quite strong, and migration as a response to natural hazards may be viewed as a last resort (Penning-Rowsell et al. 2013). Nevertheless, a modelling study by Hassani-Mahmooei and Parris (2012) suggests likely changes in population densities across

Bangladesh due to migration from drought-prone western regions and cyclone- and flood-prone areas in the south, towards northern and eastern districts. The model predicted between three and ten million internal migrants in Bangladesh by 2051, by considering drivers such as incidences of extreme poverty, socio-economic vulnerability, demography, historical drought, cyclone and flooding patterns.

19.3.4 Urbanisation

Evidence suggests that the coastal delta region is not only particularly vulnerable in terms of environmental risks but also experiences unique demographic and urbanisation dynamics (Szabo et al. 2015a). While Bangladesh as a country has undergone a rapid process of urbanisation during the last decade (2011–2011), the coastal delta region saw a slight decline both in terms of urban population size and the proportion of urban population. In 2001, the study area had approximately 2.8 million urban dwellers which accounted for around 20 per cent of the total population in coastal delta (Table 19.2). By 2011, the population declined to 2.5 million, almost 19 per cent of the overall population in the study

Table 19.2 Proportion of urban population across nine districts in the study area (Compiled using data from BBS 2012)

Division	District	Year			
		2001		2011	
		Total urban population	Per cent urban (2001)	Total urban population	Per cent urban (2011)
Barisal	Barguna	87,582	10.32	103,094	11.55
	Barisal	394,567	16.75	519,016	22.33
	Bhola	234,302	13.76	243,317	13.69
	Jhalokati	104,070	14.99	112,003	16.41
	Patuakhali	175,284	12.00	201,882	13.14
	Pirojpur	166,970	15.03	182,631	16.41
Khulna	Bagerhat	206,554	13.33	195,331	13.23
	Khulna	1,284,208	53.98	777,588	33.54
	Satkhira	171,614	9.20	197,616	9.95
Total		2,825,151	20.23	2,616,006	18.55

area. This decline was caused primarily by population dynamics in the Khulna district which experienced a large decline in its urban population, almost 40 per cent during the intercensal period (2001–2011). While without a detailed analysis it is difficult to determine why such a steep decline occurred, it is sensible to assume that both low fertility rate and high level of out-migration contributed to this pattern.

The recent district level census report (BBS 2012) confirms the findings presented in Table 19.2. While in 2001 the level of urbanisation in Khulna district was over 54 per cent, it was only 34 per cent in 2011. Comparatively, in the remaining districts of the Khulna Division (Bagerhat and Satkhira) the proportion of urban population remained at similar levels. In Barisal Division, the Barisal district saw the most rapid pace of urbanisation, with over 22 per cent of its population residing in urban areas in 2011, compared to nearly 17 per cent in 2001. These patterns of urbanisation show that overall the study area is highly diverse in terms of its internal dynamics. Further systematic qualitative or survey research is needed to elaborate the factors affecting urbanisation processes in the coastal delta.

19.4   Future Scenarios and Policy Recommendations

Population projections for the south-west coast of Bangladesh are uncertain: ongoing processes will continue towards urbanisation and demographic transitions. Szabo et al. (2015a) developed population projections based on three sets of assumptions about the excepted direction of change for different demographic components. These scenarios were aligned with the overall sustainability scenarios and named Business As Usual (BAU), Less Sustainable (LS) and More Sustainable (MS). The results of two out of three projection scenarios show that the population of the study area is likely to experience a significant decline before 2050 (Szabo et al. 2015a). Only in the MS scenario (one which presupposes lower rates of out-migration), population was expected to see a slight increase (Fig. 19.4).

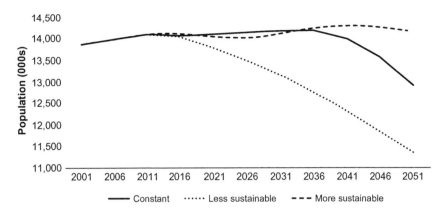

Fig. 19.4 Scenarios of population change in the GBM delta (Reproduced with permission from Szabo et al. 2015a)

It was also found that in the next decades the population structure in the coastal GBM delta will undergo considerable change regardless of the projection scenario. These future demographic trends can be explained by declining fertility, greater life expectancy and high rates of out-migration. The two key aspects of this shift will entail an ageing population and a declining proportion of younger people. More specifically, the proportion of 65 and over is likely to increase from 5.7 per cent in 2011 to 15 or 16 per cent in 2051 depending on the projection scenario. At the same time, the proportion of children and youth (ages 0–14) would decline from 34 per cent to 15 per cent under LS scenario and 16 per cent under the BAU and MS scenarios.

19.5   Summary

The demographic trends and projections presented in this chapter have important implications for the wider context of ecosystem services and sustainability in the delta region. First, they suggested the need for tailored district level policy interventions. While the demographic outcomes diverge, the policy priorities are common across the delta region: there is a need to enhance public services and public health, protect vulnerable populations and generate routes out of poverty building on sustainable resource use and mobility. Second, given the distinct demographic patterns of the study area, and the likelihood that migration will continue at some level regardless of rural interventions, the policy planning process should prioritise designing specific measures linked to population distribution at the regional and country level. The recent out-migration from the coastal GBM delta area has an important impact on other urban areas, such as Khulna and Dhaka, including the clustering of new populations in slums and informal settlements. Finally, given the interconnectedness of environmental and social factors and the impact of climate change on human well-being, it is critical to develop integrated strategies to reduce the vulnerability of disadvantaged groups. These groups may include migrants in destination areas (mostly urban centres) and those facing the challenges of ecosystem service decline and environmental degradation in rural areas.

References

Ahlburg, D.A., A.C. Kelley, and K.O. Mason. 2013. *The impact of population growth on well-being in developing countries*. Berlin: Springer Science & Business Media.

Asadullah, M.N., A. Savoia, and W. Mahmud. 2014. Paths to development: Is there a Bangladesh surprise? *World Development* 62: 138–154. https://doi.org/10.1016/j.worlddev.2014.05.013.

BBS. 2012. *Bangladesh population and housing census 2011 – Socio-economic and demographic report*. Report national series, Volume 4. Dhaka, Bangladesh: Bangladesh Bureau of Statistics (BBS) and Statistics and Informatics Division (SID), Ministry of Planning, Government of the People's Republic of Bangladesh. http://www.bbs.gov.bd/PageSearchContent.aspx?key=census%202011. Accessed 7 July 2016.

Black, R., W.N. Adger, N.W. Arnell, S. Dercon, A. Geddes, and D. Thomas. 2011. The effect of environmental change on human migration. *Global Environmental Change* 21 (Supplement 1): S3–S11. https://doi.org/10.1016/j.gloenvcha.2011.10.001.

Cash, R.A., S.R. Halder, M. Husain, M.S. Islam, F.H. Mallick, M.A. May, M. Rahman, and M.A. Rahman. 2013. Reducing the health effect of natural hazards in Bangladesh. *The Lancet* 382 (9910): 2094–2103. https://doi.org/10.1016/s0140-6736(13)61948-0.

Chowdhury, A.M.R., A. Bhuiya, M.E. Chowdhury, S. Rasheed, Z. Hussain, and L.C. Chen. 2013. The Bangladesh paradox: Exceptional health achievement despite economic poverty. *The Lancet* 382 (9906): 1734–1745. https://doi.org/10.1016/s0140-6736(13)62148-0.

Dasgupta, S., M. Huq, and D. Wheeler. 2016. Drinking water salinity and infant mortality in Coastal Bangladesh. *Water Economics and Policy* 2 (1). https://doi.org/10.1142/S2382624X1650003X.

de Sherbinin, A., D. Carr, S. Cassels, and L. Jiang. 2007. Population and environment. *Annual Review of Environment and Resources* 32 (1): 345–373. https://doi.org/10.1146/annurev.energy.32.041306.100243.

Haque, U., M. Hashizume, K.N. Kolivras, H.J. Overgaard, B. Das, and T. Yamamoto. 2012. Reduced death rates from cyclones in Bangladesh: What more needs to be done? *Bulletin of the World Health Organization* 90 (2): 150–156. https://doi.org/10.2471/blt.11.088302.

Harris, J.R., and M.P. Todaro. 1970. Migration, unemployment and development: A two-sector analysis. *The American Economic Review* 60 (1): 126–142.

Hassani-Mahmooei, B., and B.W. Parris. 2012. Climate change and internal migration patterns in Bangladesh: An agent-based model. *Environment and Development Economics* 17: 763–780. https://doi.org/10.1017/s1355770x12000290.

Hossain, M.S., and S. Szabo. 2017. Understanding the socio-ecological system of wetlands. In *Wetlands science: Perspective from South Asia*, ed. B.A. Prusty, R. Chandra, and P.A. Azeez. Berlin: Springer Nature.

Hossain, M.S., F. Amoako Johnson, J.A. Dearing, and F. Eigenbrod. 2016. Recent trends of human wellbeing in the Bangladesh delta. *Environmental Development* 17: 21–32. https://doi.org/10.1016/j.envdev.2015.09.008.

Hummel, D., S. Adamo, A. de Sherbinin, L. Murphy, R. Aggarwal, L. Zulu, J.G. Liu, and K. Knight. 2013. Inter- and transdisciplinary approaches to population-environment research for sustainability aims: A review and appraisal. *Population and Environment* 34 (4): 481–509. https://doi.org/10.1007/s11111-012-0176-2.

Jampaklay, A., K. Korinek, and B. Entwisle. 2007. Residential clustering among Nang Rong migrants in urban settings of Thailand. *Asian and Pacific Migration Journal* 16 (4): 485–510.

Joarder, M.A.M., and P.W. Miller. 2013. Factors affecting whether environmental migration is temporary or permanent: Evidence from Bangladesh. *Global Environmental Change* 23 (6): 1511–1524.

Khan, A.E., P.F.D. Scheelbeek, A.B. Shilpi, Q. Chan, S.K. Mojumder, A. Rahman, A. Haines, and P. Vineis. 2014. Salinity in drinking water and the risk of (pre) eclampsia and gestational hypertension in Coastal Bangladesh: A case-control study. *PLoS One* 9 (9). https://doi.org/10.1371/journal.pone.0108715.

Mitra, S.N., M.N. Ali, S. Islam, A.R. Cross, and T. Saha. 1994. *Bangladesh demographic and health survey, 1993–1994*. Calverton, MD, USA: National Institute of Population Research and Training (NIPORT), Mitra and Associates and Macro International Inc. http://dhsprogram.com/pubs/pdf/FR60/FR60.pdf. Accessed 25 Apr 2017.

NIPORT, b. icddr, and MEASURE Evaluation. 2011. *Bangladesh: District level socio-demographic and health care utilization indicators*. Dhaka, Bangladesh and Chapel Hill, NC, USA: National Institute of Population Research and Training (NIPORT), International Centre for Diarrhoeal Disease Research, Bangladesh (icddr,b) and MEASURE Evaluation. http://www.cpc.unc.edu/measure/resources/publications/tr-11-84. Accessed 7 July 2015.

———. 2012. *Bangladesh maternal mortality and health care survey 2010*. Dhaka, Bangladesh and Chapel Hill, NC, USA: National Institute of Population Research and Training (NIPORT), International Centre for Diarrhoeal Disease Research, Bangladesh (icddr,b) and MEASURE Evaluation. http://pdf.usaid.gov/pdf_docs/PA00K3BH.pdf. Accessed 25 Apr 2017.

NIPORT, Mitra and Associates, and ICF International. 2016. *Bangladesh demographic and health survey 2014*. Dhaka, Bangladesh and Rockville, MD, USA: National Institute of population research and training (NIPORT), Mitra and Associates, and ICF International.

Penning-Rowsell, E.C., P. Sultana, and P.M. Thompson. 2013. The 'last resort'? Population movement in response to climate-related hazards in Bangladesh. *Environmental Science & Policy* 27: S44–S59.

Rabbi, A.M.F., and M. Kabir. 2015. Explaining fertility transition of a developing country: An analysis of quantum and tempo effect. *Fertility Research and Practice* 1 (1): 1–6. https://doi.org/10.1186/2054-7099-1-4.

Ratha, D. 2013. *The impact of remittances on economic growth and poverty reduction*. Policy brief. Washington, DC: Migration Policy Institute. http://www.migrationpolicy.org/research/impact-remittances-economic-growth-and-poverty-reduction. Accessed 12 Feb 2017.

Sebesvari, Z., F.G. Renaud, S. Haas, Z. Tessler, M. Hagenlocher, J. Kloos, S. Szabo, A. Tejedor, and C. Kuenzer. 2016. A review of vulnerability indicators for deltaic social-ecological systems. *Sustainability Science* 11 (4): 575–590. https://doi.org/10.1007/s11625-016-0366-4.

Seto, K.C. 2011. Exploring the dynamics of migration to mega-delta cities in Asia and Africa: Contemporary drivers and future scenarios. *Global Environmental Change* 21: S94–S107. https://doi.org/10.1016/j.gloenvcha.2011.08.005.

Szabo, S., D. Begum, S. Ahmad, Z. Matthews, and P.K. Streatfield. 2015a. Scenarios of population change in the coastal Ganges Brahmaputra Delta (2011–2051). *Asia Pacific Population Journal* 30 (2): 51–72.

Szabo, S., S. Hossain, Z. Matthews, A. Lazar, and S. Ahmad. 2015b. Soil salinity, household wealth and food insecurity in agriculture-dominated delta. *Sustainability Science*: 1–11. https://doi.org/10.1007/s11625-015-0337-1.

Szabo, S., E. Brondizio, F.G. Renaud, S. Hetrick, R.J. Nicholls, Z. Matthews, Z. Tessler, A. Tejedor, Z. Sebesvari, E. Foufoula-Georgiou, S. da Costa, and J.A. Dearing. 2016a. Population dynamics, delta vulnerability and environmental change: Comparison of the Mekong, Ganges–Brahmaputra and Amazon delta regions. *Sustainability Science* 11 (4): 539–554. https://doi.org/10.1007/s11625-016-0372-6.

Szabo, S., R. Hajra, A. Baschieri, and Z. Matthews. 2016b. Inequalities in human well-being in the urban Ganges Brahmaputra Meghna Delta. *Sustainability* 8 (7): 608.

Tiffen, M. 1995. Population density, economic growth and societies in transition – Boserup reconsidered in a Kenyan case study. *Development and Change* 26 (1): 31–65. https://doi.org/10.1111/j.1467-7660.1995.tb00542.x.

UN. 2013. *World population prospects, the 2012 revision*. Department of Economic and Social Affairs, Population Division. https://esa.un.org/unpd/wpp/Publications/. Accessed 30 May 2014.

WHO. 2014. *Global health observatory data repository*. World Health Organization (WHO) http://gamapserver.who.int/gho/interactive_charts/mbd/life_expectancy/atlas.html

Winkels, A. 2008. Rural in-migration and global trade: Managing the risks of coffee farming in the Central Highlands of Vietnam. *Mountain Research and Development* 28 (1): 32–40.

20

Land Cover and Land Use Analysis in Coastal Bangladesh

Anirban Mukhopadhyay, Duncan D. Hornby,
Craig W. Hutton, Attila N. Lázár,
Fiifi Amoako Johnson, and Tuhin Ghosh

20.1 Introduction

In coastal Bangladesh, land use has historically been an indicator of the principal livelihood source for the resident population. The most significant land use intervention has been the development of a system of polders since the 1960s, where areas of land are enclosed by embankments

A. Mukhopadhyay (✉) • T. Ghosh
School of Oceanographic Studies, Jadavpur University, Kolkata, India

D. D. Hornby • C. W. Hutton
Geodata Institute, Geography and Environment, University of Southampton, Southampton, UK

A. N. Lázár
Faculty of Engineering and the Environment and Tyndall Centre for Climate Change Research, University of Southampton, Southampton, UK

F. Amoako Johnson
Social Statistics and Demography, Faculty of Social, Human and Mathematical Sciences, University of Southampton, Southampton, UK

© The Author(s) 2018
R. J. Nicholls et al. (eds.), *Ecosystem Services for Well-Being in Deltas*,
https://doi.org/10.1007/978-3-319-71093-8_20

and water levels are managed using drainage. The widespread development of polders encouraged agriculture but, over the long term, has degraded soil quality by preventing new fertile sedimentation from being deposited and promoting subsidence. Subsidence makes drainage more difficult and increasing potential flood depths when dikes fail (Auerbach et al. 2015). Land use change has also interacted with changes in the balance between sea water and freshwater (Clarke et al. 2015; Lázár et al. 2015) which varies seasonally, inter-annually and spatially in response to changing patterns of precipitation, sea levels, extreme events and water management (Islam et al. 2015).

In order to capture land use changes and provide underlying base for the integrated modelling, two land use maps have been created for the study area (see Chap. 4, Fig. 4.2). The area comprises one of the world's largest lowlands with an elevation up to three metres, one metre above normal high tides, which is subject to tidal exchange. Hence it is the area within Bangladesh most exposed to sea-level rise (Milliman 1991; Huq et al. 1995; World Bank 2010). There are numerous islands near the Meghna River with resident communities. It also includes the Bangladeshi portion of the Sundarbans, the largest mangrove forest in the world.

The first set of land use maps are a historical series from 1989, 2001 and 2010. These compliment the decadal census data and allow for the identification of the change in land use and land cover over a 20-year period, providing a land use trend for the generation of land use scenarios out to 2050. A second set of maps uses the most recent data from 2010; these are composed of the 2010 map used to generate the trend but with additional characteristics derived from secondary data sets available only for this year. The more detailed 2010 map allows detailed identification and analysis of the rise of the aquaculture sector as one major recent land use change.

20.2   The Development of Land Use Maps

The land use maps are based on Landsat satellite images. Four scenes were required to cover the whole study area, and scenes with the least cloud cover from January of each selected year were acquired. Four numbers of scenes were downloaded to cover the whole study area of each year (1989, 2001 and 2010 of Landsat 5 TM Satellite sensor with a resolution of 30 metres).

To rectify and standardise the classification of the Landsat images for change analysis, radiometric standardisation (Duggin and Robinove 1990; Song et al. 2001) is undertaken along with atmospheric correction to reduce the errors. In this research, Landsat visible and near-infrared (VNIR) bands are atmospherically corrected. The bands are radiometrically calibrated to transform the DN values into top of the atmosphere (TOA) radiance (L_{TOA}) using sensor calibration function (Eq. 20.1) (Chander et al. 2007). The radiance of VNIR bands are transformed to correct surface reflectance based on image-based atmospheric correction model developed by Chavez (1996) (Eq. 20.2).

$$L_{TOA} = \left(\frac{L_{max_{\lambda}} - L_{min_{\lambda}}}{QCAL_{max} - QCAL_{min}} \right) \times \left(DN - QCAL_{min} \right) + L_{min_{\lambda}} \quad (20.1)$$

Where, $L_{max_{\lambda}}$ and $L_{min_{\lambda}}$ are maximum and minimum radiance (in W/ m^{-2} sr^{-1} μm^{-1}), $QCAL_{max}$ and $QCAL_{min}$ are maximum and minimum DN value possible (255/1).

$$\rho = \frac{\left(L_{TOA} - L_p \right) \pi d^2}{ESUN_{\lambda} \cos\theta_z T_z} \quad (20.2)$$

Where

ρ = surface reflectance,
d = Earth sun distance (in Astronomical Units),
$ESUN_{\lambda}$ = band pass solar irradiance at top of the atmosphere (TOA),
θ_z = Solar zenith angle (deg).
T_Z = atmospheric transmission between ground and TOA. The values of the T_Z for band 4 and 5 were taken as 0.85 and 0.95 respectively (Chavez 1996).

L_p = radiance resulted with the interaction aerosol and atmospheric particles and estimated based on Song et al. (2001), Chavez (1996) and Sobrino et al. (2004). All related atmospheric correction was completed using ATCORE Module of ERDAS imagine software.

20.3   Image Classification for Historical Maps

Nine major land use types were considered in this study as they are the predominant land cover classes of the study area: (i) agriculture, (ii) vegetation with rural settlement, (iii) wetland/bare land, (iv) aquaculture, (v) water, (vi) mudflat, (vii) mangrove, (viii) sand and (ix) urban settlement. In order to assess the change in land classes, the Landsat imageries of 1989, 2001 and 2010 were initially analysed using an unsupervised classification which generated 100 classes. This was followed by a supervised classification using maximum likelihood classifier (MLC), one of the most recognised parametric classifiers (Lázár et al. 2015; Melesse and Jordan 2002; Otukei and Blaschke 2010), using training data gathered from the field using handheld GPS. The first five bands of Landsat TM were used in the classification. The algorithm calculates the likelihood of belonging to one of the known classes. The benefit of this method is that it takes into account the variance-covariance matrix within the class distributions (Scott and Symons 1971). It is extensively used where a pixel with the determined likelihood is classified into the consistent class. This classification process was carried out for the three years of data (Fig. 20.1), allowing for the identification of land use change over time, along with trends and rates.

Analysis of the prepared land use land cover maps (Fig. 20.2 and Table 20.1) showed vegetation with rural settlement, aquaculture and urban settlement have been increasing steadily whereas agricultural land shows a large decrease, particularly in the north-west side of the study area where it appears to have been displaced by growing areas of aquaculture. Mangrove areas show a small but steady decreasing trend over the 20-year period.

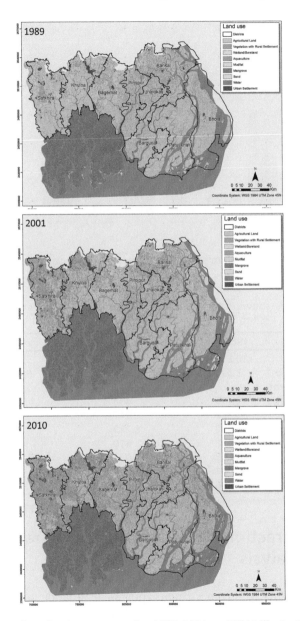

Fig. 20.1 Land use land cover maps for 1989, 2001 and 2010 illustrating changes in the seven predominant land use land cover classes

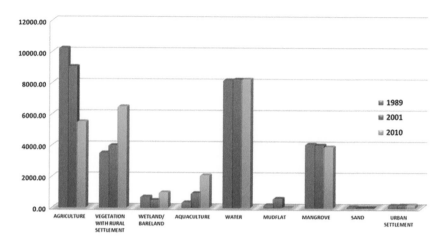

Fig. 20.2 Change in area (km²) for the land use land cover classes in the study area

Table 20.1 Table showing the changing trend of the land cover classes of the study area from 1989 to 2010

Class	1989 area (km²)	2001 area (km²)	2010 area (km²)
Agriculture	10,240	9,059	5,517
Vegetation with rural settlement	3,524	4,003	6,490
Wetland/bareland	717	494	989
Aquaculture	362	947	2090
Water	8,190	8,233	8,247
Mudflat	198	610	21
Mangrove	4,098	4,037	3,925
Sand	81	12	3
Urban settlement	196	207	215

20.4 Extraction of Aquaculture and Trend Analysis

Aquaculture in the Ganges-Brahmaputra-Meghna delta significantly impacts on livelihoods, employment and land tenure (Belton et al. 2011; Little et al. 2012; Kamruzzaman 2014) and is an important and growing land use within the study area. Since the infrared band is mostly absorbed by water, areas featuring water will be darker than surrounding areas in the satellite images. Indices used in this study for classifying water bodies were the Normalised Difference Water Index (NDWI) and Modified Normalised Difference Water Index (MNDWI) (Acharya et al. 2016). In both the water indices used, water pixels have a positive value, but differentiation between aquacultural and non-aquacultural ponds is unclear. An object-oriented image classification was therefore utilised within the study to take advantage of the fact that aquaculture ponds tend to have a specific shape and size.

Figure 20.3 indicates the spatial distribution of aquaculture and its persistence between 1989 and 2010 as calculated by the number of years identified (maximum of seven). For example, the orange colour denotes that this land has been classified as aquaculture five of the seven times. This type of analysis shows the progressive growth of aquaculture notwithstanding small areas where land has alternated in use. The growth of the shrimp industry over the last 20 years (Fig. 20.4) would appear to be associated, at least initially, with the development of the main road systems radiating out from the Khulna city south-west.

Aquaculture is replacing rice fields throughout the region that can no longer produce sufficiently in the increased saline environment. The exchange of shrimp for rice has socio-economic impacts with reductions in the localised population density, as rice farming is lost as well as the potential reduction in employment associated with the more intensive farming methodologies being associated with migration.

20.5 Differentiating Saline and Freshwater Aquaculture

Differentiating between saline and freshwater aquaculture is an important aspect of the Landsat analysis. A new method is applied to the most recent Landsat 5 TM images from January 2010 which involves field verification. The Landsat scenes are initially processed using ENVI 5.0 (Exelis Visual Information Solutions, USA). They are passed through the

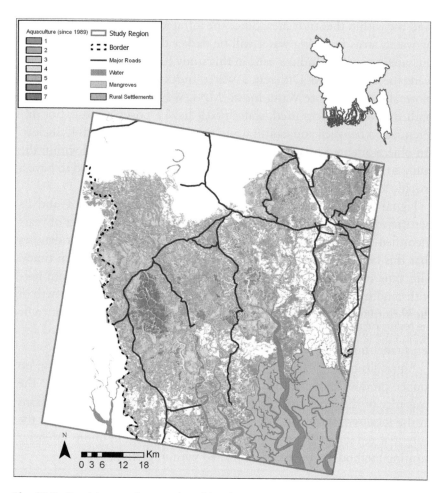

Fig. 20.3 Persistence of aquacultural land use over the period 1989–2010 in the north-west of the study area

radiometric calibration tool and then atmospherically corrected using the FLAASH tool with all downstream processing carried out in ArcGIS 10.2. Three Landsat bands (4, 5 and 3) are combined to make a composite image. This combination shows up water as a deep purple and vegetation as shades of brown and orange colours. This composite is the base layer for all subsequent processing. The 4-5-3 composite image is then

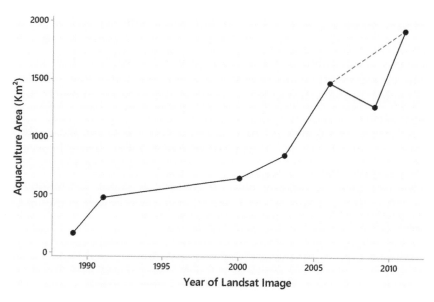

Fig. 20.4 Area under aquaculture in the study area derived from Landsat data. The reduction in area suggested by the 2009 data point is potentially due to cloud cover in the imagery

run through an isocluster tool which is an unsupervised classification technique and is set to create ten initial classes. The grid is then clipped back to the area of interest by the Union dataset to reduce the amount of data that needed processing, and a majority filter was run on the clipped image to subsume individual pixels.

The base layer is then modified to account for inaccuracies. A separate raster using the NDWI equation (McFeeters 1996) is created from the ENVI output and river pixels are extracted. As this would include flooded land, this dataset is vectorised and a manual pass of identifying rivers done. The aim of this is to distinguish between what are clearly rivers and temporarily flooded fields. The identified rivers are then rasterised and mosaicked or "stamped" back onto the majority-filtered grid. The perimeter of the urban areas Khulna and Barisal are also digitised and stamped back onto the grid. Areas of irrigated agriculture were distinguished from mangrove areas using a manually digitised mask to reclassify pixels that are erroneously classified by the isocluster tool into the mangrove class. A separate dataset that distinguished between irrigated and non-irrigated agriculture is also utilised. The newly identified irrigated agriculture polygons are also rasterised and stamped back onto the grid. These cells corresponded well to what was being classified as simply agriculture by the isocluster tool.

A further modification of the base layer uses a land use dataset for 2014 created by CEGIS in Bangladesh which differentiates between saline and freshwater aquaculture. This is supported by stakeholder evidence (see Chap. 10) and reference to available literature. The resultant map, Fig. 20.5, provides the basis for other research as well as the integrated model (e.g. Chaps. 21, 22 and 28).

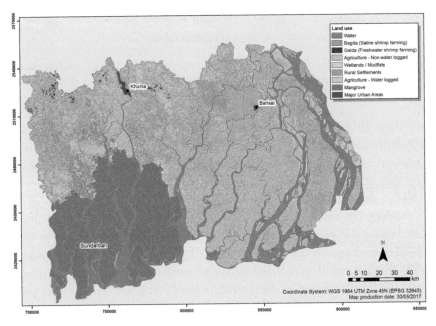

Fig. 20.5 Classification of 2010 Landsat Imagery of the study area into nine land use classes using remote sensing supported by ground truth data relating to the saline shrimp farming in the west of the study area (Reprinted under Creative Commons Attribution 4.0 International License from Amoako Johnson et al. 2016)

20.6   Scenarios of Future Land Use

Rather than simply extrapolating historical trends, future land use scenarios were developed for the study area based on stakeholder's scenario narratives for 2050 under the business as usual (BAU) scenario (e.g. brackish shrimp area slightly increased due to conversion of natural vegetation). The narratives are quantified through a process of qualitative to quantitative translation by experts and, after a final stakeholder workshop, the quantifications were projected to 2050 using agreed per cent changes from the 2011 land use baseline. The analysis does not extend beyond 2050, not least due to the large uncertainties in social and political trends more than 40 years into the future.

The land use scenarios were therefore linked to the expert demographic and socio-economic scenarios. Land use scenarios utilised in this context are a novel way to unify socio-economic and biophysical scenarios with changes in land use and land cover acting as a main driver of the scenario modelling process. Whilst such an approach allowed for land use to act as a central driver for modelling the relationship between people and ecosystem services, it also critically allows stakeholders to envisage a future in terms with which they are familiar, land use being an aspect of the day-to-day lives of the local population and decision makers.

20.7   Discussion

The land use base map and scenarios outlined here are fundamental to integrated analysis of ecosystems, environmental change and social sustainability in the delta. The land use information, for example, can be used to establish the role of tenure, or exposure to environmental stress, as a constraint on livelihoods and the spatial distribution of poverty (see Chap. 21). The maps developed show a picture of rapid change over the past two or three decades in the delta, notably the consolidation of polders and agriculture within them, and a significant increase in the prevalence of brackish shrimp farming with aquaculture mainly replacing rice fields. This is a process that might reasonably be set to continue into the future and almost certainly within the context of the development of the Bangladesh Delta Plan 2100 (BDP2100 2015), which will see substantial coastal infrastructural development raising the value of land and formalising agricultural structures. This is likely to include the further development of current polders and possibly see the introduction of further water and sediment management processes.

The trend of conversion for aquaculture has occurred concurrently with an increase in salinity which itself has been associated with a rise in poverty across the study region (Amoako Johnson et al. 2016). Indeed this region has shown a substantial growth in shrimp farming in the last 20 years as shown in the land use maps. This process is often associated with the presence of salinity and is thus highly likely to continue with future levels of salinity (see Chaps. 17 and 18). The growth of the shrimp-based aquaculture appears to be associated with both the intrusion of salt but also the proximity of a substantial settlement, in this case of Khulna, where rapid processing of the perishable product can be undertaken. This association is supported by access to main road systems which can be seen to be associated with the main hotspots of shrimp industry development.

At one level this growth might be viewed as a useful adaptation to increasing salinity intrusion in the region (Belton et al. 2011; Primavera 1997; Paul and Vogl 2011; Kamruzzaman 2014). Indeed, the high demand and perceived monetary benefits of shrimp has inspired many farmers to convert farmlands intruded by saline water into shrimp farms, whilst others have actively encouraged saline water from marine sources into their farmlands to produce shrimp (Rahman et al. 2013). However, the environmental impacts of this intense aquaculture practice (e.g. increasing soil toxicity) raise concerns over its sustainability. Intensive aquaculture, as identified in the mapping, has consequences for land tenure, livelihood displacements and income loss, food insecurity and health, rural unemployment, social unrest, conflicts and forced migration (Hossain et al. 2013; Swapan and Gavin 2011; Paul and Vogl 2011). As such the monitoring of shrimp farming by remote sensing offers a valuable insight into changes in socio-environmental context linked via land use change and offers future scenarios the trend in growth of brackish shrimp farming.

References

Acharya, T.D., D.H. Lee, I.T. Yang, and J.K. Lee. 2016. Identification of water bodies in a Landsat 8 OLI image using a J48 decision tree. *Sensors* 16 (7). https://doi.org/10.3390/s16071075.

Amoako Johnson, F., C.W. Hutton, D. Hornby, A.N. Lázár, and A. Mukhopadhyay. 2016. Is shrimp farming a successful adaptation to salinity intrusion? A geospatial associative analysis of poverty in the populous Ganges–Brahmaputra–Meghna Delta of Bangladesh. *Sustainability Science* 11 (3): 423–439. https://doi.org/10.1007/s11625-016-0356-6.

Auerbach, L.W., S.L. Goodbred, D.R. Mondal, C.A. Wilson, K.R. Ahmed, K. Roy, M.S. Steckler, C. Small, J.M. Gilligan, and B.A. Ackerly. 2015. Flood risk of natural and embanked landscapes on the Ganges-Brahmaputra tidal delta plain. *Nature Climate Change* 5 (2): 153–157. https://doi.org/10.1038/nclimate2472.

BDP2100. 2015. *Coast and polder issues: Baseline study*. Bangladesh Delta Plan. General Economic Division, Planning Commission, Government of the People's Republic of Bangladesh. http://www.bangladeshdeltaplan2100.org/publications/baseline-studies/. Accessed 19 July 2016.

Belton, B., M. Karim, S. Thilsted, K. Murshed-E-Jahan, W. Collis, and M. Phillips. 2011. *Review of aquaculture and fish consumption in Bangladesh*. Studies and reviews 2011–53. Penang: The WorldFish Center. http://pubs.iclarm.net/resource_centre/WF_2970.pdf. Accessed 8 Dec 2016.

Chander, G., B.L. Markham, and J.A. Barsi. 2007. Revised Landsat-5 thematic mapper radiometric calibration. *IEEE Geoscience and Remote Sensing Letters* 4 (3): 490–494. https://doi.org/10.1109/LGRS.2007.898285.

Chavez, P.S. 1996. Image-based atmospheric corrections revisited and improved. *Photogrammetric Engineering and Remote Sensing* 62 (9): 1025–1036.

Clarke, D., S. Williams, M. Jahiruddin, K. Parks, and M. Salehin. 2015. Projections of on-farm salinity in coastal Bangladesh. *Environmental Science-Processes and Impacts* 17 (6): 1127–1136. https://doi.org/10.1039/c4em00682h.

Duggin, M.J., and C.J. Robinove. 1990. Assumptions implicit in remote-sensing data acquisition and analysis. *International Journal of Remote Sensing* 11 (10): 1669–1694.

Hossain, M.S., M.J. Uddin, and A.N.M. Fakhruddin. 2013. Impacts of shrimp farming on the coastal environment of Bangladesh and approach for management. *Reviews in Environmental Science and Bio-Technology* 12 (3): 313–332. https://doi.org/10.1007/s11157-013-9311-5.

Huq, S., S.I. Ali, and A.A. Rahman. 1995. Sea-level rise and Bangladesh: A preliminary analysis. *Journal of Coastal Research SI* 14: 44–53.

Islam, G.M.T., A. Islam, A.A. Shopan, M.M. Rahman, A.N. Lázár, and A. Mukhopadhyay. 2015. Implications of agricultural land use change to ecosystem services in the Ganges delta. *Journal of Environmental Management* 161: 443–452. https://doi.org/10.1016/j.jenvman.2014.11.018.

Kamruzzaman, P. 2014. *Poverty reduction strategy in Bangladesh: Rethinking participation in policy making.* Bristol: Policy Press.

Lázár, A.N., D. Clarke, H. Adams, A.R. Akanda, S. Szabo, R.J. Nicholls, Z. Matthews, D. Begum, A.F.M. Saleh, M.A. Abedin, A. Payo, P.K. Streatfield, C. Hutton, M.S. Mondal, and A.Z.M. Moslehuddin. 2015. Agricultural livelihoods in coastal Bangladesh under climate and environmental change – A model framework. *Environmental Science-Processes and Impacts* 17 (6): 1018–1031. https://doi.org/10.1039/c4em00600c.

Little, D.C., B.K. Barman, B. Belton, M.C. Beveridge, S.J. Bush, L. Dabaddle, H. Demaine, P. Edwards, M.M. Haque, G. Kibria, E. Morales, F.J. Murray, W.A. Leschen, M.C. Nandeesha, and F. Sukadi. 2012. Alleviating poverty through aquaculture: progress, opportunities and improvements. In *Farming the waters for people and food: Proceedings of the Global Conference on Aquaculture 2010*, Phuket. September 22–25, 2010, ed. R.R. Subasinghe, J.R. Arthur, D.M. Bartley, S.S. De Silva, M. Halwart, N. Hishamunda, C.V. Mohan, and P. Sorgeloos, 719–783. Rome/Bangkok: FAO and NACA.

McFeeters, S.K. 1996. The use of the normalized difference water index (NDWI) in the delineation of open water features. *International Journal of Remote Sensing* 17 (7): 1425–1432.

Melesse, A.M., and J.D. Jordan. 2002. A comparison of fuzzy vs. augmented-ISODATA classification algorithms for cloud-shadow discrimination from Landsat images. *American Society for Photogrammetry and Remote Sensing* 68 (9): 905–912.

Milliman, J.D. 1991. Flux and fate of fluvial sediment and water in coastal seas. In *Ocean margin processes in global change: report of the Dahlem Workshop on Ocean Margin Processes in Global Change*, Berlin. March 18–23, 1990, ed. R.F.C. Mantoura, J.M. Martin, and R. Wollast, 69–89. Chichester: Wiley.

Otukei, J.R., and T. Blaschke. 2010. Land cover change assessment using decision trees, support vector machines and maximum likelihood classification algorithms. *International Journal of Applied Earth Observation and Geoinformation* 12: S27–S31. https://doi.org/10.1016/j.jag.2009.11.002.

Paul, B.G., and C.R. Vogl. 2011. Impacts of shrimp farming in Bangladesh: Challenges and alternatives. *Ocean and Coastal Management* 54 (3): 201–211. https://doi.org/10.1016/j.ocecoaman.2010.12.001.

Primavera, J.H. 1997. Socio-economic impacts of shrimp culture. *Aquaculture Research* 28 (10): 815–827.

Rahman, M.M., V.R. Giedraitis, L.S. Lieberman, T. Akhtar, and V. Taminskiene. 2013. Shrimp cultivation with water salinity in Bangladesh: The implications of an ecological model. *Universal Journal of Public Health* 1 (3): 131–142. http://dx.doi.org/10.13189/ujph.2013.010313.

Scott, A.J., and M.J. Symons. 1971. Clustering methods based in likelihood ration criteria. *Biometrics* 27 (2): 387. https://doi.org/10.2307/2529003.

Sobrino, J.A., J.C. Jimenez-Munoz, and L. Paolini. 2004. Land surface temperature retrieval from LANDSAT TM 5. *Remote Sensing of Environment* 90 (4): 434–440. https://doi.org/10.1016/j.rse.2004.02.003.

Song, C., C.E. Woodcock, K.C. Seto, M.P. Lenney, and S.A. Macomber. 2001. Classification and change detection using Landsat TM data: When and how to correct atmospheric effects? *Remote Sensing of Environment* 75 (2): 230–244. https://doi.org/10.1016/S0034-4257(00)00169-3.

Swapan, M.S.H., and M. Gavin. 2011. A desert in the delta: Participatory assessment of changing livelihoods induced by commercial shrimp farming in Southwest Bangladesh. *Ocean and Coastal Management* 54 (1): 45–54. https://doi.org/10.1016/j.ocecoaman.2010.10.011.

World Bank. 2010. *Economics of adaptation to climate change: Bangladesh*. Vol. 1 main report. Washington, DC: World Bank Group. https://openknowledge. worldbank.org/handle/10986/12837. Accessed 9 Jan 2017.

21

A Geospatial Analysis of the Social, Economic and Environmental Dimensions and Drivers of Poverty in South-West Coastal Bangladesh

Fiifi Amoako Johnson and Craig W. Hutton

21.1   Introduction

What are the causes of persistent poverty in the Ganges-Brahmaputra-Meghna (GBM) delta and how important are environmental dimensions and ecosystem services in explaining the uneven distribution of the observed rates of poverty? Investigation of the fine-grained linkages between the distribution of poverty and the state and health of the environment are particularly important for policy decisions and planning for the allocation of resources for development intervention. In this research, spatially aggregated population level data are used to (i) examine the extent

F. Amoako Johnson (✉)
Social Statistics and Demography, Faculty of Social, Human and Mathematical Sciences, University of Southampton, Southampton, UK

C. W. Hutton
Geodata Institute, Geography and Environment, University of Southampton, Southampton, UK

© The Author(s) 2018
R. J. Nicholls et al. (eds.), *Ecosystem Services for Well-Being in Deltas*,
https://doi.org/10.1007/978-3-319-71093-8_21

of geographical variations in poverty in the delta, (ii) identify the key socio-economic and environmental drivers of poverty and (iii) investigate how the drivers of poverty are spatially distributed and associated with spatially explicit socio-economic and environmental factors. Underlying the approach adopted is the hypothesis that if changes in the socio-economic and environmental functions in deltas have a substantial impact on the well-being of the local population, it should be reflected in the associated poverty levels measured.

Research has revealed wide social and geographical variations in poverty in Bangladesh (Amoako Johnson et al. 2016). In addition to the social, economic and cultural determinants, recent studies have shown that environmental and ecosystem services are also important associative factors of poverty (Suich et al. 2015; Amoako Johnson et al. 2016; Islam et al. 2016). Suich et al. (2015) also documented that ecosystem services act as a safety net for the poor and marginalised populations; however there is little evidence to suggest that availability of ecosystem services acts as a route out of poverty (see Chap. 2). Other studies have shown that the impact of ecosystem services on poverty are mediated by other factors including access to land, land tenure arrangements and availability of human capital (McKay and Lawson 2003; Daw et al. 2016). Yet, the ambiguity about the causes of the uneven distribution of poverty across space and across society remains unexplained. In this research, the hypothesis that the spatial dynamics of the factors associated with poverty including environmental services and human capital affect the incidence of poverty across space is explored.

Conventional approaches for measuring poverty (poverty headcount, income share and the poverty gap) rely on indicators such as income, expenditure and/or consumption which are either not covered by censuses or, where they are reported, are often not reliable (Meyer and Sullivan 2003; Nicoletti et al. 2011). To overcome these limitations, studies have used approaches based on households' ownership of assets and amenities (e.g. Filmer and Pritchett 2001). Validation of these approaches have shown that asset poverty robustly captures the multidimensionality of poverty (Filmer and Pritchett 2001) and represents chronic poverty and lack of human capital (McKay and Lawson 2003; Cooper and Bird 2012; Stein and Horn 2012; Wietzke 2015).

Most censuses in low- and middle-income countries collect detailed information on ownership of assets and amenities, which could be used to evaluate the poverty status of local communities and the associative effects of climate-related hazards and environmental stressors.

In this chapter, socio-economic data from the Bangladesh Population and Housing Census (BPHC) is linked with environmental data derived from Landsat Imagery to examine the geospatial differentials in poverty with social, economic and environmental vulnerability, including salinisation, development of shrimp and prawn farms, loss of agriculture, water logging and infrastructure development in the GBM delta. The geospatial unit of analysis is the Union, which is the lowest tier of local level administrative structure in Bangladesh and typically represent perhaps 5,000 people.

21.2 Background

Although deltas are major source of diverse ecosystem services, vital for sustaining human well-being, they remain exposed to the impacts of climate change, environmental hazards, sea-level rise and land cover changes on local ecosystems. In turn, livelihoods and survival of residents, particularly poor and vulnerable communities, are affected (Nicholls et al. 2016). Communities of the GBM delta are therefore not only marginalised by environmental dynamics but also in their social and economic development. This is reflected in the region's adult illiteracy rate, education, access to health care, nutrition, employment, transportation and gender empowerment indicators which remain very low, with high geographical inequalities (Biswas 2008). For example, a study by Szabo et al. (2016) reported large intra-urban inequalities in education and access to health care services, whilst a study by Sohel et al. (2010) identified clusters of high foetal loss and infant death in the localities of the Meghna River.

A major challenge to the ecosystem services within the delta has been the increasing salinisation of the region which has had a substantial impact on land use and land cover changes. This is illustrated by the decline of traditional (rice) agriculture and the increase in brackish shrimp farming (although freshwater prawn farming also has a 20-year history within the area). Shrimp farming, due to the high demand and perceived monetary benefits, is an economic adaptation to the impacts of the rapidly salinising delta with many farmers converting their permanently flooded farmlands into shrimp farms and others actively encouraging saline water from marine sources into their farmlands to enable shrimp production (Rahman et al. 2013). Large-scale commercial shrimp farming in the delta has also developed, leading to deforestation, loss of agricultural land and increasing soil toxicity. For the local population, this has generated issues around land tenure, livelihood displacements, income loss, food insecurity and negative health impacts, loss of rural unemployment, social unrest, conflicts and forced migration (Paul and Vogl 2011; Swapan and Gavin 2011; Hossain et al. 2013). These issues raise concerns on the benefits and sustainability of the ever-expanding shrimp farms to the vulnerable and marginalised populations of the delta.

Analysis of historical data also shows a decline in mangrove areas of about 17 per cent in the Sundarbans of the GBM delta, through both sea-level rise and deforestation (see Chap. 26 and Mukhopadhyay et al. (2015)), with projections anticipating a further decline of between 3 and 24 per cent by 2100 (Nicholls et al. 2016). Alongside salinisation and land loss, waterlogging of agricultural land is a growing phenomenon in the study area and is the result of the slow dissipation of annual flooding due to poorly maintained and overwhelmed drainage systems. These changes have important implications for provisioning ecosystem services (e.g. agriculture, fisheries) and regulating services such as the protective role of mangroves during storms.

Evidently, there is the need to examine the complex interactions between the socio-economic and environmental dynamics of the GBM delta to support policy and programmes aimed at alleviating poverty and develop sustainable approaches to preserve the regions' ecosystems and environment. In this research, multiple data sources including Census, Landsat Satellite Imagery 5 TM and Soil Salinity Survey data are used to examine the extent of geospatial clustering in poverty in the delta and their associative relationships with selected socio-economic and environmental factors in the GBM delta.

21.3 Study Area

The analysis is conducted at the Union level, which is the lowest local government administrative unit in Bangladesh (Panday 2011; MoHFW 2012). The spatial distribution of Unions is shown in Fig. 21.2. The study area focused on the south-central (Barisal, Bhola and Patuakhali districts) and south-western (Bagerhat, Barguna, Jhalokati, Khulna, Pirojpur and Satkhira districts) coastal zones of the Bangladeshi GBM delta (Fig. 21.1).

The study area covers the 653 Unions which make up the central and western coastal zones of the GBM delta, classified into 497 rural and 156 urban (cities, municipalities and *Upazila* headquarters) Unions. It is important to note that four of the nine districts in the study area (Bagerhat, Satkhira, Pirojpur and Khulna) are classified amongst the major shrimp-producing districts in Bangladesh (FAO 2015).

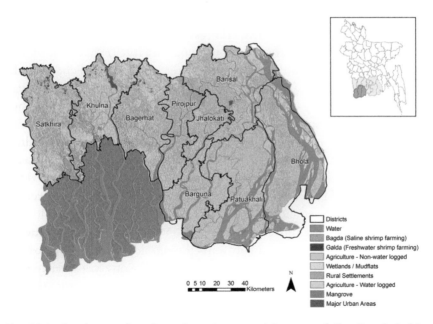

Fig. 21.1 South-central and south-western coastal zones of the Bangladeshi GBM delta. Map shows districts and land use (Reproduced from Amoako Johnson et al. 2016 under Creative Commons Attribution 4.0 International License (http://creativecommons.org/licenses/by/4.0/))

21.4   Data and Methodology

21.4.1   Socio-economic Data

Unlike environmental and climate-related data which can be generated across multiple spatial resolutions, social and economic data is more problematic as the infrastructure and resources to conduct surveys to collect data representative for small geographic areas are limited. A major source of representative local level social and economic data for most low- and middle-income countries is therefore the national Population and Housing Census (PHC).[1] In Bangladesh, although these censuses are less regular and expensive compared to population level surveys, they do provide important demographic and human capital information representative for small geographic units such as Unions.

The outcome variable of interest 'asset poverty' is a multidimensional score based on ownership of assets and amenities derived from the 2011 Bangladesh PHC (BBS 2012). The assets and amenities data include detailed information on the type of housing structure (*pucka*, *semi-pucka*, *kutcha* and *jhupri*), sources of drinking water (tap, tube well and others), type of toilet facility (water sealed, non-water sealed, non-sanitary and no toilet) and electricity connectivity. *Pucka* refers to houses built with permanent materials such as burnt bricks or concrete, *kutcha* are those built with nondurable materials such mud floors and metal sheet roofs and/or walls, whilst *semi-pucka* is a hybrid of *pucka* and *kutcha* (e.g. floors and/or walls are bricks or concrete but the rest are sheets) (Bern et al. 1994; Nenova 2010; GFDRR et al. 2014). A maximum likelihood factor analysis technique (Filmer and Pritchett 2001; Rutstein and Johnson 2004) is used to derive an asset poverty score at Union level. Maximum likelihood factor analysis is used as it circumvents the problem of multicollinearity and assigns indicator weights based on the variations in ownership of assets and amenities (Jones and Andrey 2007). The first factor score is categorised into quintiles and mapped to show the extent of spatial clustering in asset poverty.

The 2011 Bangladesh PHC also collected detailed information on demographic, economic, human capital and tenure. The socio-economic covariates derived from the 2011 Bangladesh PHC include employment status, adult literacy, school attendance, population density, dependency ratio and average household size. Information on major road density in each Union was derived from the 2011 Bangladesh Department of Roads and Highways data.

21.4.2 Environmental Data

The environmental covariates for the analysis are derived from 2009 Bangladesh Soil Salinity Survey (BSSS) (Ahsan 2010) and the 2010 Bangladesh Landsat 5 TM supplemented with the Bangladesh 2010 MODIS Terra Satellite Imagery (MODIS TSI). The 2009 BSSS is used to derive the percentage of Union area affected by soil salinity, whilst the Landsat 5 TM remote sensing images is used to extract Union area used for brackish shrimp and freshwater prawn farming. Union area affected by soil salinity is classified into four intensities: (i) low salinity (2–4 deciSiemens per metre (dS/m)), (ii) moderate salinity (4.1–8 dS/m), (iii) high salinity (8.1–12 dS/m) and (iv) very high salinity (12 dS/m or higher). The percentage of Union area used for brackish shrimp farming is classified into four categories: (i) no brackish shrimp farming, (ii) low brackish shrimp farming (less than one per cent of Union area), (iii) moderate brackish shrimp farming (one to ten per cent of Union area) and (iv) high brackish shrimp farming (greater than ten per cent of Union area). The percentage of Union area used for freshwater prawn farming was categorised into three categories: Unions with (i) no freshwater prawn farming, (ii) low freshwater prawn farming (less than one per cent of Union area) and (iii) high freshwater prawn farming (greater than one per cent of Union area). It is important to note that freshwater prawn farming is very limited in the study area, with only eight Unions where more than ten per cent of the Union area used for freshwater prawn farming. Additional environmental predictors extracted from the Landsat 5 TM include (i) the waterlogged agricultural land in a Union, (ii) Union area that is mangrove forest, (iii) permanent open water bodies and (iv) wetland and mudflats (see Amoako Johnson et al. 2016 for a more detailed discussion on the variables and their extraction). Table 21.1 shows the environmental, socio-economic and important controls variables selected for the analysis.

Table 21.1 Selected environmental and socio-economic variables

Variables and categorisation	Year and source	Type	Categorical variable coding
Outcome variable			
Asset poverty score	2011 BPHC	Categorical	1 = bottom quintile, 2 = second quintile, 3 = middle quintile, 4 = fourth quintile, 5 = top quintile
Predictor variables			
Administrative controls			
Division	2011 BPHC	Categorical	0 = Barisal, 1 = Khulna
Type of Union	2011 BPHC	Categorical	0 = Urban, 1 = Rural
Environmental predictors			
% of Union area affected by:			
2–4 dS/m salinity	2009 BSSS	Continuous	
4.1–8 dS/m salinity	2009 BSSS	Continuous	
8.1–12 dS/m salinity	2009 BSSS	Continuous	
>12 dS/m salinity	2009 BSSS	Continuous	
% of Union area used for brackish shrimp farming	2010 Landsat 5 TM	Categorical	0 = none, 1 = low (less than 1 per cent), 2 = moderate (1–10 per cent), 3 = high (greater than 10 per cent)
% of Union area used for freshwater prawn farming	2010 Landsat 5 TM	Categorical	0 = none, 1 = low (less than 1 per cent), 2 = high (greater than 1 per cent)
Presence of mangrove forest	2010 Landsat 5 TM/ MODIS TSI	Categorical	0 = no mangrove, 1 = mangrove
% of waterlogged agricultural lands in a Union	2010 Landsat 5 TM/ MODIS TSI	Continuous	

% of Union area made up of permanent open water bodies	2010 Landsat 5 TM/MODIS TSI	Continuous
% of Union area made up of wetland and mudflats	2010 Landsat 5 TM/MODIS TSI	Continuous
Socio-economic predictors		
% persons aged 15–60 years working	2011 BPHC	Continuous
% persons aged 7+ years literacy	2011 BPHC	Continuous
% of children aged 6–14 years registered or enrolled in school	2011 BPHC	Continuous
Population density	2011 BPHC	Continuous
Dependency ratio (%)	2011 BPHC	Continuous
Average household size	2011 BPHC	Continuous
Major road density	2011 BDRH	Continuous

Data sources: *BPHC* Bangladesh Population and Housing Census, *BSSS* Bangladesh Soil Salinity Survey, *MODIS TSI* MODIS Terra Satellite Imagery, *BDRH* Bangladesh Department of Roads and Highways

21.4.3 Methods

To examine the extent of spatial clustering in asset poverty, the join-count spatial autocorrelation technique is used to examine whether the observed spatial patterns of asset poverty amongst the Unions in the study area are significantly random or clustered (Cliff and Ord 1981). A Bayesian Geo-additive Semi-parametric (BGS) regression is used to examine the geospatial differentials in asset and the extent to which the socio-economic and environmental predictors are associative with the observed spatial differentials in asset poverty (Brezger et al. 2005). The BGS techniques were adopted for this analysis because it allows for the unobserved spatial heterogeneity (both spatially structured and unstructured) to be accounted for as well as the simultaneous estimation of non-linear effects of continuous covariates as well as fixed effects of categorical and continuous covariates in addition to the spatial effects (Brezger et al. 2005).

The outcome variable of interest 'asset poverty' y_i is coded 1 if a Union is in the bottom quintile of the score and 0 otherwise. The outcome variable y_i follows a binomial distribution with parameters n_i and π_i, that is $y_i \sim B(n_i, \pi_i)$, where π_i is the probability of a Union being in the bottom quintile and n_i is the number of Unions in the study area. The model linking the probabilities of a Union being in the bottom quintile with the predictors follows a logistic model of the form:

$$\pi_i = P(y_i 1 \mid \eta_i) = \frac{\exp(\eta_i)}{1 + \exp(\eta_i)} \tag{21.1}$$

where η_i is the predictor of interest. With a vector $x_i^{'} = \left(x_{i1}, \ldots, x_{ik} \right)^{'}$ of k continuous covariates and $\lambda_i^{'} = \left(\lambda_{i1}, \ldots \lambda_{id} \right)^{'}$ a vector of d categorical covariates, then predictor η_i can be specified as:

$$\eta_i = \alpha \lambda_i^{'} + \beta x_i^{'} \tag{21.2}$$

where α is a d-dimensional vector of unknown regression coefficients for the categorical covariates $\lambda_i^{'}, \beta$ is a k-dimensional vector of unknown regression coefficients for the continuous covariates x_i'. To account for non-linear effects of the continuous covariates and spatial dependence in asset wealth, the BGS framework which replaces the strictly linear predictors with flexible semi-parametric predictors was adopted. The BGS model is then specified as shown in Eq. 21.3:

$$\eta_i = \alpha\lambda_i' + f_k x_{ik}' + f^{\text{spat}(S_i)} \tag{21.3}$$

where $f_k(x)$ are non-linear smoothing function of the continuous variables x_{ik} and $f^{\text{spat}(S_i)}$ accounts for unobserved spatial heterogeneity at location i ($i=1,...,S$), some of which may be spatially structured and others unstructured. The spatially structured effects show the effect of location by assuming that geographically close areas are more similar than distant areas, whilst the unstructured spatial effect accounts for spatial randomness in the model. Then, Eq. 21.4 can be specified as:

$$\eta_i = \alpha\lambda_i' + f_k x_{ik}' + f^{\text{str}(S_i)} + f^{\text{unstr}(S_i)} \tag{21.4}$$

where f^{str} is the structured spatial effects and f^{unstr} is the unstructured spatial effects and $f^{\text{spat}(S_i)} = f^{\text{str}} + f^{\text{unstr}}$. In the case of this study, the spatially structured effects depict the extent of clustering of asset poverty and the influence of unaccounted predictor variables that themselves may be spatially clustered or random. The smooth effects of continuous factors are modelled with P-spline priors, whilst the spatial effects are modelled using Markov random field priors.

21.5 Results

The geospatial patterns in poverty derived using ownership of assets and amenities are shown in Fig. 21.2. The figure shows a strong clustering of asset poverty in the Bhola district and the Unions close to the Sundarbans. More than half (58.2 per cent) of all the Unions in the Bhola district are classified in the bottom quintile. The Pirojpur district recorded the second

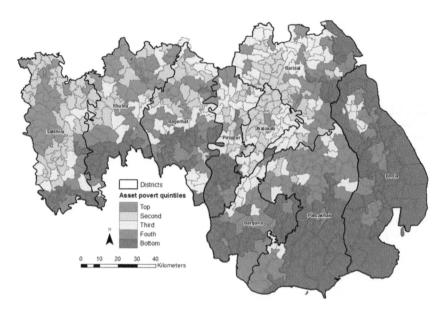

Fig. 21.2 Geospatial variations in asset poverty in the GBM delta of Bangladesh (Reproduced from Amoako Johnson et al. 2016 under Creative Commons Attribution 4.0 International License (http://creativecommons.org/licenses/by/4.0/))

highest percentage (34.9 per cent) of Unions in the bottom quintile. About a quarter of all Unions in the Barguna (25.9 per cent), Patuakhali (25.6 per cent) and Bagerhat (25.0 per cent) districts are also in the bottom quintile. Asset poverty is lowest in the Unions in Jhalokati (2.9 per cent), Barisal (5.7 per cent), Satkhira (9.5 per cent) and Khulna (9.7 per cent) districts where less than one-tenth of Unions in those districts are in the bottom quintile. A joint count spatial autocorrelation analysis showed that Unions in the bottom quintile are nearly three times more likely to be neighbours than would be expected under a random spatial pattern $(Z[BW] = -18.87, p < 0.05)$. This demonstrates that the poorest Unions are more concentrated in some parts of the study area (see Fig. 21.2).

Table 21.2 shows the posterior odds ratios and their corresponding 95 per cent credible intervals of the effect of the socio-economic and environmental predictors on poverty, after controlling for the location (the divisional administrative effect and rural/urban status of the Union) effects. Only predictors significant at the five per cent significance ($p < 0.05$) level

Table 21.2 Estimated posterior odds ratios and their corresponding 95 per cent credible intervals of the associative effects of the environmental and socio-economic predictors on poverty

Primary and control variables	Odds ratio	95% posterior CI		Significant at $p < 0.05$
		Lower bound	Upper bound	
Location effect				
Division				
Barisal	1.00			
Khulna	0.71	0.07	7.80	No
Type of Union				
Urban	1.00			
Rural	1.89	0.52	6.88	No
Environmental predictors				
Intensity and extent of salinity intrusion (% Union area)				
2.0–4.0 dS/m salinity	1.01	0.98	1.04	No
4.1–8.0 dS/m salinity	1.04	1.01	1.07	Yes
8.1–12.0 dS/m salinity	1.04	1.01	1.08	Yes
>12 dS/m salinity	1.07	1.01	1.14	Yes
Union area used for brackish shrimp farming				
None	1.00			
Low (less than 1 per cent)	1.36	0.36	5.14	No
Moderate (1–10 per cent)	1.79	0.14	23.49	No
High (greater than 10 per cent)	0.30	0.01	8.85	No
Union area for freshwater prawn farming				
None	1.00			
Low (less than 1 per cent)	0.66	0.16	2.70	No
High (greater than 1 per cent)	0.41	0.01	16.22	No
Mangroves				
Unions with no mangrove	1.00			
Unions with mangrove	6.05	1.35	27.16	Yes
Waterlogged agricultural land	1.02	1.00	1.05	Yes
Permanent open water bodies	1.03	1.00	1.06	Yes
Wetland and mudflats	Non-linear			Yes
Socio-economic predictors				
% 15–64 years employed	Non-linear			Yes
% 15 years or older who are literate	0.90	0.83	0.98	Yes
% 6–14-year-olds in school	0.84	0.73	0.98	Yes
Major road density within Union	0.01	0.00	0.45	Yes

are retained in the model, except for the shrimp farming because of its perceived economic importance and an alternative coping mechanism for salinity intrusion in the delta. The posterior odds ratios show that the divisional administrative effect and rural/urban status of the Unions are not significantly associated with poverty. For environmental predictors, the results show that whilst high levels (greater than 4 ds/m) of soil salinity in a Union are significantly associated with the probability of a Union being poor, areas used for both brackish shrimp and freshwater prawn farming are not. The estimated posterior odds ratios show that increasing intensity of soil salinity increases the odds of Union being poor. This suggests that whilst poverty is pronounced in Unions affected by high levels of salinity, large shrimp farms do not reduce the incidence of asset poverty.

The results further show that Unions with mangroves are six times more likely to be in the poorest quintile when compared with Unions with no mangrove. For permanent open water bodies, the results show that the higher percentage of permanent open water bodies within a Union, the higher the odds of the Union being in the poorest quintile. On average, if permanent water body area within a Union increases by one per cent, the odds of the Union being in the poorest quintile increases by 1.03. Similarly, the higher the wetland and mudflats area in a Union, the higher the likelihood of the Union being poor, however, the relationship is not linear.

Considering socio-economic factors, the results show that employment, literacy, school attendance and access to major road with a Union are significantly associated with poverty. The posterior odds ratios show that increase in employment rate, adult literacy and school attendance are all associated with a decline in the odds of a Union being poor (Table 21.2). Also, the higher the density of major roads within a Union, the lower the odds of the Union being poor.

The posterior mode of the structured spatial effects and their corresponding posterior probabilities at the 95 per cent nominal level are used to examine the spatial drivers of poverty. The posterior mode of the structured spatial effects could be used to identify Unions where asset poverty is high, where they are low and where they are trivial. To identify the spatial drivers of poverty, a sequential model building approach was adopted by first including the environmental predictors in the fitted model, accounting for the location effects, followed by the socio-economic predictors. The posterior probabilities are used to identify the spatial correlations of the predictors with poverty by examining Unions where the estimated posterior mode of the structured spatial effects become statistically non-significant after covariates are added to the model.

Figure 21.3 shows the spatial correlates of asset poverty for the poorest Unions in the GBM delta based on the asset score. The results show that the environmental controls exhibit significant geospatial associations with asset poverty predominantly with Unions close to the Sundarbans in the Satkhira and Khulna and Bagerhat districts, as well as those in the Barguna, Patuakhali and Pirojpur districts.

The socio-economic predictors were significantly associated with asset poverty for Unions in the Bhola and Patuakhali districts, as well as a few Unions in the Barisal district. The socio-economic factors are associated with asset poverty in 34 of the 39 Unions in the bottom quintile in the Bhola district, making up about half of all the Unions in the district. In the

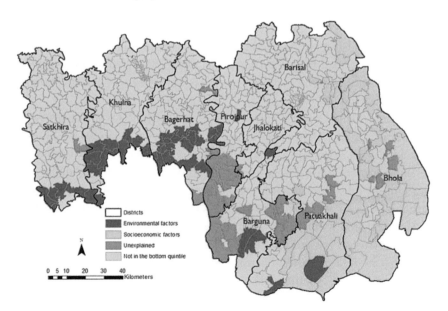

Fig. 21.3 Spatial correlates of poverty amongst Unions in the bottom quintile (Reproduced from Amoako Johnson et al. 2016 under Creative Commons Attribution 4.0 International License (http://creativecommons.org/licenses/by/4.0/))

Patuakhali district, the socio-economic factors are associated with asset poverty in 12 of the 18 Unions in the bottom quintile. The socio-economic factors are also significantly associated with asset poverty in six Unions in the Barisal district—Dhul Khola, Hizla Gaurabdi, Alimabad, Char Gopalpur, Jangalia and Bhasan Char Unions and also the Atharagashia Union in the Barguna district.

21.6   Discussion

The analysis conducted in this research illustrates that the benefits of integrating census data at lower geographic units are beneficial for designing and promoting policy-relevant research and the development of sustainable poverty-relevant programmes and targeted interventions. Using census data, this research has been able to explore the geospatial differentials in asset poverty, an important indicator and correlate of chronic poverty and lack of human capital (McKay and Lawson 2003; Cooper and Bird 2012; Stein and Horn 2012; Wietzke 2015), and provide input into the integrated modelling described in Chap. 28. Linking census and environmental data derived from Landsat 5 TM, MODIS TSI and Soil Salinity Survey, the research has explored the spatial correlates of poverty in the GBM delta.

The results show a strong clustering of poverty in the GBM delta, predominantly clustered amongst Unions in the Bhola, Bagerhat, Barguna, Patuakhali and Pirojpur districts. The multivariate analysis revealed that the environmental predictors—intensity and extent of salinity intrusion in a Union, presence of mangrove forest, waterlogged agricultural land, permanent open water bodies and wetland and mudflats—are significantly associated with Union level poverty. Whilst increasing intensity of salinity intrusion is significantly associated with poverty, the results also show that both large brackish shrimp and freshwater prawn farms do not impact on poverty in the GBM delta. This reveals that the impact of shrimp farming on alleviating poverty amongst the local population is trivial. The strong association identified between salinity intrusion and poverty could be attributed to loss of arable land, reduced agricultural productivity and income, food insecurity, rural unemployment, social unrest, conflicts and forced migration (Paul and Vogl 2011; Swapan and Gavin 2011; Hossain et al. 2013; Sá et al. 2013).

Socio-economic predictors, employment rate, adult literacy and school attendance are also significant predictors of poverty as well as access to major roads. The results show that increased employment rate, adult literacy, school attendance and access to major roads reduce the odds of a Union being poor.

The overall finding of the research is that the socio-economic and environmental drivers of poverty in the GBM delta are discernible spatially. As captured in the Bangladesh Government coastal zone policy, the coastal zone is slow in socio-economic development and lacks the resources to cope with environment deterioration and hazards (MoWR 2005). As such policy formulation aimed at improving the well-being of residents of the GBM delta should be geospatially focused and targeted. These findings provide relevant input to addressing the geospatial inequalities in poverty in the GBM delta in the government's coastal zone policy plan.

Note

1. Available at https://unstats.un.org/unsd/demographic/sources/census/censusdates.htm

References

Ahsan, M. 2010. *Saline soils of Bangladesh.* Dhaka: Soil Resource Development Institute, Ministry of Agriculture, Government of the People's Republic of Bangladesh. http://srdi.portal.gov.bd/sites/default/files/files/srdi.portal.gov.bd/publications/bc598e7a_df21_49ee_882e_0302c974015f/Soil%20salinity%20report-Nov%202010.pdf. Accessed 6 Sept 2016.

Amoako Johnson, F., C.W. Hutton, D. Hornby, A.N. Lázár, and A. Mukhopadhyay. 2016. Is shrimp farming a successful adaptation to salinity intrusion? A geospatial associative analysis of poverty in the populous Ganges–Brahmaputra–Meghna Delta of Bangladesh. *Sustainability Science* 11 (3): 423–439. https://doi.org/10.1007/s11625-016-0356-6.

BBS. 2012. Bangladesh population and housing census 2011 – Socio-economic and demographic report. Report national series. Vol. 4. Dhaka, Bangladesh: Bangladesh Bureau of Statistics (BBS) and Statistics and Informatics Division (SID), Ministry of Planning, Government of the People's Republic of Bangladesh. http://www.bbs.gov.bd/PageSearchContent.aspx?key=census% 202011. Accessed 7 July 2016.

Bern, C., J. Sniezek, G.M. Mathbor, M.S. Siddiqi, C. Ronsmans, A.M.R. Chowdhury, A.E. Choudhury, K. Islam, M. Bennish, E. Noji, and R.I. Glass. 1994. Risk factors for mortality in the Bangladesh cyclone of 1991. *Bulletin of the World Health Organisation* 71 (1): 73–78.

Biswas, A.K. 2008. Management of Ganges-Brahmaputra-Meghna system: Way forward. In *Management of transboundary rivers and lakes*, ed. O. Varis, A.K. Biswas, and C. Tortajada, 143–164. Berlin: Springer.

Brezger, A., T. Kneib, and S. Lang. 2005. BayesX: Analyzing Bayesian structured additive regression models. *Journal of Statistical Software* 14 (11): 1–22.

Cliff, A.D., and J.K. Ord. 1981. *Spatial processes, models and applications*. London: Pion.

Cooper, E., and K. Bird. 2012. Inheritance: A gendered and intergenerational dimension of poverty. *Development Policy Review* 30 (5): 527–541. https:// doi.org/10.1111/j.1467-7679.2012.00587.x.

Daw, T.M., C.C. Hicks, K. Brown, T. Chaigneau, F.A. Januchowski-Hartley, W.W.L. Cheung, S. Rosendo, B. Crona, S. Coulthard, C. Sandbrook, C. Perry, S. Bandeira, N.A. Muthiga, B. Schulte-Herbrüggen, J. Bosire, and T.R. McClanahan. 2016. Elasticity in ecosystem services: Exploring the variable relationship between ecosystems and human well-being. *Ecology and Society* 21 (2). https://doi.org/10.5751/ES-08173-210211.

FAO. 2015. *National aquaculture sector overview: Bangladesh*. Rome: Food and Agriculture Organization of the United Nations (FAO), Fisheries and Aquaculture.

Filmer, D., and L.H. Pritchett. 2001. Estimating wealth effects without expenditure data – Or tears: An application to educational enrollments in states of India. *Demography* 38 (1): 115–132. https://doi.org/10.2307/3088292.

GFDRR, UNDP, and EU. 2014. *Planning and implementation of post-Sidr housing recovery: Practice, lessons and future implications: Recovery framework case study*. Dhaka, Bangladesh: Global Facility for Disaster Reduction and Recovery (GFDRR) of the World Bank, United Nations Development Program (UNDP) and the European Union (EU).

Hossain, M.S., M.J. Uddin, and A.N.M. Fakhruddin. 2013. Impacts of shrimp farming on the coastal environment of Bangladesh and approach for management. *Reviews in Environmental Science and Bio-Technology* 12 (3): 313–332. https://doi.org/10.1007/s11157-013-9311-5.

Islam, D., J. Sayeed, and N. Hossain. 2016. On determinants of poverty and inequality in Bangladesh. *Journal of Poverty*: 1–20. https://doi.org/10.1080/1 0875549.2016.1204646.

Jones, B., and J. Andrey. 2007. Vulnerability index construction: Methodological choices and their influence on identifying vulnerable neighborhoods. *International Journal of Emergency Management* 42 (2): 269–295. https://doi.org/10.1504/IJEM.2007.013994.

McKay, A., and D. Lawson. 2003. Assessing the extent and nature of chronic poverty in low income countries: Issues and evidence. *World Development* 31 (3): 425–439. https://doi.org/10.1016/s0305-750x(02)00221-8.

Meyer, B.D., and J.X. Sullivan. 2003. Measuring the well-being of the poor using income and consumption. *Journal of Human Resources* 38: 1180–1220. https://doi.org/10.2307/3558985.

MoHFW. 2012. *Health Bulletin 2012*. Dhaka: Ministry of Health and Family Welfare, Government of the People's Republic of Bangladesh.

MoWR. 2005. *Coastal zone policy*. Dhaka: Ministry of Water Resources (MoWR), Government of the People's Republic of Bangladesh. http://lib.pmo.gov.bd/legalms/pdf/Costal-Zone-Policy-2005.pdf. Accessed 20 Apr 2017.

Mukhopadhyay, A., P. Mondal, J. Barik, S.M. Chowdhury, T. Ghosh, and S. Hazra. 2015. Changes in mangrove species assemblages and future prediction of the Bangladesh Sundarbans using Markov chain model and cellular automata. *Environmental Science-Processes and Impacts* 17 (6): 1111–1117. https://doi.org/10.1039/c4em00611a.

Nenova, T. 2010. *Expanding housing finance to the underserved in South Asia: Market review and forward agenda*. Washington, DC: The World Bank. https://openknowledge.worldbank.org/handle/10986/2475. Accessed 13 Mar 2017.

Nicholls, R.J., C.W. Hutton, A.N. Lázár, A. Allan, W.N. Adger, H. Adams, J. Wolf, M. Rahman, and M. Salehin. 2016. Integrated assessment of social and environmental sustainability dynamics in the Ganges-Brahmaputra-Meghna delta, Bangladesh. *Estuarine, Coastal and Shelf Science* 183, Part B: 370–381. https://doi.org/10.1016/j.ecss.2016.08.017.

Nicoletti, C., F. Peracchi, and F. Foliano. 2011. Estimating income poverty in the presence of missing data and measurement error. *Journal of Business and Economic Statistics* 29 (1): 61–72. https://doi.org/10.1198/jbes.2010.07185.

Panday, P.K. 2011. Local government system in Bangladesh: How far is it decentralised? *Journal of Local Self-Government* 9 (3): 205–230.

Paul, B.G., and C.R. Vogl. 2011. Impacts of shrimp farming in Bangladesh: Challenges and alternatives. *Ocean and Coastal Management* 54 (3): 201–211. https://doi.org/10.1016/j.ocecoaman.2010.12.001.

Rahman, M.M., V.R. Giedraitis, L.S. Lieberman, T. Akhtar, and V. Taminskiene. 2013. Shrimp cultivation with water salinity in Bangladesh: The implications of an ecological model. *Universal Journal of Public Health* 1 (3): 131–142. http://dx.doi.org/10.13189/ujph.2013.010313.

Rutstein, S.O., and K. Johnson. 2004. The DHS wealth index. Demographic and Health Surveys (DHS) comparative reports no. 6. Calverton: ORC Macro.

Sá, T., R. Sousa, Í. Rocha, G. Lima, and F. Costa. 2013. Brackish shrimp farming in Northeastern Brazil: The environmental and socio-economic impacts and sustainability. *Natural Resources* 4 (8): 538–550. https://doi.org/10.4236/nr.2013.48065.

Sohel, N., M. Vahter, M. Ali, M. Rahman, A. Rahman, P.K. Streatfield, P.S. Kanaroglou, and L.Å. Persson. 2010. Spatial patterns of fetal loss and infant death in an arsenic-affected area in Bangladesh. *International Journal of Health Geographics* 9 (1): 53. https://doi.org/10.1186/1476-072X-9-53.

Stein, A., and P. Horn. 2012. Asset accumulation: An alternative approach to achieving the Millennium Development Goals. *Development Policy Review* 30 (6): 663–680. https://doi.org/10.1111/j.1467-7679.2012.00593.x.

Suich, H., C. Howe, and G. Mace. 2015. Ecosystem services and poverty alleviation: A review of the empirical links. *Ecosystem Services* 12: 137–147. https://doi.org/10.1016/j.ecoser.2015.02.005.

Swapan, M.S.H., and M. Gavin. 2011. A desert in the delta: Participatory assessment of changing livelihoods induced by commercial shrimp farming in Southwest Bangladesh. *Ocean and Coastal Management* 54 (1): 45–54. https://doi.org/10.1016/j.ocecoaman.2010.10.011.

Szabo, S., E. Brondizio, F.G. Renaud, S. Hetrick, R.J. Nicholls, Z. Matthews, Z. Tessler, A. Tejedor, Z. Sebesvari, E. Foufoula-Georgiou, S. da Costa, and J.A. Dearing. 2016. Population dynamics, delta vulnerability and environmental change: Comparison of the Mekong, Ganges–Brahmaputra and Amazon delta regions. *Sustainability Science* 11 (4): 539–554. https://doi.org/10.1007/s11625-016-0372-6.

Wietzke, F.B. 2015. Who is poorest? An asset-based analysis of multidimensional wellbeing. *Development Policy Review* 33 (1): 33–59. https://doi.org/10.1111/dpr.12091.

22

Defining Social-Ecological Systems in South-West Bangladesh

Helen Adams, W. Neil Adger, Munir Ahmed,
Hamidul Huq, Rezaur Rahman, and Mashfiqus Salehin

22.1 Introduction

Deltas are diverse environments with spatially and temporally variable ecosystem services (Barbier et al. 2011). This chapter argues that different parts of the delta can be identified as distinct social-ecological systems. In doing so it proposes that recognising these distinct systems is crucial to understanding the persistence of poverty in natural resource dependent communities. In effect, the social mechanisms that have evolved to manage and govern access to diverse bundles of ecosystem services are distinctive and associated with different land uses (Rodriguez et al. 2006).

H. Adams (✉)
Department of Geography, King's College London, London, UK

W. Neil Adger
Geography, College of Life and Environmental Sciences,
University of Exeter, Exeter, UK

M. Ahmed
TARA, Dhaka, Bangladesh

© The Author(s) 2018
R. J. Nicholls et al. (eds.), *Ecosystem Services for Well-Being in Deltas*,
https://doi.org/10.1007/978-3-319-71093-8_22

405

Ecosystem services are co-produced in bundles in different social-ecological systems. A social-ecological system is the amalgamation of physical, ecological, and social phenomena into a set of recognisable and distinct systems of interaction. Social-ecological systems theories have a long and varied intellectual history, drawing on ideas of hybrid environments, human ecology, and cultural geography. Social-ecological systems as an analytical framework emphasise the interdependent nature of the human and ecological systems (Cote and Nightingale 2012).

The chapter describes the social-ecological systems (SESs) within the study area and how they have been identified and characterised. The first section shows how social-ecological systems were integrated as a guiding principle through the project. The second section presents the results of analysis of the ways in which social systems differ according to the ecological system to form distinct sub-systems of the delta. The chapter concludes by summarising some of the key findings on the relationship between poverty reduction and ecosystem services revealed through a systems perspective.

22.2 Social-Ecological Systems as an Analytical Framework

Managing natural resources, promoting sustainability, and enhancing human well-being require knowledge of populations, resources, and institutions within distinct social-ecological systems. Many studies have highlighted the benefits of understanding social-ecological interactions to promote sustainable management. Analysis of landscape change in Mediterranean Spain, for example, required modelling of both the intensity of

H. Huq
Institute of Livelihood Studies, Bangladesh University of Engineering and Technology, Dhaka, Bangladesh

R. Rahman • M. Salehin
Institute of Water and Flood Management, Bangladesh University of Engineering and Technology, Dhaka, Bangladesh

agriculture and the socio-economic context in which it was implemented (De Aranzabal et al. 2008). In the marine sector, fisheries management progressed from species-based to ecosystem-based approaches that embedded humans and their associated cultural values and different management practices within the ecosystem (Levin et al. 2009).

Social-ecological perspectives are particularly suited to incorporating diverse forms of lay and scientific knowledge in natural resource management (Hill et al. 2012), co-management practices (Cinner et al. 2012), common pool resources (Nagendra and Ostrom 2014), and governance issues (Karpouzoglou et al. 2016). As such, studies have applied system approaches to better manage and conserve ecosystem services, rather than explicitly seeking to design interventions with social goals such as alleviation of poverty. However, one major and common insight from this field which is relevant for this research is that social institutions are directly affected by the underlying characteristics of the natural resource base, including the resource fluctuations, variability, and divisibility (Ostrom 1990).

Many social mechanisms employed by poor populations to access natural resources are specific to that particular resource. In capture fisheries, for example, loans to buy equipment are paid back as a proportion of the catch because of the variable and unpredictable nature of catches and thus income (Allison and Ellis 2001). Systems of sharecropping, where landowners take significant proportions of agricultural outputs as rents, emerge as a mechanism in agricultural areas dominated by private land ownership, surplus labour, and insecure livelihoods (Wood 2003). Thus, mechanisms of rent capture in agriculture and fisheries are adapted to the specific characteristics of the bundle of ecosystem services but have a similar outcome of systematically extracting surplus value away from the poorest.

22.3   Methodology

Social-ecological systems are identified as articulated and experienced by those engaged in resource use across the study area. Categorisation was based on data generated using open-ended questions focussed on social factors that influence the ability of ecosystem services to produce well-being for the poor through livelihoods, environment, and perceptions of causal mechanisms (described in Chap. 1). Other studies have used this method for exploratory analyses of this type: respondents converge on issues of multiple causation, threshold effects, and the social dynamics and processes by which social and ecological dimensions of their lives and livelihoods are constructed (e.g. Davis 2009; Crane 2010; Fabinyi et al. 2014).

Data from 70 households purposefully sampled across the dominant land cover types of the delta region was collected between September 2012 and May 2013. Effort was made to include geographically remote areas, diverse administrative districts, different land uses, and livelihood systems in households with a range of wealth statuses. The data are available in open-access form as notes and transcripts of the interviews (see Adams and Adger 2016 to access Reshare depository). Interviews were carried out in Bengali and lasted from between 30 minutes and one hour.

Analysis of the data suggests seven distinct SESs across the study area: irrigated and rain-fed agriculture, brackish and freshwater aquaculture, char-dominated (eroding islands) areas, areas that are dependent on the Sundarbans mangrove forest, and coastal areas with easy access to offshore fisheries. The distribution of these SESs is shown in Fig. 22.1. The majority of the study area is dedicated to crop cultivation (63 percent), predominantly rice, and the Sundarbans mangrove forest, which constitutes almost a third of the land area (29 percent). Eight per cent of the land area is dedicated to freshwater prawn, brackish shrimp, and white fish aquaculture while waterways and wetlands compose around one percent. Thus, the SESs defined can correspond directly to the dominant land use (agriculture and aquaculture) or defined by proximity to a key geographical feature (Sundarbans dependence, char areas and offshore fisheries). For an examination of the ecological characteristics of these systems, please see Chaps. 21, 24, 25, and 26 and Adams et al. (2013).

Once defined, the SESs are identified and mapped using satellite imagery and GIS analysis (see Adams et al. 2016 for a full description of this process). The SESs form the basis of analysis of the ecosystem services throughout this research highlighting the diversity of delta

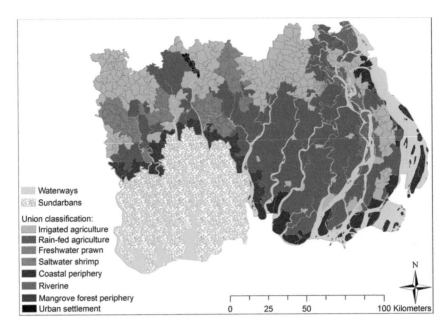

Fig. 22.1 Distribution of social-ecological systems based on survey information in the study area

environments and social trajectories across the delta region. The systems are used, for example, to analyse the influence of environmental conditions on the incidence of poverty (see Chap. 1) and form the first-level stratification of the household survey (see Chap. 23 and Adams et al. 2016). The integrated modelling approach also takes into account these different system dynamics through incorporation of the household survey data (see Chap. 28).

The remainder of this chapter presents some initial results highlighting some of the causal mechanisms linking ecosystem services and poverty in the study area and how these vary between SESs. These themes are summarised in Table 22.1, which also provides some summary statistics of the respondents. Numbers cited in brackets throughout this chapter correspond to the metadata file that accompanies the transcripts in the online database.[1]

Table 22.1 The seven social-ecological systems in the study area, summary statistics of interview data and some emerging themes

Social-ecological system	No. of respondents	Sex ratio (M:F)	Access to land (%)	Examples of respondent livelihood activities	Indicative themes emerging from interviews
Rain-fed agriculture	6	100:0	100	Agriculture Chicken farming Decorating business	Land as security and/or route out of poverty Breakdown of intra-community moral economy
Irrigated agriculture	16	56:44	56	Agricultural labourer Farmer (owner and mortgaged) Day labour Fertiliser business Remittances (overseas) Rickshaw puller White fish retailer	Shortage of off-farm livelihoods Differentiated access to irrigation Agriculture-shrimp land conflict
Freshwater prawn aquaculture	8	50:50	29	Day labourer Shrimp pond owner River fishing Sharecrops Snail crusher	Negative impacts on ecosystem service diversity Proximity to coast and cyclone impacts Influence of outside investors
Brackish shrimp aquaculture	5	60:40	40	Agriculture (owner and leased) Shrimp pond owner Day labour	Support services to shrimp
Charland	11	73:27	45	Day labour Fishing Farmer (owner and sharecropped) Tea seller Village doctor	Cultivation for subsistence Insecure land tenure Constant mobility Reliance on networks and patrons Loss of land as a cause of poverty

Sundarbans (SB) dependent	10	55:45	27	Agricultural labourer Shrimp larvae collector Fish drying Fishing (offshore and at beach) Fish trader Shop Taxi driver Farmer	Highly seasonal incomes High levels of livelihood and personal insecurity Permit systems difficult to navigate and/or ineffective
Coastal periphery	11	70:30	22	Shrimp larvae collector Crab collector (SB) Delivers shrimp Firewood collector (SB) Fishing (offshore and SB)	Highly seasonal livelihoods Long-term recovery from cyclone Sidr in 2009 Debt relations in accessing fisheries

22.4 Poverty Ecosystem Service Dynamics by Social-Ecological System

22.4.1 Agriculture

In agricultural regions, two distinct SESs emerged from the interviews based on the presence or absence of irrigation infrastructure. These two systems were kept analytically separate because (i) the cost of, and access to, irrigation serves to deepen inequalities between the landed and landless, (ii) multiple cropping seasons impact on the value of produce, labour opportunities, and associated mobility, (iii) each system has a differentiated vulnerability to external pressures such as fluctuating market prices and climate change (e.g. through an ability to withstand drought or alternate between crops), and (iv) the geographic location; irrigated areas tended to be those in direct land conflict with brackish shrimp cultivation. However, some common themes emerged across interviews in both types of agricultural zone. These include access to land for food security and/or route out of poverty, breakdown of intra-community moral economy in rural villages, and a shortage of off-farm livelihoods for the large landless population.

Large landowners interviewed had leveraged their assets to start profitable business activities such as supplying organic fertiliser, painting and decorating, intensive chicken rearing, educate their families so that they could take up professional jobs, or fund international labour migration with resulting remittances. This supports similar findings from Bangladesh (Tourfique 2002). However, there is another kind of landowner, constrained by the status in society conferred by land ownership but lacking sufficient assets to access profitable off-farm activities: *"Being part of the middle class society has closed the path of asking help from someone. My father was elected member [local politician] twice, this has given us an illusion of aristocracy and also prevents me from taking any small jobs."* (56).

Nearly half the rural population in Bangladesh is landless (Saha 2002). Landless interviewees mentioned that the landed could use loans to further consolidate their wealth. Agricultural banks and NGO loans require land as collateral: *"if we give them our land documents temporarily"* (65).

Thus, loans are not so easily accessible for the landless. Furthermore, they felt landowners could use their loans on economic activities rather than subsistence agriculture because they can grow their own food. This is supported by statements from land owners who avoid selling their land to maintain the subsistence security it confers. One farmer said, he *"would no more sell his land than kill a crazy son"* (48); meaning although agricultural activities are not always profitable, he would be reluctant to sell land.

Some respondents mention that a strong intra-community moral economy is breaking down: *"Everyone is guided by his or her own judgement. All of us think why shall we be guided by others?"* (41), but the role of large landowners as patrons to the poor continues. The poorest landless households, when in trouble, *"seek help from other rich people. If we tell them that we're in big trouble, they help as much as they can"* (42). Sometimes this assistance to the poor also takes the form of a place to stay, in return for work and political support: *"Since we live on their land, we must do as they say"* (45). Respected village members such as teachers and war veterans still play a role in mediator of disputes, even if the role of the informal village head has diminished (56, 5).

Seasonality remains a driving factor. The landless respondents described the wet season as the most difficult because reduced opportunities for agricultural labour combined with fewer off-farm activities and opportunities to buy food. One respondent said: *"(We) have needs all year round- but the rainy season is the worst time for us…at that time we don't have work like loading-unloading [at the border with India]"* (42). Another said: *"Rainwater wets everything. They can't sow paddy and can't sell that"* (57). Although there is a year-round shortage of opportunities for the landless: *"Do your sons go outside for work?"* *"Of course. Otherwise what will we eat?"* (57).

22.4.2   Aquaculture

Aquaculture ponds (*ghers*) can be used to cultivate shrimp, prawn, white fish, or rice (or a combination of them all) depending on the elevation of the plot, the salinity of surface water, access to irrigation, precipitation, and season. An economy exists supporting the *ghers* by collecting shrimp post-larvae from natural waters (31), providing fertiliser and other inputs such as snail meat (35), and delivering supplies to the pond owners (7, 8). This goes some way to remedying the loss of farm labouring opportunities, caused by low labour requirements on *ghers*, particularly shrimp farming (Swapan and Gavin 2011).

Aquaculture is associated with a loss of open-access resources. In brackish water shrimp areas, salinity precludes other ecosystem services. However, freshwater prawn can be cultivated in rotation with other crops, vegetables can be grown on pond walls, and there is sufficient plant life for animal fodder. Freshwater prawn cultivation is also associated with a loss of open-access wetland areas which has provided security of income for those with land but reduced open-access resources for the landless (5): "*Earlier we had the freedom for fishing in the beel [a wetland], now we earn more [from aquaculture]*" (32). Furthermore, nets used to collect the shrimp larvae to supply the ponds are perceived almost universally to have led to a decrease in fish stocks due to the by-catch. These nets are illegal but bans against them are not followed or well enforced.

Distinct differences can be observed between areas that are dominated by brackish shrimp aquaculture and areas that are dominated by freshwater prawn that, in turn, differ from those of farmers who concentrate solely on rice cultivation. Therefore, two different systems were defined, one for dominantly brackish shrimp cultivation, another for freshwater prawn dominated areas. The systems are separated based on: (i) forms of investment and issues of land rights, (ii) impacts on the poorest, and (iii) geographic location of the areas and associated stresses.

Forms of investment in, and access to, aquaculture practices differ. Expansion of brackish water shrimp aquaculture has been driven by external investors: "*The owners of those large ghers were people coming from outside the area*" (39). Thus, benefits tend to accrue outside the area to absentee landlords. Land conflicts arise, as rice farmers adjacent to shrimp areas have no choice but to convert to shrimp due to the negative effects of saline water intrusion on the productivity of their crops (see Faruque et al. 2017). In freshwater prawn areas, external investors and absentee landlords were not mentioned. What did arise in conversation were the large debts to set up the ponds for those who own land and different prices of land inside the polder (where cultivation can occur) and outside.

The issues raised by the most marginalised households differ between the two systems. In brackish shrimp areas the poorer and landless respondents stated that their well-being would be improved by the return of a local landlord cultivating crops to whom they could go in emergences for food, loans, and help with medical costs (62). The return of agriculture would also provide bare subsistence for the ultra-poor: one destitute respondent mentioned that if there were still agriculture in the surrounding areas she could collect rice that had fallen to the ground during harvesting (9). More of the *gher* owners in the freshwater shrimp areas are owner-cultivators, and a wider range of crops can be cultivated (including through sharecropping opportunities), perhaps why such issues were not raised by respondents from this area.

Brackish shrimp areas are more remote from health services, markets, and alternative livelihood sources and more exposed to risks of storm surges and cyclones. However, they are also very close to the Sundarbans mangrove forest and the coast with easy access to offshore fisheries. This means that the poor are able to diversify into these alternative but precarious livelihoods of fishing, shrimp larvae collection, and forest collection. The case is different in the freshwater prawn areas; there are few open-access resources to support those without access to private property. Canals belong to the government or are on private land so the landless are unable to fish in them (35). Thus job opportunities are found in the nearby city of Khulna.

22.4.3   Riverine Areas and Charlands

Riverbank erosion occurs across the study area, often exacerbated by man-made river diversion or by malfunctioning systems of dykes and sluice gates due to siltation. By contrast in the eastern part of the delta, adjacent to the Meghna River, high levels of erosion and accretion lead to highly dynamic char islands. These stretches of land are, in effect, sandbanks in the river, attached to or detached from river banks, but with their own highly mobile and sometimes marginal populations.

The physical dynamics of chars has led to them being an important SES in their own right, recognised throughout the south Asian region and described as having "livelihoods defined by water" and populations living constantly with risk of displacement (Lahiri-Dutt and Samanta 2013). The principal characteristics of Charland SESs, as articulated by interview respondents, are (i) highly seasonal income and seasonal shifts in income between fishing and rain-fed agriculture, (ii) loss of land resulting in sudden changes in material well-being, (iii) high mobility of landless households, (iv) high reliance on richer patrons to help landless households, and (v) high insecurity of land tenure for those whose land has been submerged with the constant threat of land-grabbing.

The data from interviews with char dwellers reveal a high seasonality of income between crops in the wet season and fishing in the dry season. Cultivation is principally for subsistence. Charlands share characteristics common to agriculture and fisheries SESs. For example, sharecropping for the landless and mortgaging land to others when in financial difficulty and loans for fishing equipment that are paid back through a percentage of the profit on the catch.

Insecure land tenure was perceived by char dwellers as a critical issue in accessing ecosystem services. When land is eroded (and thus submerged) property rights remain with the owner if sediment accretes above the water level within 30 years. If not, the land returns to government ownership and can be redistributed to the landless. However, the interviewees mentioned that land can be appropriated by more influential people before it is reclaimed by the family (75), or names are changed on title deeds within government offices.

Associated with the constant erosion of riverbanks and islands is constant mobility. Households living on unclaimed strips of land on dykes and riverbanks find themselves continually moving as the riverbank erodes. In turn, constant mobility and the constant search for new land on which to build a household can create a dependence on wealthier neighbours and relations for support and patron-client relationships. One landless respondent who had been forced to move multiple times was able to generate income by raising cattle for a wealthier resident (73). Another respondent was allowed to live on a relative's land in return for work (75).

Loss of land is associated with change in livelihood. A person who has lost all their agricultural land to erosion must find alternative sources of income; respondents often mentioned that people become fishermen (e.g. 63). Loss of land through river erosion is also associated with a fall in income. When asked who were the poorest in the village, one respondent replied it was those who had lost their land to river erosion (27).

22.4.4 Sundarban Mangrove Dependence

This SES takes into account the people living directly adjacent to the Sundarbans mangrove forest. People fish on its margins and interior, and enter on a daily or weekly basis to collect firewood, honey, fish, crabs, and thatching among other resources. The Sundarbans is a nature reserve so people are prohibited from living within its boundaries. Households therefore live on its border among the brackish water shrimp ponds. Three themes emerged from the interviews with mangrove collectors: (i) highly seasonal incomes, (ii) high levels of livelihood and personal insecurity, and (iii) systems of permits that are difficult to navigate and/or ineffective.

Interviews with these natural resource users revealed livelihoods highly affected by seasonality. Wet season rains make collection of firewood difficult (9); in the dry season there is a scarcity of freshwater and people have to drink salty water (9); fishermen fish in different locations in different seasons (8, 14); resource collection is banned during certain periods (68); and the quality of resources (e.g. the size of crabs) changes with the season (11).

Some respondents move between different resource and day labour opportunities, while others exhibit extremely low livelihood mobility due to a strong livelihood-based identity (e.g. traditional fishermen), or a lack of human capital. For example, despite collection of firewood being prohibited, a woman of around 50 still went into the forest every day. In doing so, she faced the threat of both being detected and physical punished by forest guards as well as exposure to extreme weather and natural hazards (9).

This theme of insecurity commonly arose in interviews. Not just in terms of a stable income source but also in terms of the potential of physical harm from encounters with pirates, forest guards or wild animals, physically demanding working conditions, and the periodic threat of cyclones. This physical insecurity is partly a result of a lack of alternative livelihood sources outside the forest. One resource collector said: "*You know in his heart nobody wants to go to the Sundarbans*" because of fear (68). The loss of agriculture in the area to make way for shrimp *ghers* (described in Sect. 22.4.2) has removed a key alternative livelihood (10) increasing reliance on the forest resources to fill income gaps.

Another theme frequently occurring in this set of interviews is the system of permits and moratoriums on resource collection in the forest and the efficacy of the government forest regulators in enforcing them. Some interviewees collected wood despite it being illegal while others paid for the permits and respected periods when resource collection is prohibited. Respondents perceive a decrease in the quantity and diversity of fish catch and blame people using fine nets (14) as well as organised gangs collecting fish illegally in ways that are destructive to other species (e.g. poison—15). That is to say, because they perceive that, forest regulators are not enforcing rules effectively.

22.4.5   Offshore Fisheries and the Coastal Periphery

It is difficult to geographically define an SES based on fisheries. Reliance on local inland fisheries (in *beels*, canals, and rivers) is ubiquitous across the study area. People will also travel from inland areas to access offshore fishing activities. There are fishing villages (often majority Hindu), where fishing is a traditional livelihood and closely linked to identity, where men will fish from nearby rivers during the wet season, and offshore during the dry season; these villages can be found across the study area. Also, Sundarbans fishermen often live between the brackish shrimp *ghers*, and coastal fishermen live among agricultural land to take advantage of the subsistence agricultural opportunities it offers.

However, those living adjacent to the coast have easier access to the resources the ocean offers. Thus, this research defined a SES based on those areas with direct access to the Bay of Bengal. Three aspects of this system are commonly highlighted by interviewees: (i) seasonal livelihoods, (ii) the long-term recovery from Cyclone Sidr in 2009, and (iii) the role of debt relations in accessing fisheries.

This SES is characterised by seasonal changes in livelihoods. Some fishermen alternative between species with different profitabilities: Hilsa (*Tenualosa ilisha*), in summer, and less profitable fish in other seasons or Hilsa in summer and shrimp post-larvae in the winter (31). Others alternate between livelihoods: one household collected forest resources from small patches of mangroves when food stores from subsistence agriculture were low (30).

Cyclone Sidr in 2009, and its continued impact on livelihoods, is a key feature for this system. People are still recovering from loss of assets and land damaged by salinity. One household had invested profits from fishing into cattle that were lost during Cyclone Sidr (54), a businessman lost his stock of dried fish that was on the beach when the cyclone hit (28).

Another key feature of this system is loans as a means to access fisheries. Loans are taken to access equipment, boats, and supplies, and are paid back as a proportion of the profit on the catch made. There is a more complex system for larger boats where groups of up to eight men travel offshore. These loans are accessed in advance per season and as a group through cooperation of the crew of the boat. Profits are subsequently apportioned to the crew and money lender (29, 68). These loans are associated with the accumulation of debt, and debt bondage as the catch is often insufficient to pay back the advance (54, 68).

However, while continuation of the loan system is not perceived as a positive aspect of household economies, some respondents felt there was no alternative (e.g. 29). Respondents also mentioned that loans have allowed people to access fisheries more easily as, whereas in the past men would wade off the beach up to their necks to catch fish, now they have boats, nets, and diesel and as such, access to more profitable species of fish that are found further offshore (29).

22.5   Conclusion

The timing and nature of ecosystem services across the study area give rise to different livelihood opportunities, means of access to ecosystem services, and coping mechanisms. This research demonstrates how these patterns of ecosystem service and human-environment interactions form seven distinct SESs. Each of these systems has, in effect, different barriers and enablers of access to ecosystem services and hence potential pathways out of poverty. The seven social-ecological systems demonstrate a large spatial range of ecosystem services as well as high variation throughout the year. Given the prevalence of diverse livelihoods by individuals and households described within these systems, this analysis suggests that policy interventions need to take account of that diversification and also be tailored to the dominant mechanisms and institutions within diverse social-ecological systems.

The research has highlighted that ecosystem service access is critical for subsistence of poor parts of delta populations across all the seven systems and that these ecosystem services are critical in poverty prevention for marginal sections of society. Yet there are potential trade-offs between ecosystem services and the need for subsistence and for finding routes out of poverty. Income from shrimp post-larvae collection, for example, is a key safety net for the poorest when they lose land or livelihoods. However, the collection of this shrimp has negative impacts on fish catch. Fishing communities retain solidarity and identity from the traditions and institutions of fishing. Yet with successive generations entering fishing, the resource is stressed. The situation is further complicated by the management of floods via sluice gates for the benefit of agriculture. This limits the ability of fish fry to enter the polder early in the breeding season, and thus maintain fish stocks.

Across all seven social-ecological systems highlighted here, some social trends and mechanisms are common. A crucial example is that the part of each population most directly connected to the ecosystem services are least able to benefit from, their presence. For fishers, for example, surplus accrues to the money lenders and traders. For landless labourers, profits accrue to landowners through rent or sharecropping. Hence, a common issue across all seven systems is to design mechanisms by which the poorest populations retain value and benefits as the ecosystem service travels up the commodity chain. Such interventions would be a significant step in enabling allowing ecosystem services to alleviate poverty and building the sustainability of the diverse social and ecological circumstances within delta regions.

Note

1. http://reshare.ukdataservice.ac.uk/852356/

References

Adams, H., and W.N. Adger. 2016. *Mechanisms and dynamics of wellbeing-ecosystem service links in the south-west coastal zone of Bangladesh.* UK Data Service Reshare. https://doi.org/10.5255/UKDA-SN-852356, https://doi.org/10.5255/UKDA-SN-852356.

Adams, H., W.N. Adger, H. Huq, M. Rahman, and M. Salehin. 2013. *Transformations in land use in the south-west coastal zone of Bangladesh: Resilience and reversibility under environmental change.* Proceedings of Transformation in a Changing Climate International Conference, University of Oslo, Oslo.

Adams, H., W.N. Adger, S. Ahmad, A. Ahmed, D. Begum, A.N. Lázár, Z. Matthews, M.M. Rahman, and P.K. Streatfield. 2016. Spatial and temporal dynamics of multidimensional well-being, livelihoods and ecosystem services in coastal Bangladesh. *Scientific Data* 3: 160094. https://doi.org/10.1038/sdata.2016.94.

Allison, E.H., and F. Ellis. 2001. The livelihoods approach and management of small-scale fisheries. *Marine Policy* 25 (5): 377–388. https://doi.org/10.1016/S0308-597X(01)00023-9.

Barbier, E.B., S.D. Hacker, C. Kennedy, E.W. Koch, A.C. Stier, and B.R. Silliman. 2011. The value of estuarine and coastal ecosystem services. *Ecological Monographs* 81 (2): 169–193. https://doi.org/10.1890/10-1510.1.

Cinner, J.E., T.M. Daw, T.R. McClanahan, N. Muthiga, C. Abunge, S. Hamed, B. Mwaka, A. Rabearisoa, A. Wamukota, E. Fisher, and N. Jiddawi. 2012. Transitions toward co-management: The process of marine resource management devolution in three east African countries. *Global Environmental Change* 22 (3): 651–658. https://doi.org/10.1016/j.gloenvcha.2012.03.002.

Cote, M., and A.J. Nightingale. 2012. Resilience thinking meets social theory: Situating social change in socio-ecological systems (SES) research. *Progress in Human Geography* 36 (4): 475–489. https://doi.org/10.1177/0309132511425708.

Crane, T.A. 2010. Of models and meanings: Cultural resilience in socio-ecological systems. *Ecology and Society* 15 (4): 19. www.ecologyandsociety.org/vol15/iss4/art19/.

Davis, P. 2009. Poverty in time: Exploring poverty dynamics from life history interviews in Bangladesh. In *Poverty dynamics: Interdisciplinary perspectives*, ed. T. Addison, D. Hulme, and R. Kanbur, 154–182. Oxford: Oxford University Press.

De Aranzabal, I., M.F. Schmitz, P. Aquilera, and F.D. Pineda. 2008. Modelling of landscape changes derived from the dynamics of socio-ecological systems – A case of study in a semiarid Mediterranean landscape. *Ecological Indicators* 8 (5): 672–685. https://doi.org/10.1016/j.ecolind.2007.11.003.

Fabinyi, M., L. Evans, and S.J. Foale. 2014. Social-ecological systems, social diversity, and power: Insights from anthropology and political ecology. *Ecology and Society* 19 (4): 28. https://doi.org/10.5751/ES-07029-190428.

Faruque, G., R.H. Sarwer, M. Karim, M. Phillips, W.J. Collis, B. Belton, and L. Kassam. 2017. The evolution of aquatic agricultural systems in Southwest Bangladesh in response to salinity and other drivers of change. *International Journal of Agricultural Sustainability* 15 (2): 185–207. https://doi.org/10.108 0/14735903.2016.1193424.

Hill, R., C. Grant, M. George, C.J. Robinson, S. Jackson, and N. Abel. 2012. A typology of indigenous engagement in Australian environmental management: Implications for knowledge integration and social-ecological system sustainability. *Ecology and Society* 17 (1): 23. https://doi.org/10.5751/es-04587-170123.

Karpouzoglou, T., A. Dewulf, and J. Clark. 2016. Advancing adaptive governance of social-ecological systems through theoretical multiplicity. *Environmental Science and Policy* 57: 1–9. https://doi.org/10.1016/j.envsci.2015.11.011.

Lahiri-Dutt, K., and G. Samanta. 2013. *Dancing with the river: People and life on the chars of South Asia*. New Haven: Yale University Press.

Levin, P.S., M.J. Fogarty, S.A. Murawski, and D. Fluharty. 2009. Integrated ecosystem assessments: Developing the scientific basis for ecosystem-based management of the ocean. *PLoS Biology* 7 (1): 23–28. https://doi.org/10.1371/journal.pbio.1000014.

Nagendra, H., and E. Ostrom. 2014. Applying the social-ecological system framework to the diagnosis of urban lake commons in Bangalore, India. *Ecology and Society* 19 (2): 67. https://doi.org/10.5751/es-06582-190267.

Ostrom, E. 1990. *Governing the commons: The evolution of institutions for collective action*. Cambridge: Cambridge University Press.

Rodriguez, J.P., T.D. Beard, E.M. Bennett, G.S. Cumming, S.J. Cork, J. Agard, A.P. Dobson, and G.D. Peterson. 2006. Trade-offs across space, time, and ecosystem services. *Ecology and Society* 11 (1): 28. www.ecologyandsociety.org/vol11/iss1/art28/.

Saha, B.K. 2002. Rural development trends: What the statistics say. In *Hands not land: How livelihoods are changing in rural Bangladesh*, ed. K.A. Toufique and C. Turton. Dhaka: Bangladesh Institute of Development Studies, Dhaka and Department for International Development, London.

Swapan, M.S.H., and M. Gavin. 2011. A desert in the delta: Participatory assessment of changing livelihoods induced by commercial shrimp farming in Southwest Bangladesh. *Ocean and Coastal Management* 54 (1): 45–54. https://doi.org/10.1016/j.ocecoaman.2010.10.011.

Tourfique, K.A. 2002. Agricultural and non-agricultural livelihoods in rural Bangladesh: A relationship in flux. In *Hands not land: How livelihoods are changing in rural Bangladesh*, ed. K.A. Toufique and C. Turton. Dhaka: Bangladesh Institute of Development Studies, Dhaka and Department for International Development, London.

Wood, G. 2003. Staying secure, staying poor: The "Faustian bargain". *World Development* 31 (3): 455–471. https://doi.org/10.1016/s0305-750x(02) 00213-9.

23

Characterising Associations between Poverty and Ecosystem Services

Helen Adams, W. Neil Adger, Sate Ahmad, Ali Ahmed,
Dilruba Begum, Mark Chan, Attila N. Lázár,
Zoe Matthews, Mohammed Mofizur Rahman,
and Peter Kim Streatfield

23.1 Introduction

In order to develop insight and potential actions to co-manage ecosystem services for both healthier ecosystem functioning and to alleviate poverty in natural resource-dependent communities within deltas, it is necessary to understand how poverty is manifest and the level of dependence of populations on the ecosystems and social-ecological systems in which they live and work.

One strategy to develop this insight involves the direct observation of how people live, their own management of the resources around them,

H. Adams (✉) • M. Chan
Department of Geography, King's College London, London, UK

W. Neil Adger
Geography, College of Life and Environmental Sciences,
University of Exeter, Exeter, UK

S. Ahmad
Faculty of Agricultural and Environmental Sciences, University of Rostock,
Rostock, Germany

© The Author(s) 2018
R. J. Nicholls et al. (eds.), *Ecosystem Services for Well-Being in Deltas*,
https://doi.org/10.1007/978-3-319-71093-8_23

the outcomes of that interaction with ecosystem services in terms of material well-being and their health and their perceptions of those relationships. Hence this research uses social science survey techniques to generate extensive data on the ways in which households use ecosystem services to generate well-being as part of diverse rural livelihoods. In doing so, the survey also provides a quantitative baseline understanding that is also essential to the integrated model (see Chap. 28). Simulations of winners and losers of future interventions on ecosystem services are thus based on real-life starting conditions.

There is significant value in generating primary observational data on ecosystem service use. Alternative sources of social data on life and livelihood include national census data and generalised livelihood surveys, such as the standard Household Income and Expenditure Surveys (HIES) carried out in low-income countries throughout the world (Deaton 1997). Census data are limited to demographic variables, typically with the main occupations of adults within households, and as such give an economic picture of populations. They are less useful to demonstrate where and how people interact with their local environments—the objective of this work (see Chap. 1). Household surveys typically provide detailed analysis of consumption patterns, expenditure patterns and economic activities, as well as demographic

A. Ahmed • D. Begum
Climate Change and Health, International Center for Diarrheal Disease
Research, Bangladesh (icddr,b), Dhaka, Bangladesh

A. N. Lázár
Faculty of Engineering and the Environment and Tyndall Centre for Climate
Change Research, University of Southampton, Southampton, UK

Z. Matthews
Social Statistics and Demography, Faculty of Social, Human and Mathematical
Sciences, University of Southampton, Southampton, UK

M. M. Rahman
International Center for Diarrheal Disease Research, Bangladesh (icddr,b),
Dhaka, Bangladesh

P. K. Streatfield
Formerly of the International Center for Diarrheal Disease Research,
Bangladesh (icddr,b), Dhaka, Bangladesh

variables. They do so using nationally representative samples of households from which inferences concerning national trends can be drawn, and the patterns of poverty and well-being are developed. Use of the Bangladesh HIES by Szabo et al. (2016) showed how food security at the household level is negatively associated with creeping salinity in the study area. Yet these data sources are limited in information about the direct benefits people derive from their ecosystems, about their mobility and other responses, and on the health and well-being of populations. Hence the survey here provides a unique set of insights on human-environment relations in this delta.

As highlighted in Chap. 21, ecosystem services are highly variable in space and time. This bespoke survey therefore builds in temporal variability by repeat interviews in three waves over a full calendar year: the analysis constructs detailed livelihood calendars. A further challenge for human-environment models is the multi-dimensional and contested nature of poverty, both as manifest in lack of material assets, an absence of health and also as a lived experience (Baulch 1996). Hence the survey is comprehensive in collecting specific variables that facilitate interdisciplinary analyses and consideration of material and subjective measurements of well-being alongside use of ecosystem services and livelihood diversity. It allows multilevel analysis and intra-household analyses: variables relate to individual men and women and to whole households.

This chapter first briefly outlines the survey methodology and implementation (Sect. 23.2). Section 23.3 summarises the data available from the household survey; it highlights unique variables and aspects of the survey and those that are comparable with other standard datasets. Section 23.4 describes each of the publicly available datasets associated with the household survey, illustrated with selected descriptive statistics. The publicly available datasets are land cover data by Union for the field area in Khulna and Barisal Divisions, household listing data of 9,300 households, a household roster dataset that presents separately the basic data of the 8,000 people living in the households surveyed and three rounds of household survey data for approximately 1,500 separate households taking into account attrition between the three rounds. The chapter closes with a reflection on the reuse potential of the dataset.

All data are available to download from the ReShare UK-based online data repository.[1] The data are accompanied by English and Bengali versions of the questionnaire, as well as a glossary of terms used in the questionnaire. The survey design process itself is described in more detail in Adams et al. (2016).

23.2   Methodology

This section explains how the survey was carried out. First, it describes the sampling strategy and how the concept of social-ecological systems was operationalised and used to stratify the sample; social-ecological systems were integrated from the beginning. Second, it describes the household listing process, required to ensure a random sampling from within villages. Third, it describes the process of implementing the survey.

23.2.1   Sampling Strategy

Households were identified through systematic random sampling. The sample was stratified initially by social-ecological system (SES): rain-fed agriculture, irrigated agriculture, freshwater prawn aquaculture, brackish shrimp aquaculture, coastal aquaculture, riverine areas including eroding islands, and mangrove dependence. These were identified through expert elicitation and land cover maps and verified through semi-structured interviews (see Chap. 22). The sample was not driven by livelihood or ecosystem service use as these changed throughout the year and thus were not compatible with a seasonal approach.

In order to create a sampling frame, a land cover map is overlaid with an administrative map and Unions assigned to a SES. Agricultural and aquaculture systems could be directly assigned based on 80 per cent minimum land coverage per Union. Riverine, marine and Sundarbans systems were assigned based on contiguous boundaries with the associated feature. Some Unions do not have a clearly dominating land use. These Unions were excluded from the sampling process. Table 23.1 shows the number of Unions included in the sampling belonging to each of the strata.

Systematic random sampling was used to select three Unions from each SES and three *Mouzas* from each Union. A segment of approximately 125 households was listed in each *Mouza* to randomly select the 21 households that were interviewed. Households were eligible if both a man (aged 18–54) and women (aged 15–49) were present. The target

Table 23.1 Number of Unions sampled included in each social-ecological system

Social-ecological system	No. of Unions
Rain-fed agriculture	223
Irrigated agriculture	29
Freshwater prawn aquaculture	11
Brackish water shrimp aquaculture	31
Charland (riverine)	17
Sundarban dependent	24
Coastal periphery	11

respondent for the survey was the main earner, who completed the structured questionnaire. Information on global satisfaction of life, anthropometry (height and weight) and blood pressure was collected from both a male and a female member of the selected household.

The sample size was calculated based on a head count ratio of poverty prevalence, poverty defined by the inability of households to meet the costs of basic food needs (BBS 2011). In Barisal 27 per cent of people are below this poverty line, and in Khulna 15 per cent and so a population weighted average of 22 per cent was calculated. Ten per cent was added for potentially non-responses, and an additional ten per cent was added to take into account attrition between rounds (although actual attrition rates were below five per cent). Further information on sampling strategy can be found in Adams et al. (2016).

23.2.2   Survey Implementation

The survey was administered to selected households three times: first in June 2014, then over October to November 2014 and finally in March 2015, each time with a four-month recall period. Thus the data covers the period from February 2014 to February 2015. Attrition rates were low: 1,586 households were initially selected; this fell to 1,516 in the second round, and came back up to 1,531 in the third round. However, when all three rounds are considered, 1,478 households have consistent and complete records across the three surveys.

23.3   Survey Data

The survey contains up to nearly 3,000 potential variables corresponding to hundreds of different survey questions contained within 15 different sections. Table 23.2 describes the type of variables that are contained in the survey.

Reuse of the data is facilitated by the interdisciplinary nature of the questions, the number of standard measures of multi-dimensional well-being that can be recreated from the data, and the multilevel nature of the data: it can be disaggregated by season, by social-ecological system, by Union, or by individual.

Questions were also included to allow the creation of standard measures to facilitate direct comparison with other surveys. This includes the variables required to recreate the Progress out of Poverty index[2], the Multi-Dimensional Poverty Index[3] and the FANTA III food diversity score.[4] The survey also contains multiple questions that are also contained within national censuses in order to facilitate comparison and validation.

The items on the asset list (Sect. 2 of the survey) and expenditure (Sects. 13 and 14) were taken from the Bangladesh Household Income and Expenditure 2010 (BBS 2011), with a few additions to take into account the ecosystem service-focus of this research. The global satisfaction with life scale is the same one that is applied in Gallup surveys and

Table 23.2 Metadata summary of the survey data

Design type	Systematic random sampled longitudinal household questionnaire survey
Factor types	Assets, income type and livelihood diversification, agricultural and aquaculture output and expenditure, fisheries and mangrove activities and expenditure, migration, loans, livestock and poultry, homestead forestry, landholdings, shocks and migration strategies, place attachment, perception of environmental quality, household food diversity, household food consumption, non-food expenditure, impacts of oil spill, women's empowerment, height, weight, blood pressure, global satisfaction with life
Sample characteristics	Bangladesh, social-ecological system, Khulna Division, Barisal Division, household, male aged 18–54, female aged 15–49, child aged <5

the UK well-being survey (Evans 2015; Gallup 2015). The set of nine questions answered on Likert scales to measure place attachment has been previously developed, tested and applied within social psychology and human geography (Lewicka 2011; Devine-Wright 2013).

23.4   Datasets Available for Reuse

Each of the datasets described here is openly available for reuse and analysis; some illustrative descriptive statistics are discussed here.

23.4.1   Land Cover Database

In addition to the main survey, an Excel spreadsheet is also available[5] that provides land cover data for each of the Unions in the study area. While this information is used to create a sampling strategy based on SESs, it could also be used to create other land cover-based sampling strategies or to better understand the land use characteristics of Unions in Khulna and Barisal Divisions.

23.4.2   Household Listing

A listing was carried out at the start of the data collection process as a rapid census to determine eligibility of households for the full survey. Thus the data comprises a very limited set of variables. However, the value of this dataset lies in the sample size: 9,327 households were surveyed. Data is available for these households on sex and age of the main earner; primary, secondary and tertiary occupation of the earning member; estimated total monthly income in Bangladesh Taka (BDT); and floor, wall and roof materials.

Figure 23.1 shows the average income by social-ecological system disaggregated by sex of the main household earner. Men consistently earn more than women. The coastal periphery SES is worst for income parity, and the freshwater prawn aquaculture zone is the most favourable. However, the difference in average earnings across the systems is twice as great for women than for men. Average income for male household heads varies

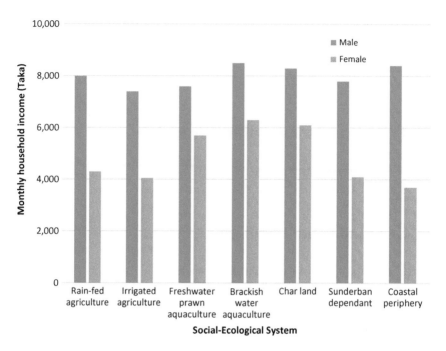

Fig. 23.1 Mean monthly income by social-ecological system and sex from listing data

from 8,508 BDT a month in the riverine system to 7,207 BDT a month in irrigated agricultural SES (a difference of 1,301 BDT). For women, average income varies from 6,447 in the brackish shrimp aquaculture SES to 3,532 in the coastal periphery zone (a difference of 2,915 BDT).

Figure 23.2 shows the most common income sources disaggregated by SES. This includes primary, secondary and tertiary incomes. The five most important income types are shown using two metrics: contribution to total income of the system and the proportion of households involved in this livelihood.

The most common income types vary in ways expected by SES; fishing as an income source is more common in the coastal and riverine zones, pond-based aquaculture is only present in the aquaculture zones (freshwater prawn and brackish shrimp), and agriculture (rain-fed or irrigated) exists across all zones. Interesting findings emerge in the differences between the most common and the most lucrative income types.

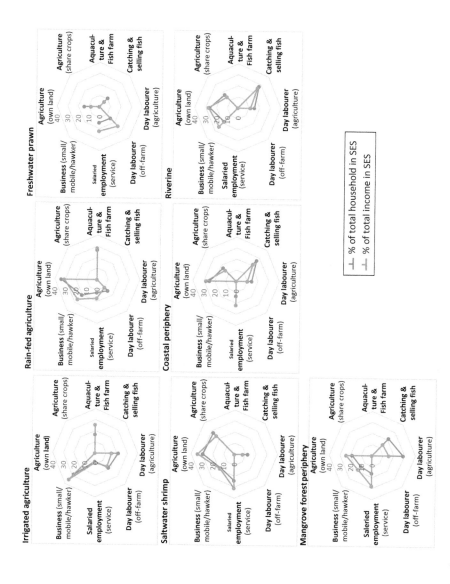

Fig. 23.2 Five most common income sources in terms of contribution to household income, by social-ecological system

In the Mangrove dependent SES, for example, professional salaried jobs are most important economically, but most households are engaged as day labourers.

23.4.3 The Household Roster

The household roster information is available as a separate dataset so the basic characteristics of all individuals included in the survey can be easily analysed. Although the survey was administered to 1,586 households, this comprises 7,993 different women, men and children. Information on whether the person was a visitor or permanent household member, relationship to household head, age, marital status, school attendance, highest level of education reached, number of times the person has attended school, employment status, whether the person is working away and birth place of person (by *Upazila* and urban/rural) is collected for all three rounds. Figure 23.3 shows the average household size in the survey,

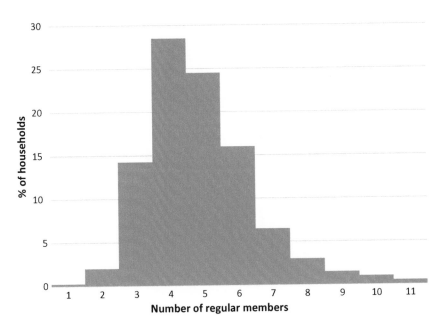

Fig. 23.3 Household size distribution

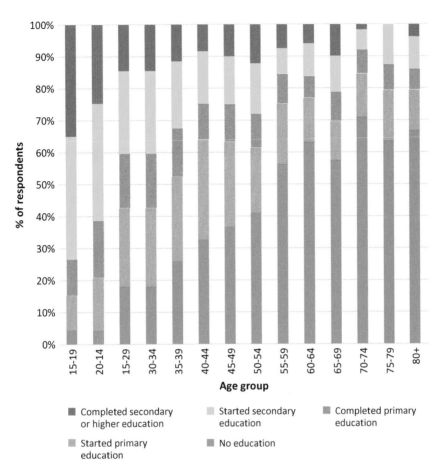

Fig. 23.4 Educational attainment by age group

and Fig. 23.4 shows educational attainment of individuals by age group. Most households had four or five members and educational attainment increases with younger age groups, with people aged 55 and over least likely to have had any education.

Information on place of birth was included in the roster in addition to standard questions on relation to household head, age, education level and income. Analysis of the responses on place of birth show that almost everyone surveyed (90 per cent) was born in the same *Upazila* in which they were now living; therefore this is not an area of in-migration. The approximately ten per cent of the population born outside the *Upazila*, was made up of women (613 people born outside the *Upazila* compared to 132 men), suggesting that in-migration, where it does occur, is predominantly a result of marriage.

As the household roster was carried out for each round of the survey, a window into changing intra-household dynamics is provided. Table 23.3, for example, shows changes in the percentage of men and women working and the proportion of those working who are migrating outside the village to access these opportunities. A tenth of the number of women are working compared to men. However, in two of the rounds, a very similar proportion of working women as working men were seeking livelihood opportunities outside the village.

23.4.4   Seasonal Household Survey Variables

There are three main categories of variable types in the household survey: those with an aim of measuring multi-dimensional poverty, those measuring ecosystem service use and those recording coping strategies used by the household to cope with variability in income.

23.4.4.1   Measuring Multi-dimensional Poverty

Much of the survey instrument is dedicated to the measurement of multi-dimensional well-being. Various sections of the survey measure material poverty at the household level: Section 2 of the survey records household

Table 23.3 Change in the percentage of men and women working within and outside the village by survey round

Round/ season	Household members working within the village	Household members working outside the village
First	54% men	16% men
	7% women	9% women
Second	54% men	20% men
	4% women	19% women
Third	54% men	18% men
	4% women	19% women

assets, Sect. 3 income type and livelihood diversification, Sect. 9D collects information on landholdings, Sect. 13 on household food diversity and food expenditure and Section 14 non-food expenditure.

Information is also collected to record health status at the individual level. Height, weight and blood pressure of a man, woman and child under five was measured in each household. Chapter 27 focuses on the health outcomes of the survey. Questions on subjective well-being, through a ten-point scale on satisfaction with life, were also asked to men and women separately in each household (Sect. 15 of the survey). Finally, in the third round of the survey, a section was included to measure levels of perceived empowerment of women in the household.

23.4.4.2   Ecosystem Use and Quality

Section 4 of the survey contains four sections on agriculture, aquaculture, fisheries and mangrove activities, including species or variety, productivity, price and profit made. Section 9 contains questions on livestock and poultry and homestead forestry. In addition, questions on perceptions of different dimensions of the quality of the natural environment are asked (Sect. 12). Figures 23.5 and 23.6 provide two examples from this section: on perceptions of water quantity and quality by SES and season. Perceptions change with season and with different SES. While the aquifers from which most water is drawn do not follow the SES boundaries, perceptions of poor water quality are highest where problems of salinity intrusion are most acute. Due to an oil spill occurring in the Sundarbans between the second and third rounds, a module on the impacts of the oil spill was added to the final round.

23.4.4.3   Coping Strategies

The survey instrument contains various sections dedicated to understanding coping strategies. Three sections are dedicated to understanding mobility: past migration and current migration strategies of the household (Sects. 5 and 6 of the survey) and place attachment of the main respondent (Sect. 11). Section 7 is devoted to understanding the use of loans and Sect. 10 examines costs and responses to specific shocks. Table 23.4 shows the source of loans by season. Formal loans from agricultural banks

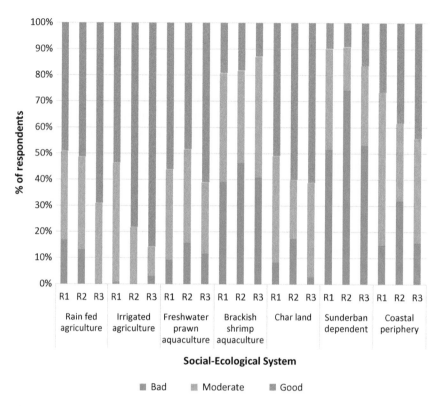

Fig. 23.5 Respondent perceptions of availability of drinking water by social-ecological system throughout the seasons (i.e. rounds—R1–R3)

and non-governmental organisations are most commonly stated as sources of loans, reported approximately five times as often as loans from friends and family are reported. Loans from money lenders are reported slightly less, although people may not want to share the degree to which they owe money in this form.

23.4.4.4   Seasonal Changes in Productivity of Ecosystems

The seasonal nature of the survey and four-month recall period allows for the reconstruction of seasonal calendars for the different ecosystem services. For example, Figs. 23.7 and 23.8 show the total production of

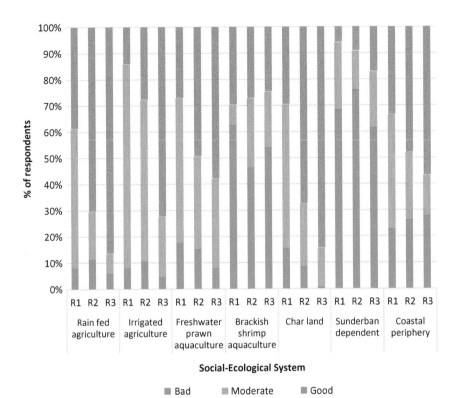

Fig. 23.6 Respondent perceptions of salinity of drinking water by social-ecological system in each season (i.e. rounds – R1–R3)

Table 23.4 Number of households taking loans from different sources by survey round

Loan type	Number of households		
	Round 1	Round 2	Round 3
Formal loans only	305	251	392
Informal loans only	42	26	41
Kinship loans only	52	48	66
Loans from patrons	0	0	1
Formal and Informal loans	5	4	16
Formal and Kinship loans	7	5	15
Informal and Kinship loans	5	1	5

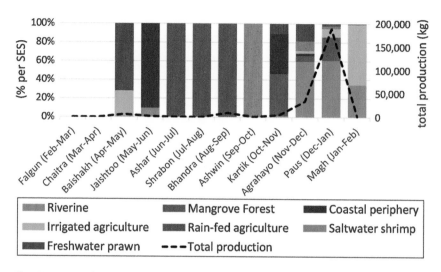

Fig. 23.7 Total *Aman* rice production in kilograms by social-ecological system and month

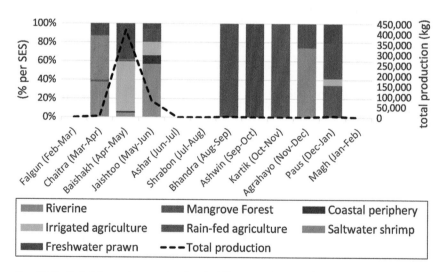

Fig. 23.8 Total *Boro* rice production in kilograms by social-ecological system and month

Aman and *Boro* rice varieties by month for each of the SESs. *Aman* rice is typically grown during the wet, monsoon season and harvested during the winter; however, it can be grown in small quantities in other months as well. Surprisingly, the Charland SES produces the most *Aman* rice, followed by the mangrove forest and finally the rain-fed agriculture SES. *Boro* rice requires irrigation during the dry season and typically harvested during spring. Not surprisingly, the irrigated agriculture SES produces the most *Boro* rice. *Boro* rice often produced together with freshwater aquaculture crops, such as fish and prawn. *Boro* rice is dominantly produced in the irrigated agriculture and freshwater prawn SESs, but the riverine SES also harvests significant quantities of *Boro* rice.

Figure 23.9 shows seasonality in total fish catch by month for each of the SESs. Fish is generally available all year round. Easy access to the Bay of Bengal fisheries seem to invite fishers in larger quantities: caught fish is most important in the coastal periphery and Charland SESs although during the monsoon months and during the early dry season months (July to January), fishing has some importance in the Sundarban dependent and rain-fed irrigation SESs. Collected data generally refers to Bengali months

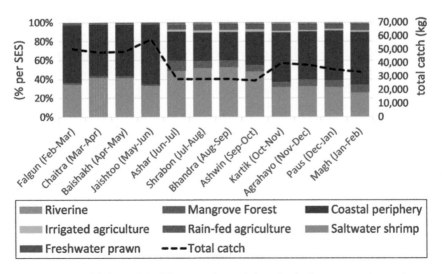

Fig. 23.9 Total fish catch in kilograms by social-ecological system and month

that are slightly offset compared to Western months (e.g. Falgun runs from mid-February to mid-March), figures provides both Bengali and English month names for clarity.

23.5 Conclusion

The survey described in this chapter provides key insights into the complexity of rural livelihoods in the coastal zone and helps us to understand the characteristics of households benefitting from ecosystem services and those of households using them as a last resort in the absence of other safety nets. Analysis of the data shows that having any proportion of household income originating in provisioning ecosystem services within farming, aquaculture, fisheries or forest-based livelihoods increases the likelihood of that household being above the poverty line. However, being above a poverty line does not necessarily mean the household is on a positive well-being trajectory, or that they will not fall under the poverty line again. High levels of ecosystem services are associated with high levels of well-being only in those with significant land assets and associated social capital to enter into agriculture-based business opportunities.

The data reported here for the GBM delta demonstrates a wide variety of levels of income and trajectory: while significant proportions of the population would not be classified as being under the poverty line, incomes and the multiple dimensions of well-being are limited throughout the populations surveyed. Further analysis of the survey results shows how these households combine ecosystem services with off-farm livelihoods and risk diversification strategies such as loans and migration, to maintain more secure rural livelihoods.

Notes

1. Available at https://doi.org/10.5255/UKDA-SN-852179 from the file: *Spatial and temporal dynamics of multidimensional well-being, livelihoods and ecosystem services in coastal Bangladesh.*
2. http://www.progressoutofpoverty.org/ppi-construction
3. http://www.ophi.org.uk/resources/online-training-portal

4. http://www.fantaproject.org/research/comparing-household-food-consumption-indicators-acute-food-insecurity
5. https://doi.org/10.5255/UKDA-SN-852356

References

Adams, H., W.N. Adger, S. Ahmad, A. Ahmed, D. Begum, A.N. Lázár, Z. Matthews, M.M. Rahman, and P.K. Streatfield. 2016. Spatial and temporal dynamics of multidimensional well-being, livelihoods and ecosystem services in coastal Bangladesh. *Scientific Data* 3: 160094. https://doi.org/10.1038/sdata.2016.94.

Baulch, B. 1996. Neglected trade-offs in poverty measurement. *IDS Bulletin* 27 (1): 36–42. https://doi.org/10.1111/j.1759-5436.1996.mp27001004.x.

BBS. 2011[2010]. *Report of the household income and expenditure survey.* Dhaka: Bangladesh Bureau of Statistics.

Deaton, A. 1997. *The analysis of household surveys: a microeconometric approach to development policy.* Washington, DC: World Bank and Johns Hopkins University Press. http://documents.worldbank.org/curated/en/593871468777303124/pdf/multi-page.pdf. Accessed 10 Apr 2017.

Devine-Wright, P. 2013. Explaining NIMBY objections to a power line: The role of personal, place attachment and project-related factors. *Environment and Behavior* 45 (6): 761–781. https://doi.org/10.1177/0013916512440435.

Evans, J. 2015. *Measuring national well-being: Life in the UK, 2015.* London: Office for National Statistics. https://www.ons.gov.uk/peoplepopulationandcommunity/wellbeing/articles/measuringnationalwellbeing/2015-03-25. Accessed 10 Apr 2017.

Gallup. 2015. *Gallup world poll.* www.gallup.com/services/170945/world-poll.aspx. Accessed 17 July 2016.

Lewicka, M. 2011. On the varieties of people's relationships with places: Hummon's typology revisited. *Environment and Behavior* 43 (5): 676–709. https://doi.org/10.1177/0013916510364917.

Szabo, S., E. Brondizio, F.G. Renaud, S. Hetrick, R.J. Nicholls, Z. Matthews, Z. Tessler, A. Tejedor, Z. Sebesvari, E. Foufoula-Georgiou, S. da Costa, and J.A. Dearing. 2016. Population dynamics, delta vulnerability and environmental change: Comparison of the Mekong, Ganges–Brahmaputra and Amazon delta regions. *Sustainability Science* 11 (4): 539–554. https://doi.org/10.1007/s11625-016-0372-6.

Part 5

Present and Future Ecosystem Services

24

Prospects for Agriculture Under Climate Change and Soil Salinisation

Derek Clarke, Attila N. Lázár, Abul Fazal M. Saleh, and Mohammad Jahiruddin

24.1 Agriculture as an Ecosystem Service in Bangladesh

Agriculture is the largest and most important provisioning ecosystem in the world (Zhang et al. 2007), covering 38 per cent of the global land area (DeClerck et al. 2016). In addition to providing important ecosystem services, agriculture also relies on, is impacted by and alters many other ecosystem services (Fig. 24.1). As the population of the world and

D. Clarke (✉) • A. N. Lázár
Faculty of Engineering and the Environment and Tyndall Centre for Climate Change Research, University of Southampton, Southampton, UK

A. F. M. Saleh
Institute of Water and Flood Management, Bangladesh University of Engineering and Technology, Dhaka, Bangladesh

M. Jahiruddin
Department of Soil Science, Bangladesh Agricultural University, Mymensingh, Bangladesh

© The Author(s) 2018
R. J. Nicholls et al. (eds.), *Ecosystem Services for Well-Being in Deltas*,
https://doi.org/10.1007/978-3-319-71093-8_24

447

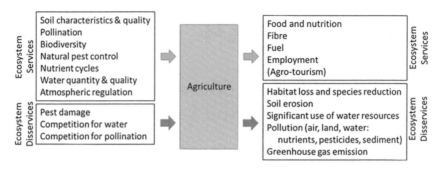

Fig. 24.1 The role of the agriculture ecosystem within ecosystem services

urbanisation increases, agriculture is becoming more and more important and the trade-off between the natural ecosystem services and provisioning ecosystem services from agriculture has to be clearly identified and managed with care (DeClerck et al. 2016; Gordon et al. 2010; Power 2010; Zhang et al. 2007). For example, in 2010, approximately 20 per cent of the world's population was employed in agriculture, but in certain countries, such as Bangladesh, this is significantly higher at 47.5 per cent (World Bank 2017). Changes in agricultural productivity can therefore have implications for both health and well-being.

Agricultural production in the coastal areas of Bangladesh faces many challenges including climatic and environmental change, water shortages and natural hazards (pest, diseases, floods) (FAO and MA 2013). Constraints to crop production are heavy soil consistency, low soil fertility, flooding in the monsoon season, poor soil structure causing delayed draining, high osmotic pressure causing a reduction in the ability of plants to absorb water and nutrients and cyclonic storm surges (see Chap. 18). In addition, soil salinisation is a major concern to farmers in these coastal regions (Baten et al. 2015). Salt naturally occurs in both soil and water; however the amount of salt present depends on both the soil characteristics and the hydrological settings. A high level of salt in the soil limits the ability of the plant to take up water and nutrients. Under normal conditions, the root cells have higher concentration of solutes than the soil water, and the difference allows a free and efficient movement of water into the plant root (i.e. osmotic effect). Increased salinity in the soil water lowers the rate of water transfer to the root; therefore, the plant needs to adapt osmotically by either accumulating salts or synthesising organic compounds such as sugars and organic acids (Hanson et al. 2006). These extra processes use energy, and thus the plant will be sub-optimally developed. In addition salts, like chloride, boron and sodium ions, can also cause toxicities for the plants when these are accumulated in the stems and leaves causing leaf-burns and twig-die-backs (Brown and Shelp 1997). Chloride, boron and sodium ions can be also absorbed through the leaves, thus irrigation water quality is even more important. Crop tolerance to salinity varies widely: some crops are very salt tolerant, but most crops are sensitive to salinity. This sensitivity also depends on the plant growth stage (i.e. germination, vegetative growth or reproductive growth). Many crops are more sensitive to salinity during the early vegetative stage (Mondal et al. 2015).

The accumulation of salts in the soils of agricultural regions occurs progressively when water evaporates from irrigated or flooded fields, leaving residual minerals. In Bangladesh, this problem is reduced during the annual monsoon which brings fresh rainwater to the system and displaces accumulated salts vertically down through the soil profile. As a result, monsoon season rice crop yields are reasonably good and adequate to support a subsistence level of farming in the coastal regions (see Chap. 23). However, evaporation in the dry season causes the salinity to re-develop each year (see Fig. 24.3, lower panel) which results in reduced soil fertility.

Irrigation in the dry season can counteract some of these problems, but the quality of the irrigation water quality is important. Many famers use groundwater or river water which can be partially contaminated with dissolved minerals, and these minerals add to the soil salinity when the applied water evaporates. Extreme events such as cyclones cause inundation of polders with sea water. This can take several years to flush out and typically requires two to three monsoon seasons to return the soils to their pre-inundation salinity levels (Rabbani et al. 2013). The salinity problem is further exacerbated by the move from rice production to brackish water shrimp and fish farming in coastal areas of Bangladesh. Although shrimp production is economically valuable to the owners, the deliberate inundation of polders with brackish or salt water contaminates soils and groundwater in adjacent areas and causes detrimental effects on both biodiversity and crop production of the region (Kartiki 2011).

The aim of this chapter is to assess the effect of the present and future climate and irrigation practices on the crop growth potential in the south-west coastal zone of Bangladesh. Thus, Sect. 24.2 provides an overview of the crops and cropping systems in coastal Bangladesh, Sect. 24.3 introduces the methods, and Sect. 24.4 presents and discusses the results and implications.

24.2   Current Conditions and Crops in the Study Area

The agricultural sector of Bangladesh constitutes an important component of the national economy accounting for around 21 per cent of the national gross domestic product (GDP) (Clarke et al. 2015). The study area is located in the south-west coastal zone of Bangladesh and covers an area of approximately 19,000 km^2 dominated, in 2010, by agriculture (45 per cent), followed by natural vegetation (12 per cent), aquaculture (11 per cent), water (8 per cent) and wetland (8 per cent). This area represents some 30 per cent of the cultivated lands of Bangladesh (Karim et al. 1990; Lázár et al. 2015) and is mostly cultivated by subsistence farmers (Fig. 24.2). The warm humid climate of the region with relatively high annual rainfall (FAO CLIMWAT) provides farming communities excellent opportunities to grow crops on the silty clays of the delta (FAO and MA 2013).

Fig. 24.2 Subsistence farmers (left) and agricultural landscape (right) of the south-west coastal region of Bangladesh (Photographs: Derek Clarke)

24.2.1   Crops, Seasonality and Irrigation

The region experiences two contrasting seasons each year—wet (*Kharif*) season, between June and October, when agriculture concentrates on wet foot rice production and the dry (*Rabi*) season, between November and May, where water availability is reduced (Fig. 24.3). The agriculture calendar allows up to three crops per year during the monsoon (*Kharif*-1), post-monsoon (*Kharif*-2) and dry season (*Rabi*). Rice is the dominant crop, and there are three broad categories of rice cultivated

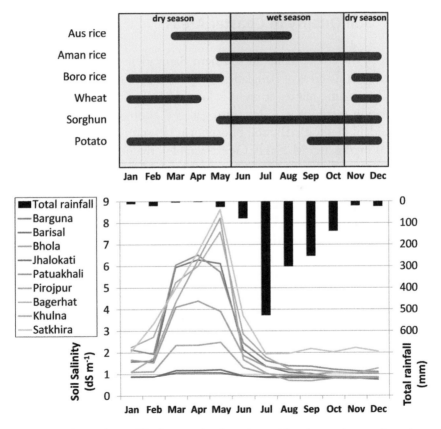

Fig. 24.3   (Upper) Simplified crop calendar adapted for the study area (based on FAO/WFP 2008); (lower) Rainfall distribution and soil salinity in southern Bangladesh (after Lázár et al. 2015)

in different 'agriculture seasons': *Aman* rice (May to December), *Boro* (November to May) and *Aus* (March to August) (Murshid 2012; Banglapedia 2012). *Aman* rice is the most important staple food, and due to the monsoon rains, it requires minimal irrigation. *Boro* rice on the other hand requires irrigation during the dry winter period. Increasing soil salinity in the dry season has encouraged farmers to diversify production to include salt-tolerant crops such a mustard seed and watermelon (Ibrahim et al. 2009). The lack of irrigation water, however, prevents all-year-round agriculture in many places of the south-east part of the study area.

24.2.2   Soil Quality in Coastal Areas

The soils of the coastal lands in southern Bangladesh are relatively poor compared to other areas of Bangladesh, requiring the use of fertiliser (FRG 2012). The organic matter content is low to medium (1.0–3.4 per cent, wet oxidation method) with low nitrogen (N) content (0.091–0.18 per cent, Kjeldahl method). The soil phosphorus (P) status is very low to low (1.0–15.0 mg/kg, 0.5 M $NaHCO_3$ extractable), the potassium (K) status is medium to optimum (0.181–0.36 c mol/kg, 1 M NH_4OAc extractable) and the sulphur (S) status is also medium to optimum (15.1–30.0 mg/kg, CaH_2PO_4 extractable). Micronutrients especially the zinc (Zn) status is low to medium (0.451–1.35 mg/kg, DTPA extractable), and the boron (B) status is medium to optimum (0.31–0.60 mg/kg, $CaCl_2$ extractable).

Many of these soils are classified as alkaline soils. In the Ganges tidal floodplain and the Young Meghna Estuarine Floodplains, 25 per cent of the agricultural land is highly saline (soil electrical conductivity (EC) above 12.0 deciSiemens per metre (dS/m)), and the remaining 75 per cent suffers from slight to moderate soil salinity (EC value 2.0–6.0 dS/m) (Ahsan 2010). The relationship between irrigation water salinity and soil salinity is complex and depends on the volumes of water used, the effectiveness of the monsoon rainfalls and water management practices such as land drainage. Mondal et al. (2015) describe series of experiments on the impacts of irrigation water quality on rice production. Salinity stress on rice crops starts if the irrigation water quality exceeds 3 dS/m, and crop yields will fall by 80 per cent if the irrigation water quality is greater than 10 dS/m.

24.2.3 Historical Changes in Crop Varieties

Farmers are always willing to adopt new varieties of crops if they perceive the benefits. However, farmers in the coastal areas are generally more conservative and reluctant to use balanced doses of fertiliser application. With assistance from plant breeders and extension workers from the Bangladesh Department of Agricultural Extension (DAE),[1] farmers have gradually moved from traditional low-yielding rice varieties and broadcast planting to higher-yielding varieties (HYV) and transplanted rice which, if kept well-watered and fertilised, can more than double production. In the 1990s, local crop varieties dominated the cropping patterns (e.g. broadcast *Aman*), but after 2000, the crop production shifted towards high-yielding and more resilient varieties (Fig. 24.4). At the same

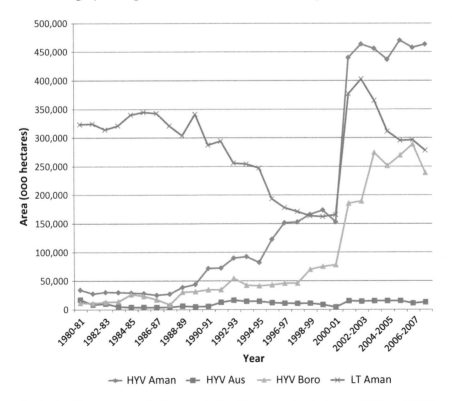

Fig. 24.4 Rice crop area (ha) grown in the Khulna region between 1980 and 2008 (Based on data from the Center for Environmental and Geographic Information)

Fig. 24.5 Dry season salinity effects: (left) rice failing to germinate, (centre) maize yield reduction and (right) melon damaged by salinity and drought (Photographs: Derek Clarke)

time, diversity of the crops also increased (e.g. potato, wheat, chilli, etc.) thanks to the dissemination of modern technologies and knowledge by various agencies and Non-Governmental Organisations (NGOs).

However, despite these changes in the dry season, agriculture remains difficult and large areas of land are sometimes incapable of being used productively until the next monsoon rainfall dilutes the accumulated salts (Fig. 24.5). More salt-tolerant crop types and varieties, for example, mustard, chilli and rapeseed, have been adopted, but the more marginal areas continue to be constrained by soil conditions, high dry season soil salinity and lack of finance to invest in alternate crops.

24.3   Estimating Changes in Climate and Soil Salinity on Agriculture

Research on the dry season salinity problem was undertaken to investigate the climate variability within coastal Bangladesh, to determine whether inter-annual variability is more important than the longer-term climatic trends associated with future climate change, whilst also identifying the key salt mechanism and its likely relationship with reductions in agricultural yield.

CROP RESPONSE MODELS

EXTERNAL FACTORS

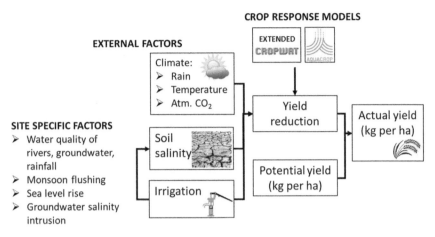

Fig. 24.6 Information used to model and evaluate future salinity and impacts on crop production

24.3.1 Methodology

The modelling-based assessment used both soil moisture and salt balance models (e.g. Andales et al. 2011; Clarke et al. 2015), run using future estimates of rainfall and climate factors that affect evaporation (Fig. 24.6). The soil moisture model is based on the FAO CROPWAT methodology which is a vertical water balance model (Clarke et al. 1998). This was extended by the authors to include a calculation of salt mass transfer through a typical irrigated farm. Salt inputs are from irrigation water from groundwater pumps or river channels at known salinity concentrations (Clarke et al. 2015). The extended CROPWAT model calculates the volume of water needed to irrigate crops in the dry season together with the mass of salt accumulated from minerals in the slightly salty irrigation water. The model calculates the soil leaching requirement and the volume of fresh rainwater to remove the salts in the monsoon season. The models require information on the crop types grown, their tolerances to water and salt stress and soil water storage characteristics. These were obtained from the FAO guidelines for crop evapotranspiration (Allen et al. 1998). The models are run at a daily time step using daily values of rainfall of the 1980–2100 period (HADRM3P climate model, UK Met Office, pers.

comm, described in Chap. 11 and Caesar et al. 2015). Evapotranspiration was calculated using the Penman-Monteith equations for potential evapotranspiration using climatic data for Barisal obtained from the FAO CLIMWAT climatic database (CLIMWAT 2017). Estimates of future potential evapotranspiration were made using the baseline FAO CLIMWAT climate data set perturbed according to temperature increases proposed by Agrawala et al. (2003).

The model runs include calculations of the soil moisture deficit (SMD),[2] the masses of salts accumulated in the soils in the dry season and the masses removed by leaching by fresh rainwater in the wet season over a typical one hectare field near Barisal. The overall aim was to calculate changes in and draw conclusions on the future seasonal and inter-annual variability of evaporative demands, the amount and schedule of irrigation required in the dry season for growing vegetables, as well as determining the accumulation of annual unleached salts and its effect on potential crop yields.

24.3.2 Modelling Results

24.3.2.1 Timing and Length of the Monsoon Season

Using the rainfall and calculated SMD in each year of the simulations, it is possible to determine the start and length of the monsoon season, that is, when the soil is sufficiently wet to grow main season *Aman* rice. This is shown in Fig. 21.7. Inter-annual (i.e. year to year) variability remains the dominant forcing in terms of dry season climate throughout the twenty-first century, despite the long-term climate trends also present within the future climate (see Chap. 11). This variability is most evident within the changes to the length of the seasons, whereas changes to both the onset and magnitude exhibit smaller trends. The source of this variability is not yet fully understood, but research suggests a significant part is due to the monsoon circulation and its interconnection with both the El Niño-Southern Oscillation (ENSO) and Madden-Julian Oscillation (MJO—the largest element of the intra-seasonal variability in the tropical atmosphere).

The simulations indicate that the monsoon season is expected to shorten by two to three weeks, and the onset of the season will be four to five days later by the end of the twenty-first century. However, despite the reduced season length, the total annual rainfall is likely to remain the same or increase by five to ten per cent. This implies that the monsoon rainfall is expected to intensify by the end of the century. This shortening of the wet season, when combined with the year to year variability could result in some monsoon seasons being too short for a successful rice crop by the end of the twenty-first century (e.g. Fig. 24.7 upper—2055,

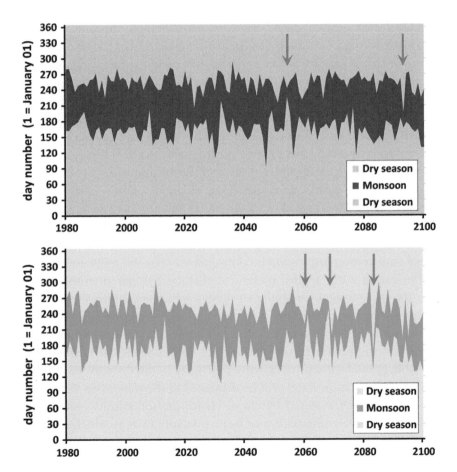

Fig. 24.7 (Upper) Calculated changes in the start and duration of the Monsoon based on rainfall duration (Q8 realisation of the HadRM3P regional climate model); (lower) Calculated changes in the start and duration of the Monsoon based on soil moisture deficit.

2093; Fig. 24.7 lower—2061,2068,2083, as indicated by arrows). This phenomenon is also visible in other climate realisations (e.g. Q0 and Q16; see Chap. 11).

24.3.2.2 Dry Season Water Requirements and Salinity Accumulation in Soils

The dry season exhibits long-term climate forcing with the length of the season expected to increase by 14 days towards the end of the twenty-first century. This is due to the fact that higher temperatures will result in more evapotranspiration and a longer period when the soils are dry, requiring more initial rainfall to re-wet the soil at the start of the monsoon season. Potential evapotranspiration was found to be less variable year to year and tended to rise slightly with increasing temperature. The dry season is thus expected to become drier with the SMD increasing by 25 mm of soil water and, with the onset of the maximum SMD occurring four to five days earlier, an earlier onset of the dry season is suggested. However, dry season irrigation requirements will not increase because the growing period of crops will remain unchanged. Although higher temperatures increase potential evapotranspiration (by five to ten per cent), actual crop evapotranspiration is controlled more by soil moisture storage and irrigation water applied by farmers then by changes in temperature. It is the inter-annual variability of rainfall which will cause changes to both the amount and timing in the application of irrigation water.

Human water management as well as natural environmental changes both have impact on future soil salinity accumulation. If farmers use available groundwater or river water to irrigate in the dry season, the salt load in the irrigation water will be deposited in the soils when the water is evaporated or transpired by crops. Hence the soil salinity conditions will depend on a combination of human factors such as which crop to grow, crop salinity tolerance, irrigation or no irrigation and maintenance of drainage systems to remove leached salts.

Additional environmental factors also exist, such as duration of the dry season, salinity of the irrigation water used, shallow water table problems, magnitude of the next monsoon season and its ability to naturally leach salts out of the soil profile. Typical 'good' irrigation water may contain one to two parts per thousand (ppt) of minerals. Medium-quality water contains 4–6 ppt, and low-quality water may contain 8–12 ppt or higher. Simulations explored the effect of irrigation water quality on salt accumulation in the soil.

The extended CROPWAT model results show that at present (2015) in the Barisal region, if the dry season irrigation water contains less than 4 ppt salts, the monsoon rains in the following season will be able to flush these salts away (the amount of rain is higher than leaching required). However in drier years or if farmers use lower-quality irrigation water, the residual salts will start to accumulate to be carried over to the next dry growing season, worsening soil salinity and decreasing crop yields. The models were run into the future (to 2100) with a range of assumed values of irrigation water quality (Fig. 24.8). Irrigation water qualities of less than 4 ppt result in a small impact of salinity causing less than ten per cent reduction in expected crop yields of dry season vegetable crops. If the irrigation water quality in the dry season remains at 4 ppt, then the monsoon in the following season is able to flush away the accumulated salts in almost every year (there is one exception in this simulation in year 2070. Note that this simulation uses stochastic methods, so the date is not absolute).

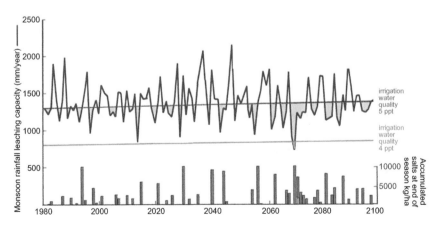

Fig. 24.8 Changes in the leaching capacity of the monsoon season rains compared with salts introduced by dry season irrigation with water qualities of 4 ppt and 5 ppt dissolved mineral content

If the regional water quality was to worsen (e.g. caused by various factors which might include sea-level rise, cyclones, use of lower-quality groundwater for irrigation), then irrigating crops results in the salts not being removed during the following monsoon season (see 5 ppt irrigation quality on Fig. 24.8) in 48 of the 120 years of the simulation. The residual salts are carried over to the next dry crop season with consequences for crop yields; in this case the anticipated crop yield loss due to salinity is 25 per cent. Note that this is a cumulative process, and, if this occurs in consecutive years, soil salinity levels increase over time and even lower potential crop yields are produced.

With the saline intrusion along river estuaries and into the shallow groundwater along the coast associated with anticipated sea-level rise (Kay et al. 2015; Bricheno et al. 2016), the decrease in the quality of irrigation water sources will be a significant factor determining future dry season agricultural productivity within the coastal zone. These findings align with the work of the World Bank (Dasgupta et al. 2014) which indicated that the coastal regions with close proximity to the coast and/or major rivers are expected to see crop yields reduced by more than 50 per cent by the end of the twenty-first century. In this situation farmer's incomes will be reduced, livelihoods threatened and they will have to be more and more reliant on the main season rice as a subsistence crop.

24.3.2.3   Temperature and Carbon Dioxide

In addition to water and salinity stress, other effects on plants will develop due to anticipated climatic change. Crop response to changes in atmospheric carbon dioxide (CO_2) and temperature differs between crop types and individual varieties of these crops. Changes in CO_2 concentrations will assist plant development and biomass accumulation. Parry et al. (2004) suggest that grain crop production will increase by between two to five per cent if atmospheric CO_2 rises from the 2016 value of 400 parts per million (ppm) to 500 ppm, which is anticipated to occur in the 2040s.

Fig. 24.9 shows a possible trajectory of CO_2 (SRES A1B scenario—KNMI 2016) and the calculated CO_2 fertilisation (i.e. atmospheric fertigation) response factor. This crop growth response factor is not a direct a multiplier of yield, rather just one factor of many that influences the crop

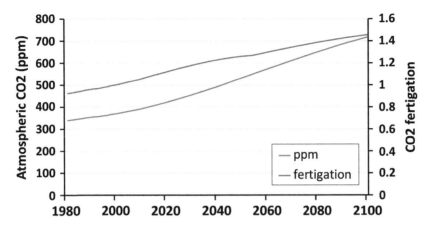

Fig. 24.9 Projected changes in carbon dioxide in the study area and changes in the crop fertigation

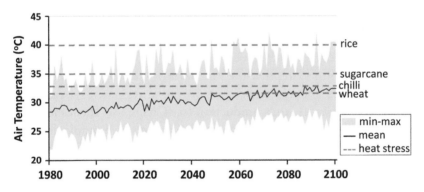

Fig. 24.10 Projected changes in air temperature in the study area compared with crop tolerance to heat stress

physiology. The overall crop yield is affected by numerous factors including crop type and variety, water availability, soil salinity, tolerance to heat, good farming practices and the appropriate use of adequate fertilisers on the poor soils.

Increased crop yields due to CO_2 fertigation are likely to be counteracted by heat stress on plants as global temperatures increase (Sánchez et al. 2014; Peng et al. 2004). Figure 24.10 shows the projected changes in temperatures used in this study and the values which will cause the onset of reduced crop yields. Rice is not likely to be affected by heat limitations in the near future. However other key crops are already potentially constrained and this effect will increase as the twenty-first century progresses. This suggests that the staple rice crop will remain viable, but relatively recently adopted crops such as chilli may not be so successful and new non-rice crop types may have to be adopted in the future.

24.4 Discussion

Agriculture is the single most important provider of ecosystem services to the people of south-west Bangladesh. Monsoon rains supply adequate water to grow a main season rice crop. However, income from agriculture is currently constrained by the limited availability of good quality irrigation water in the dry season, access to markets and the cost of fertilisers. In Bangladesh, the market prices for rice are low so farmers have little or no surplus income to develop coping strategies. Compounding these problems is the fact that the coastal farming communities are also most at risk of flooding due to cyclones.

Farmers attempt to improve their incomes by growing dry season crops such as oil seeds and vegetables, but in coastal regions, dry season soil salinity severely limits crops production (see Chap. 28). The responses to these stressors by low-income farmers are limited, for example, localised irrigation with water of low quality or flooding some areas with brackish river water for aquaculture. These desperate actions, however, serve to increase the salt loading introduced to the fields, and a successful rice crop in the following wet season will depend on removal of accumulated salts. Therefore, although the development of soil salinity is an environmental process, soil salinisation is closely linked to farmers' behaviour and land use practices. This, in turn, not only reduces the agriculture productivity but also is closely associated with the decline in other ecosystem services associated with water regulation (coastal wetlands, mangroves, etc.).

These issues are exacerbated by projections of regional changes in sea levels, higher temperatures and reduced dry season river flows. Higher sea levels will increase mean water levels and the potential overtopping of polders by sea water (Kay et al. 2015), reducing the effectiveness of drainage systems to remove saline water to the sea. Higher temperatures will increase evaporation rates which will increase the need for irrigation in the dry season. Reduced river flows (see Chap. 13 and Whitehead et al. (2015)) mean that brackish water will penetrate further up river channels (Chap. 17 and Bricheno et al. (2016)), and when this water is used for dry season irrigation, this will accelerate the salt loading in the fields. Impacts on groundwater quality have not been fully quantified, but there are strong linkages between river water quality and groundwater salinity (Chap. 18 and Salehin et al. (2014)), so increased use of groundwater for agriculture may accelerate the deterioration of soil productivity due to soil salinity.

These findings reveal that both climate change, inter seasonal climate variability and saline intrusion are important drivers determining the productivity of dry season agriculture in the coastal regions of Bangladesh. There is a tipping point of water quality around 4 ppt beyond which soil salinity development accelerates. In effect salinity is an impact resulting from the trade-offs between ecosystems system functioning, natural environmental change and human responses that lead in turn to trade-offs between provisioning and regulating ecosystems (see Chap. 2).

Notes

1. Agricultural extension workers provide knowledge to farmers on good management practices including eco-friendly, safe, climate resilient and sustainable methods and new crop varieties. In Bangladesh the extension system is run by the Department of Agricultural Extension http://www.dae.gov.bd/

2. Soil moisture deficit (SMD) is a measure of the dryness of the soil. SMD is defined as the amount of water (in mm, the same units as rainfall) that is needed to return the soil to field capacity (close to saturation). A fully wet soil has an SMD = 0 mm, and a soil with SMD greater than approximately 80 mm will cause crop stress and start to reduce yields.

References

Agrawala, S., T. Ota, A.U. Ahmed, J. Smith, and M.V. Aalst. 2003. *Development and climate change in Bangladesh: Focus on coastal flooding and the Sundarbans.* Paris: Organisation for Economic Co-operation and Development. www.oecd.org/env/cc/21055658.pdf. Accessed 28 June 2016.

Ahsan, D.A. 2010. Saline Soils of Bangladesh. Soil Resource Development Institute, Ministry of Agriculture. http://srdi.portal.gov.bd/sites/default/files/files/srdi.portal.gov.bd/publications/bc598e7a_df21_49ee_882e_0302c974015f/Soil%20salinity%20report-Nov%202010.pdf.

Allen, R.G., L.S. Pereira, D. Raes, and M. Smith. 1998. *Crop evapotranspiration – Guidelines for computing crop water requirements.* FAO Irrigation and drainage paper 56. Rome: Food and Agriculture Organization of the United Nations (FAO). http://www.fao.org/docrep/x0490e/x0490e00.htm. Accessed 6 Mar 2017.

Andales, A.A., J.L. Chavez, and T.A. Bauder. 2011. *Irrigation scheduling: The water balance approach.* Colorado: Extension, Colorado State University. http://www.etgage.com/04707.html. Accessed 6 Mar 2017.

Banglapedia. 2012. *National Encyclopedia of Bangladesh,* 2nd ed. Asiatic Society of Bangladesh. www.http://en.banglapedia.org/

Baten, M., L. Seal, and K. Lisa. 2015. Salinity intrusion in interior coast of Bangladesh: Challenges to agriculture in South-Central Coastal Zone. *American Journal of Climate Change* 4: 248–262. https://doi.org/10.4236/ajcc.2015.4302.

Bricheno, L.M., J. Wolf, and S. Islam. 2016. Tidal intrusion within a mega delta: An unstructured grid modelling approach. *Estuarine Coastal and Shelf Science* 182: 12–26. https://doi.org/10.1016/j.ecss.2016.09.014.

Brown, P.H., and B.J. Shelp. 1997. Boron mobility in plants. *Plant and Soil* 193 (1): 85–101. https://doi.org/10.1023/A:1004211925160.

Caesar, J., T. Janes, A. Lindsay, and B. Bhaskaran. 2015. Temperature and precipitation projections over Bangladesh and the upstream Ganges, Brahmaputra and Meghna systems. *Environmental Science-Processes and Impacts* 17 (6): 1047–1056. https://doi.org/10.1039/c4em00650j.

Clarke, D., M. Smith, and K. El-Askari. 1998. New software for crop water requirements and irrigation scheduling. *Irrigation and Drainage* 47 (2): 45–58.

Clarke, D., S. Williams, M. Jahiruddin, K. Parks, and M. Salehin. 2015. Projections of on-farm salinity in coastal Bangladesh. *Environmental Science-Processes and Impacts* 17 (6): 1127–1136. https://doi.org/10.1039/c4em00682h.

CLIMWAT. 2017. *Databases and software*. Food and Agriculture Organisation of the United Nations (FAO). http://www.fao.org/land-water/databases-and-software/climwat-for-cropwat/en/. Accessed 11 May 2017.

Dasgupta, S., F.A. Kamal, Z.H. Khan, S. Choudhury, and A. Nishat. 2014. *River salinity and climate change: Evidence from coastal Bangladesh*. Policy working paper series 6817. Washington, DC: The World Bank. http://documents.worldbank.org/curated/en/522091468209055387/River-salinity-and-climate-change-evidence-from-coastal-Bangladesh. Accessed 11 Apr 2014.

DeClerck, F.A.J., S.K. Jones, S. Attwood, D. Bossio, E. Girvetz, B. Chaplin-Kramer, E. Enfors, A.K. Fremier, L.J. Gordon, F. Kizito, I. Lopez Noriega, N. Matthews, M. McCartney, M. Meacham, A. Noble, M. Quintero, R. Remans, R. Soppe, L. Willemen, S.L.R. Wood, and W. Zhang. 2016. Agricultural ecosystems and their services: the vanguard of sustainability? *Current Opinion in Environmental Sustainability* 23: 92–99. https://doi.org/10.1016/j.cosust.2016.11.016.

FAO, and MA. 2013. *Master plan for agricultural development in the Southern Region of Bangladesh*. Food and Agriculture Organization of the United Nations (FAO) and Government of the People's Republic of Bangladesh, Ministry of Agriculture (MA).

FAO/WFP. 2008. *FAO/WFP crop and food supply assessment mision to Bangladesh*. Annex 2. Rome: Food and Agriculture Organisation of the United Nations (FAO). http://www.fao.org/docrep/011/ai472e/ai472e00.htm#29. Accessed 11 May 2017.

FRG. 2012. *Fertilizer recommendation guide-2012*. Soils publication no. 32. Dhaka: Bangladesh Agricultural Research Council (BARC). https://archive.org/details/FertilizerRecommendationGuide2012_201509. Accessed 6 Mar 2017.

Gordon, L.J., C.M. Finlayson, and M. Falkenmark. 2010. Managing water in agriculture for food production and other ecosystem services. *Agricultural Water Management* 97 (4): 512–519. https://doi.org/10.1016/j.agwat.2009.03.017.

Hanson, B.R., S.R. Grattan, and A. Fulton. 2006. *Agricultural salinity and drainage*. Division of Agriculture and Natural Resources Publication 3375. Davis: University of California. http://hos.ufl.edu/sites/default/files/faculty/gdliu/HansonGrattan2006_0.pdf. Accessed 21 Mar 2017.

Ibrahim, M., A.A. Hassan, I. Huque, and A.U. Ahmed. 2009. *Adaptive crop agriculture including innovative farming practices in the coastal zone of Bangladesh*. Technical report. Dhaka: Center for Environmental and Geographic Information Services (CEGIS). https://www.researchgate.net/publication/271210558_Adaptive_Crop_Agriculture_Including_Innovative_Farming_Practices_in_the_Coastal_Zone_of_Bangladesh

Karim, Z., S.G. Hussain, and M. Ahmed. 1990. *Salinity problems and crop intensification in the coastal regions of Bangladesh.* Soils publication no. 33. Dhaka: Bangladesh Agricultural Research Council (BARC).

Kartiki, K. 2011. Climate change and migration: a case study from rural Bangladesh. *Gender and Development* 19 (1): 23–38. https://doi.org/10.108 0/13552074.2011.554017.

Kay, S., J. Caesar, J. Wolf, L. Bricheno, R.J. Nicholls, A.K.M.S. Islam, A. Haque, A. Pardaens, and J.A. Lowe. 2015. Modelling the increased frequency of extreme sea levels in the Ganges-Brahmaputra-Meghna delta due to sea level rise and other effects of climate change. *Environmental Science-Processes and Impacts* 17 (7): 1311–1322. https://doi.org/10.1039/c4em00683f.

KNMI. 2016. CO_2 Time Series Data. Royal Netherlands Meteorological Institute (KNMI) Climate Explorer. http://climexp.knmi.nl/getindices. cgi?WMO=CDIACData/A1B&STATION=A1B&TYPE=i&id=someone@ somewhere&NPERYEAR=1. Accessed 11 May 2017.

Lázár, A.N., D. Clarke, H. Adams, A.R. Akanda, S. Szabo, R.J. Nicholls, Z. Matthews, D. Begum, A.F.M. Saleh, M.A. Abedin, A. Payo, P.K. Streatfield, C. Hutton, M.S. Mondal, and A.Z.M. Moslehuddin. 2015. Agricultural livelihoods in coastal Bangladesh under climate and environmental change – A model framework. *Environmental Science-Processes and Impacts* 17 (6): 1018–1031. https://doi.org/10.1039/c4em00600c.

Mondal, M.S., A.F.M. Saleh, M.A. Razzaque Akanda, S.K. Biswas, A.Z. Md, S. Zaman Moslehuddin, A.N. Lazar, and D. Clarke. 2015. Simulating yield response of rice to salinity stress with the AquaCrop model. *Environmental Science: Processes and Impacts* 17 (6): 1118–1126. https://doi.org/10.1039/ C5EM00095E Licence 10.1039/C5EM00095E.

Murshid, S.M. 2012. *Impact of sea level rise on agriculture using groundwater in Bangladesh.* MSc thesis CoMEM programme, University of Southampton. https://repository.tudelft.nl/islandora/object/uuid:e484b9b8-e1d1-40b1-99ee-5c7a274ef500/?collection=research. Accessed 22 May 2017.

Parry, M.L., C. Rosenzweig, A. Iglesias, M. Livermore, and G. Fischer. 2004. Effects of climate change on global food production under SRES emissions and socio-economic scenarios. *Global Environmental Change-Human and Policy Dimensions* 14 (1): 53–67. https://doi.org/10.1016/j.gloenvcha.2003.10.008.

Peng, S., J. Huang, J.E. Sheehy, R.C. Laza, R.M. Visperas, X. Zhong, G.S. Centeno, G.S. Khush, and K.G. Cassman. 2004. Rice yields decline with higher night temperature from global warming. *Proceedings of the National Academy of Sciences of the United States of America* 101 (27): 9971–9975. https://doi.org/10.1073/pnas.0403720101.

Power, A.G. 2010. Ecosystem services and agriculture: Tradeoffs and synergies. *Philosophical Transactions of the Royal Society B: Biological Sciences* 365 (1554): 2959. Licence. https://doi.org/10.1098/rstb.2010.0143.

Rabbani, G., A. Rahman, and K. Mainuddin. 2013. Salinity-induced loss and damage to farming households in coastal Bangladesh. *International Journal of Global Warming* 5 (4): 400–415. https://doi.org/10.1504/IJGW.2013.057284.

Salehin, M., M.S. Mondal, D. Clarke, A. Lazar, M. Chowdhury, and S. Nowreen. 2014. *Spatial variation in soil salinity in relation to hydro-climatic factors in southwest coastal Bangladesh.* Deltas in times of climate change II, September 24–26, Rotterdam.

Sánchez, B., A. Rasmussen, and J.R. Porter. 2014. Temperatures and the growth and development of maize and rice: A review. *Global Change Biology* 20 (2): 408–417. https://doi.org/10.1111/gcb.12389.

Whitehead, P.G., E. Barbour, M.N. Futter, S. Sarkar, H. Rodda, J. Caesar, D. Butterfield, L. Jin, R. Sinha, R. Nicholls, and M. Salehin. 2015. Impacts of climate change and socio-economic scenarios on flow and water quality of the Ganges, Brahmaputra and Meghna (GBM) river systems: Low flow and flood statistics. *Environmental Science-Processes and Impacts* 17 (6): 1057–1069. https://doi.org/10.1039/c4em00619d.

World Bank. 2017. *Employment in Agriculture.* World Bank Open Data. http://data.worldbank.org/indicator/SL.AGR.EMPL.ZS. Accessed 9 Feb 2017.

Zhang, Wei, Taylor H. Ricketts, Claire Kremen, Karen Carney, and Scott M. Swinton. 2007. Ecosystem Services and Dis-services to Agriculture. *Ecological Economics* 64 (2): 253–260. https://doi.org/10.1016/j.ecolecon.2007.02.024.

25

Marine Ecosystems and Fisheries: Trends and Prospects

Manuel Barange, Jose A. Fernandes, Susan Kay,
Mostafa A. R. Hossain, Munir Ahmed,
and Valentina Lauria

25.1   Introduction

Bangladesh is a country deeply connected to water, dominated by extensive flood plains, a complex network of rivers, and a dynamic system of estuaries and islands, and a major coastal sea. A significant dependence on aquatic resources also results in growing pressures on coastal ecosystems and added risks from environmental change: the population of Bangladesh

M. Barange (✉)
Food and Agriculture Organization of the United Nations, Rome, Italy

Plymouth Marine Laboratory, Plymouth, UK

J. A. Fernandes • S. Kay • V. Lauria
Plymouth Marine Laboratory, Plymouth, UK

M. A. R. Hossain
Department of Fisheries Biology and Genetics, Bangladesh Agricultural University, Mymensingh, Bangladesh

M. Ahmed
TARA, Dhaka, Bangladesh

© The Author(s) 2018
R. J. Nicholls et al. (eds.), *Ecosystem Services for Well-Being in Deltas*,
https://doi.org/10.1007/978-3-319-71093-8_25

in its coastal region has doubled since the 1980s. The dependence is further challenged by significant environmental variability, often experienced as a set of cyclones, floods and other hazards.

The dependence on aquatic ecosystems extends to its resources, such as fish and fisheries, which contribute significantly to Bangladeshi culture, economy and tradition. Fisheries supplies more than 60 per cent of the country's protein intake (DoF 2013) at an average per capita consumption of approximately 20 kg/year (DANIDA-DFID 2003). The sector contributes approximately four and a half per cent to the national gross domestic product (GDP), and provides direct employment to two million people on a full-time basis (DoF 2002) and indirect employment to a further 13 million people (Alam et al. 2012; World Bank 2015). Such is its fisheries dependence that Bangladesh is both the fourth largest inland capture producer in the world (2003–2012) and the sixth largest aquaculture producer (FAO 2016). Since independence in 1971 the fisheries industry has seen steady growth with production tripling in the last two decades (Alam et al. 2012). Between 1984 and 2009 annual average growth in fish production was more than five per cent largely driven by the expansion in inland aquaculture fisheries, which grew at a rate of almost ten per cent per year (Golub and Varma 2014).

This chapter explores how climate change will affect marine fisheries in Bangladesh, and how management interventions can minimise or exacerbate such impacts. It will demonstrate that the marine ecosystem of Bangladesh will remain productive but that the species benefiting from this production may change. This is illustrated by investigation of the specific impacts of climate change on Hilsa shad (*Tenualosa ilisha*) and Bombay duck (*Harpadon nehereus*), two of the fish that contribute the most to food security and nutrition in Bangladesh.

25.2   Structure of Fisheries Sector

The fishery of Bangladesh includes marine, inland and aquaculture sectors (Fig. 25.1), each one with its own specific stressors and production trends. In 2012–2013 the full fisheries sector produced 3.4 million tonnes of fish. Marine and inland fisheries are connected because both target in particular

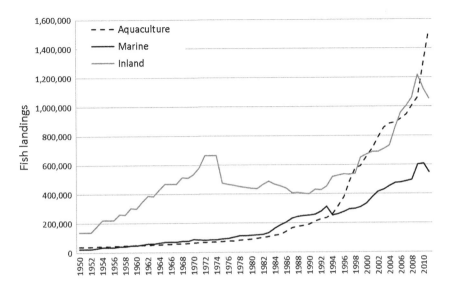

Fig. 25.1 Fisheries production (expressed in tonnes) in Bangladesh 1950–2012: marine catches, inland catches and aquaculture production (Based on data from the FAO Global database)

Hilsa shad, a fish that spends part of its life cycle in the sea and part of it in rivers. Aquaculture activities are also connected with capture fisheries because they depend on the provision of larvae and juveniles from rivers and marine ecosystems (Kathun 2004). However, there are many processes involved in the development of aquaculture as an industrial activity, beyond the availability of natural fish and shellfish resources. Fisheries has been continuously growing in Bangladesh since 1950 (Fig. 25.1), with aquaculture providing the majority of production in recent years.

The marine-based fishery in particular, which relies entirely on natural production processes, is subdivided into subsistence, artisanal and industrial. The artisanal sector is the most productive with circa 500,000 tonnes of fish landings in 2012 (94 per cent of volume of landings). Subsistence fisheries are of greatest importance for national food provision, and focuses on catches of species of lower commercial value. The only industrial fishing developed in Bangladesh operates out of Chittagong on the east coast and comprises two distinct industrial fisheries: shrimp trawl and bottom trawl (FAO 2006).

25.3 Marine Fish Catches and Their Management

Freshwater habitats in Bangladesh are inhabited by 260 fin fish species, 25 species of prawn and 25 species of turtles (Ali 1999), as well as 23 exotic fin fish species that were introduced for aquaculture purposes (Hossain 2010). Marine waters are habitat to 475 species of fin fish, of which 65 are of economic importance, and host 38 species of marine shrimps (Ali 1999).

Hilsa shad is the biggest fishery and the national fish of Bangladesh. Locally known as *ilish* or *ilisha*, it is a euryhaline anadromous fish found in marine, coastal and freshwater environments. A significant part of its catch is exported to India, where it is especially consumed in religious holidays and by Bangladeshis living in a number of outside countries. In 2012–2013, Hilsa contributed singlehandedly about ten per cent of the total fish production of Bangladesh, with a market value of US$2250 million, about one per cent of Bangladesh's GDP. In the last two decades, Hilsa production from inland waters has declined by about 20 per cent, whereas marine water yield has increased threefold (Kathun 2004).

Bombay duck supplies the second largest fish catch in the Bangladesh coastal region and is the most popular subsistence fishery, representing over 12 per cent of the catch (1.7 million tonnes), and consumed fresh or dried. In addition to its food provision role, it is a lucrative fishery in the Bay of Bengal despite its price being six times lower than Hilsa (65 BDT. kg-1 versus 430 BDT.kg-1; 1US$=79 Bangladesh Taka (BDT)).

Bangladesh has a centralised fisheries management system under the Department of Fisheries of the Ministry of Livestock and Fisheries. The present Fisheries Policy was adopted in 1998 to enhance resources and production, reduce poverty through self-employment in the sector, supply the need for animal protein, achieve economic growth and earn foreign exchange while maintaining ecological balance, biodiversity and public health. In 2006, the Ministry of Fisheries and Livestock adopted a Fisheries Strategy pushing further towards poverty reduction, co-management and conservation of resources. But even though the Hilsa fishery is regulated through some technical management measures, such as reduction in mesh size and fishing closures, the overall management of fishery resources in the country is currently insufficient to control effort. This is primarily due to the lack of funding and resources committed to sustainable fisheries management (FAO 2003).

25.4   Estimating the Effects of Climate Change on the Marine Ecosystems and Fisheries of Bangladesh

In fish-dependent economies such as Bangladesh (Toufique and Belton 2014), the combination of climate change, population growth and protein consumption patterns is a significant amplifier of pressures on critical national resources (Delgado et al. 2003; Merino et al. 2012). Yet there are no major studies that have projected the structure and availability of aquatic resources and ecosystem services into the future accounting for the impacts of climate change. This research analyses the prospects for fisheries maintaining its critical role in providing employment and provisioning ecosystem services and nutrition across the delta region.

Previous analyses have predicted a decrease in the productive potential of fisheries in South and Southeast Asia as a result of climate change (Barange et al. 2014). Understanding how this impact translates into the future provision of fish products is crucial for the sustainability of fisheries-dependent communities in coming decades (see Chap. 23). In previous work based on simpler models, Merino et al. (2012) concluded that marine fisheries production in Bangladesh may be reduced by five per cent by 2050, leaving aquaculture to produce between 0.3 and 3.8 million tonnes of fish to achieve national fish consumption targets of between 13 and 31 kg fish/capita, accounting for expected human population growth. In order to better quantify how climate change will change the marine fish production potential in Bangladesh, a combination of atmospheric, hydrological, ocean circulation and ocean biogeochemical models is used (Fig. 25.2). A full description of these models is detailed in Fernandes et al. (2016): here, a summary of these models is provided.

Climate data were taken from the UK Met Office regional climate model (RCM) HadRM3P, which is dynamically downscaled from the

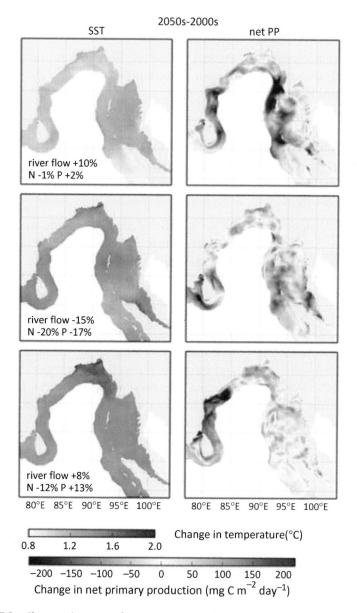

Fig. 25.2 Changes in sea surface temperature (SST) and net primary production (net PP) projected for the Bay of Bengal by 2050s, compared to baseline of values for the 2000s. The rows show results for the three climate ensemble runs Q0 (top row), Q8 (middle) and Q6 (bottom row). Changes in river flow volume, nitrate (N) and phosphate (P) loads of the GBM rivers are shown in the left-hand panel of each pair (Reproduced with permission from Fernandes et al. 2016 – Published by Oxford University Press)

global circulation model HadCM3 (see Chap. 11 and Caesar et al. 2015). As delta regions are particularly sensitive to precipitation and river run-off, outputs from an integrated catchment model (see Chap. 13 and Whitehead et al. 2015a, b) were used to determine run-off and associated nutrient loadings from the delta rivers into the Bay of Bengal. The model took account of both climatic scenarios (Q0, Q8, Q16) and patterns of upstream water use according to three socio-economic scenarios (Less Sustainable (LS), Business As Usual, (BAU), More Sustainable (MS)) scenarios (Whitehead et al. 2015a, b). The results of the Q0-BAU, Q8-LS and Q16-MS were used in further simulations.

A regional POLCOMS-ERSEM coupled model (Holt et al. 2009; Butenschön et al. 2016) was used to project both the physical state of the ocean (temperature, salinity, currents, light level), as well as the biogeo-chemistry and plankton production in the Bay of Bengal (see Chap. 14). The model simulates the dynamics of nutrients (C, P, N and Si), phyto-plankton, zooplankton and bacteria, as well as particulate and dissolved organic matter. The model domain covers the entire coastal area of Bangladesh up to 200 km beyond the edge of the continental shelf, with a horizontal resolution of 0.1° and 42 vertical levels. For each climate dataset, the model was run continuously for 1971–2099.

Outputs from the POLCOMS-ERSEM model were then used to drive a dynamic marine ecosystem model that predicts potential production by size class of fish, taking into account food availability, predation effects and temperature effects on feeding and mortality (Blanchard et al. 2012). Size-based methods like this are useful in that they capture the basic met-abolic properties and dynamics of marine food webs, describing energy flux and production by size class, independent of species' ecology (Blanchard et al. 2012). To make specific projections for key species, however, a second model, called SS-DBEM, was used (Fernandes et al. 2013, 2016, 2017), which projects changes in species distribution and abundance over time while explicitly considering changes in production, dispersal and physiology as a result of changing ocean conditions, as well as species interactions (Fernandes et al. 2013). The size-spectrum model was applied to explore potential changes in the total productivity of the Bangladesh Exclusive Economic Zone under both climate change and fishing scenarios and the SS-DBEM for the two target species (Hilsa shad and Bombay duck), for up to 2099.

In order to estimate potential fish catches not only need to project production potential is required but also management interventions. Fisheries management in Bangladesh is limited to technical measures that are insufficient to control and limit overfishing practices. In order to understand the importance of management interventions, three plausible fisheries management scenarios were defined, based on specific levels of fishing pressure in relation to the species' maximum sustainable yield (MSY). MSY is defined as the highest average theoretical equilibrium catch that can be continuously taken from a stock under average environmental conditions (Hilborn and Walters 1992). It aims to maintain the population size at the point of maximum growth rate by harvesting the individuals that would normally be added to the population, allowing the population to continue to be productive indefinitely. The three scenarios are based on adjusting fishing mortality, the parameter that determines the amount of fish that is removed from the stock through fishing. These scenarios are named 'Sustainability', 'Business As Usual', and 'OverFishing'. The sustainability scenario (MSY) involves adjusting the fishing mortality (F_{MSY}) to ensure the population remains at levels of biomass consistent with their maximum Sustainable Yield. This is a theoretical value that results in maximum catches while maintaining the population at their productivity peak. The Business As Usual scenario (BAU) represents fishing mortality that is consistent with the average estimates of fishing mortality in the country (F_{BAU}), which as illustrated later are below long-term sustainability. The average fishing mortality for Hilsa shad in recent years, the largest fishery in the country, is 1.86, or three times the fishing mortality consistent with MSY (F_{MSY}). Hilsa shad and many of the brackish species in the Bay of Bengal are considered to be being exploited at that rate for the purpose of the BAU scenario. The OverFishing scenario (F_{OF}), on the other hand, involves the highest level of exploitation and corresponds to a scenario where management is not a constraint to the fishery. Initial runs indicate collapse of catches and biomass at 4 * MSY (F_m of 2.4). The three scenarios are therefore consistent with status quo (BAU), inaction (OF) and active management (MSY).

25.5   Climate Change Impacts on the Bay of Bengal Ecosystem

Projections of change by the middle of the twenty-first century show a steady rise in sea surface temperature throughout the Bay of Bengal but a more mixed picture for river flows, nutrients and net primary production (Fig. 25.2). The temperature rise is greatest for the Q16 climate run, which has the highest climate sensitivity. River flows vary between increases of ten per cent to decreases of 15 per cent, depending on the climate projection. Similar variability is observed in the nutrient loading. There is a consistent fall in net primary production across most of the northern Bay of Bengal under Q0. Primary production in the Exclusive Economic Zone (EEZ) of Bangladesh rises in both of the warmer runs, Q8 and Q16, and is greatest for Q8. The EEZ is the zone prescribed by the UN Convention on the Law of the Sea over which a state has special rights regarding the exploration and use of marine resources.

Figure 25.3 shows how these changes in environmental conditions and primary production affect the overall fish production potential of the Bangladeshi EEZ. The results indicate that all three climate runs result in declines in fish productivity. Averaging the results per decade shows that

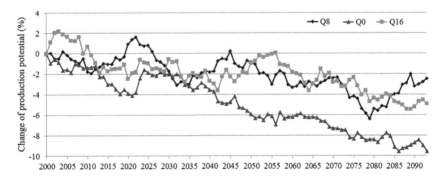

Fig. 25.3 Time series of changes in projected total fisheries production potential in the Bangladesh EEZ according to the three climate ensemble runs (Q0, Q8 and Q16) and in the absence of fisheries extractions. Values are expressed as a percentage deviation from the 2000 production for each ensemble run (Reproduced with permission from Fernandes et al. 2016 – Published by Oxford University Press)

the total fish productivity would decline between 1.3 per cent and 4.9 per cent by 2050, and between 2.6 per cent and 8.3 per cent by 2100, depending on the climatic ensemble run considered. These are not very significant changes in potential production, and close to the error margins of such a modelling exercise.

25.6   The Role of Management Interventions

In order to explore the impact of particular fisheries, management scenarios model runs using the species-based model for Hilsa shad and Bombay duck were undertaken. For this exercise, averaged outputs of the three climate ensembles were calculated to enable a focus specifically on the relative impacts of fisheries management decisions. The results are presented relative to the year 2000 model outputs (not the actual catches), but the different management interventions are implemented at year 1980, as the model needs to settle and reflect the interventions before the period of analysis. For this reason, by 2000 the three management trajectories already diverge from the actual recorded catches. Results conclude that both Hilsa shad and Bombay duck catches will decline over time regardless of the fisheries management regime, but to different degrees (Fig. 25.4). For Hilsa the decline stabilises under MSY management at 175,000 t by 2035, while it virtually collapses around the same period under the OF scenario. A significant inter-annual variability is also observed (Fig. 25.4). By the 2050s the decline in catches is between 39 per cent (under MSY) and 87 per cent. For Bombay duck inter-annual variability is reduced (Fig. 25.4), and while catch potential declines continuously under all management scenarios, they do not lead to biological collapse. By the 2050s this decline is around 35 per cent for all management scenarios. Potential catches are on average higher in the more sustainable scenarios (MSY) than in the BAU scenario for both species (91 per cent in Hilsa Shad and 37 per cent in Bombay duck by the 2050s).

It is useful to compare the outcomes of the BAU scenario in the 2000s decade with those of a more sustainable scenario in coming decades. The results indicate that a change to sustainable management would result in a very minor decline in potential catches by the 2020s but a still significant

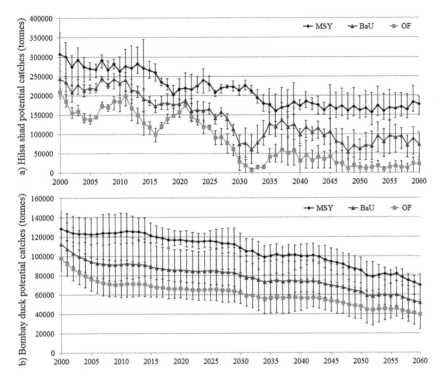

Fig. 25.4 Time series of catch potential projections (t) in the Bangladesh EEZ for Hilsa shad (**a**) and Bombay duck (**b**) under different fisheries management scenarios (MSY or Sustainable exploitation, BAU or Business As Usual and OF or OverFishing scenario). Error bars indicate variability between the three climate ensemble runs (Q0, Q8 and Q16) and associated river run-off and nutrient loadings (Reproduced with permission from Fernandes et al. 2016 – Published by Oxford University Press)

(25 per cent) decline by the 2050s (Table 25.1). Conversely, a future that follows BAU with an OF scenario will bring catches by 2050s almost 95 per cent lower than in the BAU scenario for the start of the twenty-first century (Table 25.1). For Bombay duck, potential catches by 2020s under an MSY scenario will produce over 20 per cent more fish than under BAU at the start of the century, with a decline of less than 20 per cent by the end of the projection period (Table 25.1). Conversely, maintaining BAU fishing to the end of century would result in 40 per cent decline in catch potential. What this demonstrates is that management

Table 25.1 Change in catch potential for Hilsa shad and Bombay duck in the 2020s and 2050s, referenced to the 2000s decade (in per cent), and average catch in each of the three decades (tonnes/year), according to three fisheries management scenarios. The reported values are the mean and standard deviation across Q climate scenarios (Reproduced with permission from Fernandes et al. 2016 – Published by Oxford University Press)

		2020s–2000s	2050s–2000s	2000s	2020s	2050s
		Δ catch (%)	Δ catch (%)	Average catch ('000 tonnes/year)	Average catch ('000 tonnes/year)	Average catch ('000 tonnes)
Hilsa shad	MSY	−20.6 ± 14.5	−39.0 ± 20.1	283.0 ± 30.0	221.7 ± 14.9	168.5 ± 35.3
	BAU	−30.7 ± 12.7	−42.1 ± 7.6	227.9 ± 14.0	156.7 ± 19.7	101.3 ± 11.9
	OF	−29.3 ± 8.8	−87.2 ± 11.1	165.6 ± 3.0	116.9 ± 13.0	15.7 ± 13.6
Bombay duck	MSY	−7.4 ± 4	−37.2 ± 4.0	124.2 ± 15.6	114.7 ± 10.2	77.8 ± 11.4
	BAU	−13.9 ± 1.9	−32.4 ± 7.0	98.2 ± 23.7	84.4 ± 19.5	57.3 ± 15.4
	OF	−19.2 ± 1.7	−33.5 ± 7.3	80.3 ± 20.7	65.0 ± 17.3	43.5 ± 13.6

interventions have the capability to mitigate or exacerbate the effects of climate change on ecosystem productivity.

25.7   Implications and the Prospects for Fisheries

In spite of the slight increase in primary production in Bangladeshi waters by the mid twenty-first century, projections indicate some decreases in the potential fish production. The complex relationship between temperature and primary production change may result in a higher proportion of fish biomass of smaller size, despite lower total fish biomass. Marine and inland fish catches in Bangladesh have doubled since 1995 reaching 1.6 million tonnes per year (Fig. 25.1), of which Hilsa has contributed ca. 350,000 tonnes. The analysis in this chapter indicates no evidence of increased productivity, and hence it is likely that the reported increase in catches is driven directly by growing demand for fish products rather than increase productivity of the Bangladesh EEZ. Increases in Bangladesh population from 120 million to 158 million people since 1995 (World Bank 2015) and growing protein consumption (from 42.1 in 2005 to 49.5 in 2010, grammes per capita per day, BBS 2011) are putting additional pressure on resources whose productivity is limited. Managing this growth in demand, including better use of not just marine but also inland and aquaculture resources (Bene et al. 2015) would be important as climate change impacts start to make themselves felt.

The analysis here suggests larger decreases are projected for the two main species compared to the total fisheries productivity change: decreases in Hilsa and Bombay duck may provide ecological space for other species to replace the main targets of the fishery. Possible replacement could include Chacunda Gizzard Shad (*Anodontostoma chacunda*) and Toli Shad (*Tenualosa toli*) for Hilsa Shad. It is well known that the complexity of ecological interactions in the marine food web makes it difficult to extrapolate studies on individual species to community or ecosystem level (Walther et al. 2002). Thus, it is quite possible to observe differential impacts at the community level compared to species patterns.

The analysis also demonstrates that the management options to be followed in coming decades are crucial for the sustainability of fisheries and their role as a nutritional and economic resource for the country (see Chaps. 23 and 27). The implementation of sustainable management practices for Hilsa would stabilise the marine catch to approx. 170,000 tonnes by the 2050s, but that continuing with current fishing mortalities would reduce this to around 100,000 tonnes. A decrease in Hilsa shad catch would have important consequences for a species with intrinsic cultural value, and may cause a shift in the workforce from fishing to other livelihood options (Hossain et al. 2015; Nicholls et al. 2015).

Bombay duck productivity is around half that of Hilsa shad, but the catches and biomass are more stable. The projections here indicate that catches of Bombay duck may not collapse as a result of unsustainable fishing practices and climate effects by 2060, contrary to projections for Hilsa. The reason for the different response compared to Hilsa relies on its different sensitivity to environmental and climate conditions and a broader range of feeding options (Fernandes et al. 2016; Zhang and Jin 2014).

The impact that changes in the extension and composition of mangroves would have on fish production are not included in the model. Although the importance of mangroves as fish habitats and nursery grounds is well recognised (Islam and Haque 2004; Hutchison et al. 2014), the exact impacts of changes on fish populations are still uncertain and unquantified.

Although this chapter focuses explicitly on marine environments and their resources, climate change impacts do include coastal zones and their associated inland fisheries. Drought coupled with siltation and lowering water level is reducing overwintering habitat for indigenous fish species, resulting in reduced fish recruitment. Lower dry season flows in major rivers have and will continue to deplete riverine fisheries. Due to a decrease in groundwater and surface water, extreme pressure has been exerted on floodplains to convert them to crop fields, brick kiln and other infrastructures, resulting in worrying declines in fish diversity and production. Erratic and irregular rainfall as well as temperature change will affect the readiness, maturity and gonad development of fishes in the breeding season. Higher water temperature brings changes in physiology and sex ratios of fish species, altered timing of spawning, migrations, and peak abundance, changes in timing and levels of productivity across marine and freshwater systems and increased invasive species, diseases and algal blooms. Thus, the effects quantified here are not comprehensive for other inland fisheries.

Since fishing is the second most important source of livelihood in coastal Bangladesh, exploring plausible future trajectories of the resources will allow the exploration of alternatives futures comparing the use of all natural ecosystem services of Bangladesh as a tool to aid sustainable resource management and regional development planning.

25.8 Conclusions

Results show that the impacts of climate change, under greenhouse emissions scenario A1B, are likely to decrease the potential total fish production in the Bangladesh EEZ by less than ten per cent. However, these impacts are larger for the two major species (Hilsa shad and Bombay duck), even under sustainable management practices.

It is demonstrated that sustainable management practices would stabilise the marine catch of Hilsa around 170,000 tonnes by the 2050s decade, approximately 70 per cent of current catches. However, failure to implement sustainable management practices, combined with climate effects could reduce Hilsa catches by 2050 by up to 90 per cent of current catches. On the other hand, catches of Bombay duck may not collapse as a result of unsustainable fishing practices and climate effects by 2060. Sustainable management measures would maintain current catches to a large extent. The reasons for the different response rely on the fact that Hilsa has slightly higher estimated intrinsic population growth rates and adult movement rate compared to Bombay duck, allowing it to track environmental changes more closely.

Hilsa shad is the largest fishery by volume in Bangladesh and a species with very significant economic and cultural value. Bombay duck is the second highest catch in Bangladesh and is a much cheaper fish than Hilsa Shad. The analyses indicate that environmental and climate change would impact negatively Bangladesh fisheries. However, it is demonstrated that good management can mitigate catches losses, while bad or no management can exacerbate the impacts of climate change.

The results of this research should allow for the exploration of future sustainable trajectories of the fishery sector through testing plausible scenarios. These could be used as a tool to aid sustainable resource management and regional development planning.

Fish is critically important to food security and good nutrition (Beveridge et al. 2013), and fisheries a crucial provisioning ecosystem service in Bangladesh. Fisheries not only provide food but also livelihoods and economic trade, particularly important for coastal communities who are often among the poorest in society. Ensuring the sustainability of the underlying marine and inland resources, in the face of climate change is crucial (see Chap. 2). This research has demonstrated that this ecosystem service can be maximised through responsive and effective management, for the purpose of feeding and supporting coastal communities. Inaction has a cost, and in the context of climate change, this cost would be too high in terms of food provision alone to be forgotten.

References

Alam, M.F., M.S. Palash, M.I.A. Mian, and M.M. Dey. 2012. *Marketing of major fish species in Bangladesh: A value chain analysis.* 'A value-chain analysis of international fish trade and food security with an impact assessment of the small-scale Sector' – Project report. Rome: Food and Agriculture Organization of the United Nations.

Ali, M.Y. 1999. Fish resources vulnerability and adaptation to climate change in Bangladesh. In *Vulnerability and adaptation to climate change for Bangladesh,* ed. S. Huq, Z. Karim, M. Asaduzzaman, and F. Mahtab, 113–124. Dordrecht: Springer Netherlands.

Barange, M., G. Merino, J.L. Blanchard, J. Scholtens, J. Harle, E.H. Allison, J.I. Allen, J. Holt, and S. Jennings. 2014. Impacts of climate change on marine ecosystem production in societies dependent on fisheries. *Nature Climate Change* 4 (3): 211–216. https://doi.org/10.1038/nclimate2119.

BBS. 2011. *Foreign trade statistics of Bangladesh, 2010–2011.* Dhaka: Bangladesh Bureau of Statistics (BBS).

Bene, C., M. Barange, R. Subasinghe, P. Pinstrup-Andersen, G. Merino, G.-I. Hemre, and M. Williams. 2015. Feeding 9 billion by 2050-putting fish back on the menu. *Food Security* 7 (2): 261–274. https://doi.org/10.1007/s12571-015-0427-z.

Beveridge, M.C.M., S.H. Thilsted, M.J. Phillips, M. Metian, M. Troell, and S.J. Hall. 2013. Meeting the food and nutrition needs of the poor: The role of fish and the opportunities and challenges emerging from the rise of aquaculturea. *Journal of Fish Biology* 83 (4): 1067–1084. https://doi.org/10.1111/jfb.12187.

Blanchard, J.L., S. Jennings, R. Holmes, J. Harle, G. Merino, J.I. Allen, J. Holt, N.K. Dulvy, and M. Barange. 2012. Potential consequences of climate change for primary production and fish production in large marine ecosystems. *Philosophical Transactions of the Royal Society B-Biological Sciences* 367 (1605): 2979–2989. https://doi.org/10.1098/rstb.2012.0231.

Butenschön, M., J. Clark, J.N. Aldridge, J.I. Allen, Y. Artioli, J. Blackford, J. Bruggeman, P. Cazenave, S. Ciavatta, S. Kay, G. Lessin, S. van Leeuwen, J. van der Molen, L. de Mora, L. Polimene, S. Sailley, N. Stephens, and R. Torres. 2016. ERSEM 15.06: A generic model for marine biogeochemistry and the ecosystem dynamics of the lower trophic levels. *Geoscientific Model Development* 9 (4): 1293–1339. https://doi.org/10.5194/gmd-9-1293-2016.

Caesar, J., T. Janes, A. Lindsay, and B. Bhaskaran. 2015. Temperature and precipitation projections over Bangladesh and the upstream Ganges, Brahmaputra and Meghna systems. *Environmental Science-Processes and Impacts* 17 (6): 1047–1056. https://doi.org/10.1039/c4em00650j.

DANIDA-DFID. 2003. *The future for fisheries*. Findings and recommendations from the fisheries sector review and future development Study, FAO Representation – Bangladesh, 65p. Available from http://www.lcgbangladesh.org/

Delgado, C.L., N. Wada, M.W. Rosegrant, S. Meijer, and M. Ahmed. 2003. *Fish to 2020: Supply and demand in changing global markets*. Worldfish Center Technical Report 62. Washington, DC/Penang: International Food Policy Research Institute.

DoF. 2002. *Fisheries resources survey system (2001–2002)*. Dhaka: Department of Fisheries (DOF), Government of the People's Republic of Bangladesh.

———. 2013. *National fish week 2013 compendium (in Bengali)*. Dhaka: Department of Fisheries (DOF), Government ofthe People's Republic of Bangladesh.

FAO. 2003. *Subregional review: Eastern Indian Ocean.* Review of the state of world marine capture fisheries management: Indian Ocean. FAO Fisheries Technical Paper. No. 488. Rome: Food and Agriculture Organization of the United Nations (FAO). http://www.fao.org/docrep/009/a0477e/a0477e05. htm. Accessed 12 Aug 2015.

———. 2006. *Review of the state of world marine capture fisheries management: Indian Ocean.* Rome: Fishery and Aquaculture Economics and Policy Division, Food and Agricultural Organization of the United Nations (FAO).

———. 2016. *The state of the world fisheries and aquaculture – 2016. Contributing to food security and nutrition for all.* Rome: Food and Agricultural Organization of the United Nations (FAO).

Fernandes, J.A., W.W.L. Cheung, S. Jennings, M. Butenschoen, L. de Mora, T.L. Froelicher, M. Barange, and A. Grant. 2013. Modelling the effects of climate change on the distribution and production of marine fishes: Accounting for trophic interactions in a dynamic bioclimate envelope model. *Global Change Biology* 19 (8): 2596–2607. https://doi.org/10.1111/gcb.12231.

Fernandes, J.A., S. Kay, M.A.R. Hossain, M. Ahmed, W.W.L. Cheung, A.N. Lázár, and M. Barange. 2016. Projecting marine fish production and catch potential in Bangladesh in the 21st century under long-term environmental change and management scenarios. *ICES Journal of Marine Science* 73 (5): 1357–1369. https://doi.org/10.1093/icesjms/fsv217.

Fernandes, J.A., E. Papathanasopoulou, C. Hattam, A.M. Queirós, W.W.W.L. Cheung, A. Yool, Y. Artioli, E.C. Pope, K.J. Flynn, G. Merino, P. Calosi, N. Beaumont, M.C. Austen, S. Widdicombe, and M. Barange. 2017. Estimating the ecological, economic and social impacts of ocean acidification and warming on UK fisheries. *Fish and Fisheries* 18 (3): 389–411. https://doi.org/10.1111/faf.12183.

Golub, S., and A. Varma. 2014. *Fishing exports and economic development of least developed countries: Bangladesh, Cambodia, Comoros, Sierra Leone and Uganda.* Paper prepared for United Nations Conference on Trade and Development (UNCTAD). Swarthmore: Swarthmore College. http://www.swarthmore. edu/profile/stephen-golub. Accessed 4 May 2016.

Hilborn, R., and C.J. Walters. 1992. *Quantitative fisheries stock assessment: Choice, dynamics and uncertainty.* New York: Chapman & Hall.

Holt, J., J. Harle, R. Proctor, S. Michel, M. Ashworth, C. Batstone, I. Allen, R. Holmes, T. Smyth, K. Haines, D. Bretherton, and G. Smith. 2009. Modelling the global coastal ocean. *Philosophical Transactions of the Royal Society a-Mathematical Physical and Engineering Sciences* 367 (1890): 939–951. https://doi.org/10.1098/rsta.2008.0210.

Hossain, M.A.R. 2010. Inland fisheries resource enhancement and conservation in Bangladesh. In *Inland fisheries enhancement and conservation in Asia (RAP Publication 2010/22)*, ed. M. Weimin, S. De Silva, and B. Davy, 1–17. Bangkok: Regional Office for the Asia and the Pacific, Food and Agriculture Organisation (FAO).

Hossain, M.S., L. Hein, F.I. Rip, and J.A. Dearing. 2015. Integrating ecosystem services and climate change responses in coastal wetlands development plans for Bangladesh. *Mitigation and Adaptation Strategies for Global Change* 20 (2): 241–261. https://doi.org/10.1007/s11027-013-9489-4.

Hutchison, J., M. Spalding, and P. zu Ermgassen. 2014. *The role of mangroves in fisheries enhancement.* Arlington/Wageningen: The Nature Conservancy and Wetlands International.

Islam, M.S., and M. Haque. 2004. The mangrove-based coastal and nearshore fisheries of Bangladesh: Ecology, exploitation and management. *Reviews in Fish Biology and Fisheries* 14 (2): 153–180. https://doi.org/10.1007/s11160-004-3769-8.

Kathun, F. 2004. *Fish trade liberalization in Bangladesh: Implications of SPS measures and eco-labelling for the export-oriented shrimp sector.* Policy Research – Implications of liberalization of fish trade for developing countries. Project PR26109. Rome: United Nations Food and Agriculture organisation (FAO).

Merino, G., M. Barange, J.L. Blanchard, J. Harle, R. Holmes, I. Allen, E.H. Allison, M.C. Badjeck, N.K. Dulvy, J. Holt, S. Jennings, C. Mullon, and L.D. Rodwell. 2012. Can marine fisheries and aquaculture meet fish demand from a growing human population in a changing climate? *Global Environmental Change* 22 (4): 795–806. https://doi.org/10.1016/j.gloenvcha.2012.03.003.

Nicholls, R.J., P. Whitehead, J. Wolf, M. Rahman, and M. Salehin. 2015. The Ganges-Brahmaputra-Meghna delta system: Biophysical models to support analysis of ecosystem services and poverty alleviation. *Environmental Science-Processes and Impacts* 17 (6): 1016–1017. https://doi.org/10.1039/c5em90022k.

Toufique, K.A., and B. Belton. 2014. Is aquaculture pro-poor? Empirical evidence of impacts on fish consumption in Bangladesh. *World Development* 64: 609–620. https://doi.org/10.1016/j.worlddev.2014.06.035.

Walther, G.-R., E. Post, P. Convey, A. Menzel, C. Parmesan, T.J.C. Beebee, J.-M. Fromentin, O. Hoegh-Guldberg, and F. Bairlein. 2002. Ecological responses to recent climate change. *Nature* 416 (6879): 389–395. Licence https://doi.org/10.1038/416389a.

Whitehead, P.G., E. Barbour, M.N. Futter, S. Sarkar, H. Rodda, J. Caesar, D. Butterfield, L. Jin, R. Sinha, R. Nicholls, and M. Salehin. 2015a. Impacts of climate change and socio-economic scenarios on flow and water quality of the Ganges, Brahmaputra and Meghna (GBM) river systems: Low flow and flood statistics. *Environmental Science-Processes and Impacts* 17 (6): 1057–1069. https://doi.org/10.1039/c4em00619d.

Whitehead, P.G., S. Sarkar, L. Jin, M.N. Futter, J. Caesar, E. Barbour, D. Butterfield, R. Sinha, R. Nicholls, C. Hutton, and H.D. Leckie. 2015b. Dynamic modeling of the Ganga river system: Impacts of future climate and socio-economic change on flows and nitrogen fluxes in India and Bangladesh. *Environmental Science-Processes and Impacts* 17 (6): 1082–1097. https://doi.org/10.1039/c4em00616j.

World Bank. 2015. *Total population for Bangladesh.* World Bank Group. http://data.worldbank.org/indicator/SP.POP.TOTL/countries/BD?display=graph. Accessed 11 Aug 2015.

Zhang, B., and X. Jin. 2014. Feeding habits and ontogenetic diet shifts of Bombay duck, Harpadon nehereus. *Chinese Journal of Oceanology and Limnology* 32 (3): 542–548. https://doi.org/10.1007/s00343-014-3085-7.

26

Dynamics of the Sundarbans Mangroves in Bangladesh Under Climate Change

Anirban Mukhopadhyay, Andres Payo,
Abhra Chanda, Tuhin Ghosh,
Shahad Mahabub Chowdhury, and Sugata Hazra

26.1 Introduction

Mangroves are the third most productive and bio-diverse ecosystem in the world, followed by the tropical rain forests and the coral reefs (Lang'at and Kairo 2008). They are mostly confined to the tropics and subtropics covering approximately 75 per cent of the world's coastline between 25° N and 25° S (Borges et al. 2003). Mangroves provide a wide range of ecosystem services ranging from mitigation of global climate change by carbon capture to the sustenance of local communities, especially locally poor people (e.g. fishermen, honey collectors, etc.) whose livelihood depends upon the mangrove forest products (see Chap. 23). Operating as a natural barrier to hazards such

A. Mukhopadhyay (✉) • A. Chanda • T. Ghosh • S. Hazra
School of Oceanographic Studies, Jadavpur University, Kolkata, India

A. Payo
British Geological Survey, Keyworth, Nottingham, UK

S. M. Chowdhury
International Union for Conservation of Nature, Dhaka, Bangladesh

© The Author(s) 2018
R. J. Nicholls et al. (eds.), *Ecosystem Services for Well-Being in Deltas*,
https://doi.org/10.1007/978-3-319-71093-8_26

as tropical cyclones and tsunamis, mangroves also help in the protection of shoreline areas and prevent coastal erosion (Das and Vincent 2009). Additionally, they act as biological filters, maintaining water quality in coastal regions and providing nursing grounds for a number of diverse flora and fauna species particularly fish. However, mangroves are extremely sensitive to changing environmental conditions which make these ecosystems highly endangered around the globe (Hutchison et al. 2014). Around one-third of the world's mangroves have been lost in the last 50 years (Alongi 2002), and some scientists have suggested that the entire mangrove community might be lost by the end of the twenty-first century (Duke et al. 2007). The need for assessment of mangrove forest trends in the context of global change is therefore paramount (Polidoro et al. 2010; Duke et al. 2007; Spencer et al. 2016).

The Sundarbans, situated between 88°55′ E to 89° E and 21°30′ N to 23°30′ N, comprise the world's largest single block of mangrove forest. The region is estimated to have come into existence 4000 years ago (Ali 1998) and is shared between India (4,000 km²) and Bangladesh (6,000 km²). In 1996, the entire Bangladesh Sundarbans mangroves were declared a 'reserve forest' and, in 1997, were designated as a 'UNESCO World Heritage Site'.[1] The Sundarbans Reserve Forest (SRF) is home to around 700 species of flora and fauna giving shelter to a number of endangered species, notably the Royal Bengal Tiger (*Panthera tigris*) (Rahman 2009). Income generated from the forest ecosystem services is extremely important, accounting for 74 per cent and 48 per cent of total household income for lower- and middle-income households residing in some areas of the Sundarbans (Abdullah et al. 2016). This implies that the forest produces play a major role in the poverty alleviation of the local communities, particularly those people who are exclusively dependent on the forest derived products for their livelihood.

This chapter highlights the potential impacts that the Bangladesh Sundarbans ecosystem (Fig. 26.1) faces in the future due to ongoing sea-level rise and climate change and suggests responses that should be taken into account when framing suitable policy options. These options are based on assessing the extent of forest cover and changing species composition associated with physical change and its consequences for forest carbon stocks and ecosystem services that would affect the livelihood of the local communities.

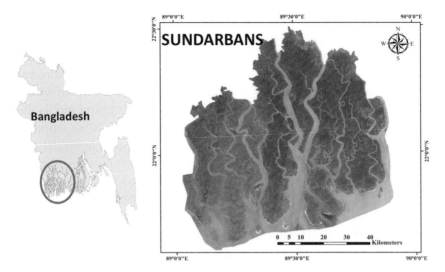

Fig. 26.1 Map showing the extent of the Bangladesh Sundarbans study area

26.2 Climate Change and Mangroves

In order to assess the effect of climate change on the mangroves of Bangladesh Sundarbans, two principal approaches were adopted for the present research. One discusses the Bangladesh Sundarbans after a century of geomorphological change under three different sea-level rise scenarios using the Sea Level Affecting Marshes Model (SLAMM) model (Sect. 26.2.1). This focusses on physical land loss due to inundation and erosion associated with projected sea-level rise. Secondly, potential changes in species assemblage during this century, and hence the blue carbon stock and ecosystem services, are identified under the Business As Usual (BAU) scenario using a hybrid model of cellular automata and Markov chain (Sects. 26.2.2, 26.2.3 and 26.3). This research emphasises that mangrove loss is not exclusively associated with extent but also due to changes in the species composition and loss.

26.2.1 Effects of Regional Sea-Level Rise

With global mean sea-level rise projected as up to 0.98 m or greater by 2100 relative to the baseline period (1985–2005), the Sundarbans—mean elevation currently approximately 2 m above mean sea level—is at risk from inundation and subsequent wetland loss; however the magnitude of loss remains unclear. Huq et al. (1995) suggested that 'one meter rise in mean sea level will probably lead to the destruction of the Sundarbans by its complete inundation'. Loucks et al. (2010) estimated that most of the Sundarbans in Bangladesh will be overwhelmed with a rise in sea level of 0.28 m over the next 50–90 years assuming no change in the local net rate of sea-level rise of 4–7.8 mm/year. More recently, based on the concept of the loss of 'elevation capital' (the potential of a mangrove ecosystem to remain within a suitable inundation regime between highest astronomical tide (HAT) and mean sea level (MSL)), Lovelock et al. (2015) suggested that the Sundarbans will be able to sustain itself beyond the year 2100 even under high rates of sea-level rise (1.4 m by 2100). Thus it is clear there is no consensus on the future of the Bangladesh Sundarbans under sea-level rise by the end of this century.

To address this uncertainty remote sensing data and field measurements, geographic information systems and simulation modelling are applied to investigate the potential effects of different sea-level rise scenarios on the Sundarbans (Fig. 26.2). Assuming that MSL is a better proxy for mangrove area delineation, the SLAMM model is able to reproduce the observed area losses for the period 2000–2010 (Payo et al. 2016). Using this calibrated model, the estimated mangrove area net losses (relative to year 2000) are estimated to be 81–178 km^2 (two to five per cent), 111–376 km^2 (three to ten per cent) and 583–1393 km^2 (15–37 per cent) for sea-level rise scenarios to 2100 of 0.46 m, 0.75 m and 1.48 m, respectively, with net subsidence of ±2.5 mm/year. Where relative sea-level rise is less than 0.75 m, these area losses are very small (less than ten per cent of present day mangrove area of 3,778 km^2) and significantly smaller than Lovelock et al. (2015) has suggested. Simulations also suggest that erosion rather than inundation may remain the dominant loss driver to 2100 under certain scenarios. Only under the highest scenarios of relative sea-level rise does inundation due to sea-level rise become the dominant loss process, suggesting that the mangrove system will persist until 2100.

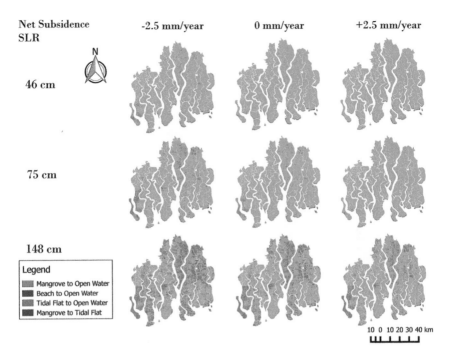

Fig. 26.2 Simulated mangrove area losses by the year 2100 under nine different relative sea-level rise scenarios (see also Payo et al. (2016))

26.2.2   Species Assemblage Change

One of the most salient aspects of the Bangladesh Sundarbans mangrove ecosystem is its rich species diversity. The wide variety of floral species in this ecosystem largely accounts for the overall high biodiversity of this region. This high heterogeneity makes characterising the species assemblage and predicting future trends a challenging task.

Analysis of the spatial distribution of mangrove species assemblages is carried out by implementing a hybrid model of Markov chain and cellular automata in business as usual scenario (detailed methodology can be found in Mukhopadhyay et al. 2015). Results show that species distribution will alter substantially within a hundred years alongside a significant decrease in the forest cover area (Fig. 26.3). The areal distribution

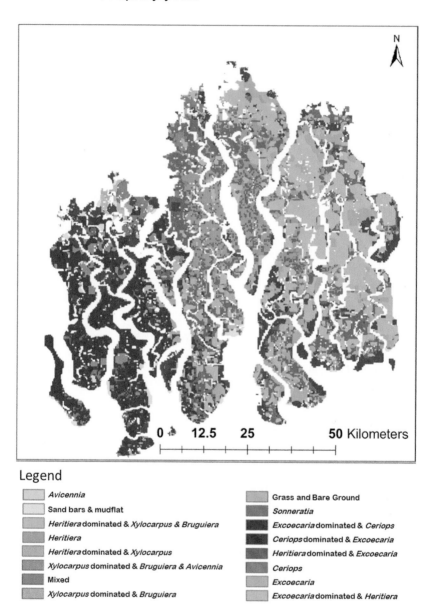

Fig. 26.3 Species assemblages of Bangladesh Sundarbans after 2100 (see also Mukhopadhyay et al. (2015))

of species assemblages with the following dominant species decrease in this order: (1) Goran (*Ceriops*) (2) Sundari (*Heritiera*), (3) Passur (*Xylocarpus*) and (4) Baen (*Avicennia*). On the other hand, assemblages with the following dominant species are predicted to increase: (1) Gewa (*Excoecaria*), (2) Keora (*Sonneratia*) and (3) Kankra (*Bruguiera*). This is potentially related to changes in salinity patterns. Karim (1988) and Hoque et al. (2006) previously differentiated the salinity zones of the Sundarbans into low saline (Oligohaline), moderate saline (Mesohaline) and high saline (Polyhaline) zones. Based on these categories and within the overall decline of 17 per cent of forest area, it is observed that the freshwater-dependent mangroves like that of *Heritiera* (Oligohaline) would diminish in abundance while there is an increase in the moderate to high salt-tolerant species such as *Excoecaria*, *Avicennia* and *Bruguiera*.

As mentioned earlier, within the next 100 years, almost one-fifth (17 per cent) of the existing forest would disappear from this ecosystem, under the BAU scenario. Extreme episodic events (e.g. cyclonic disaster, tsunami, etc.), which are not taken into account in this study, might make the scenario worse. Apart from the forest cover loss, the modelled output also highlights that low saline species along with their characteristic produce and ecosystem services would decline in future.

It is important to note that the predictions, as discussed in Mukhopadhyay et al. (2015), assume that the trend of environmental factors such as salinity, temperature, rainfall and sea-level rise leading to these changes will remain the same for the next few decades. However, with the anticipated accelerating changes in the previously mentioned controlling factors, these impacts may be underestimates.

26.2.3   Blue Carbon Stock

Forests, including mangroves, are acknowledged to play a crucial role in mitigating global climate change. Mangroves constitute rich soil carbon content up to several metres depth (Donato et al. 2011) and a high below-ground carbon content in their root system in comparison to other tropical

forests (Lovelock 2008). They therefore have high rates of carbon sequestration in both above- and below-ground live biomass (Alongi 2012).

Chanda et al. (2016) assessed the changes in the Bangladesh
Sundarbans's blue carbon stock (i.e. the carbon captured by ocean and
coastal ecosystems in the form of biomass and sediments) over the last 30
years and, based on observed trends, generated a plausible scenario of the
future carbon stock (after a hundred years). Applying this scenario with a
hybrid model of Markov chain and cellular automata (a similar approach
can be found in Mukhopadhyay et al. 2015), the change in blue carbon
stock is assessed. The magnitude of the above- and below-ground biomass of the various prominent species assemblages at present was acquired
from Rahman et al. (2015), and the carbon concentration of the various
species was measured by sampling in selected points. Combining these
datasets, the total blue carbon stock of all the species composition classes
was estimated (see Mukhopadhyay et al. (2015) for details). At present,
36.24 Tg C^2 is stored above ground and 54.95 Tg C below ground in this
forest resulting in a total blue carbon stock of 91.19 Tg C. According to
the modelling, 15.88 Tg C (17.4 per cent of current capacity) would be
lost from the area by the year 2115.

The low saline *Heritiera*-dominated species composition classes currently account for the major portion of the carbon sock (45.60 Tg C;
almost 50 per cent of the total carbon stock), while the remainder is
locked up in the moderate to high saline species. The predictions revealed
that almost 22.42 Tg C would be lost from low saline regions accompanied by an increase of 8.20 Tg C in the high saline regions dominated
mainly by *Excoecaria* and *Avicennia*. Low saline mangrove species are
capable of higher carbon sequestration and photosynthetic carbon
assimilation compared to those that thrive in high saline zones (Rahman
et al. 2015; Nandy and Ghose 2001). As already shown, due to the
anticipated changes in salinity, this class of species would be most
affected leading to a net forest carbon loss of 15.88 Tg C, and the
increase in the moderate to high saline mangroves is insufficient to
counterbalance the blue carbon loss from the low saline regions
(Fig. 26.4). Hence, a substantial net forest area loss along with the

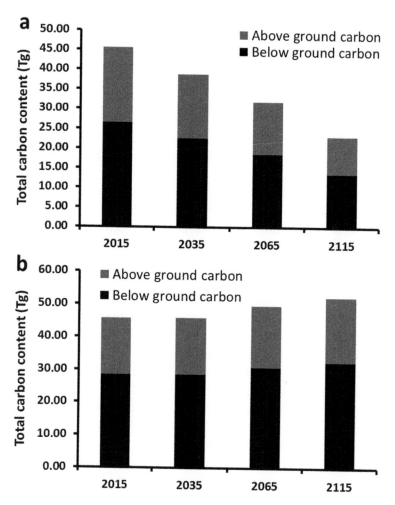

Fig. 26.4 Present and predicted trend of total carbon content in the (*a*) low saline and (*b*) moderate to high saline zones

changes in species assemblage accompanied by carbon dioxide (CO_2) emission of 1121 Mg CO_2 ha/year is anticipated over the next 100 years. At the current cost of new atmospheric carbon (Polidoro et al. 2010), almost US$2.26 billion worth of carbon will be lost from the Bangladesh Sundarbans by the year 2115.

26.3   Sundarbans Mangrove Ecosystem Services

Although mangroves are known for a wide range of ecosystem services, there have been few attempts to quantify and predict the future benefits provided by these ecosystems to both human beings and the climate as a whole. The Millennium Ecosystem Assessment classification of ecosystem services is probably the most well-known classification scheme (MEA 2005). The Sundarbans mangrove ecosystem services include provisioning, for example, food, fuel wood, protection and a nursery for fish and crustaceans, and regulating/supporting, for example, by providing security to coastal communities from cyclones, wind action, wave action and currents and reducing damages to housing, infrastructure and food sources by reducing saline water intrusion. The mangroves also help to stop erosion as roots hold sediments in place and maintain the biotic water quality of the region, which in turn helps the fishing community. Another service extended by these mangroves include items like timber, fuel wood, thatching materials, honey and waxes, thus supporting the local as well as the national economy (see Chaps. 2 and 23). More than 200 tonnes of honey and 50 tonnes of beeswax are harvested per year from the Bangladesh Sundarbans under the supervision of Forest Department. The Sundarbans accounts for about 50 per cent of the honey produced in Bangladesh (Gani 2001). By also providing cultural services and having recreational values in the form of eco-tourism, the Sundarbans mangrove ecosystem is one of the most valued ecosystems of the world.

Based on the changes predicted in species assemblages (Sect. 26.2.2), it can be observed that species like *Heritiera* and *Ceriops* are likely going to reduce in future, whereas the abundance of species like *Excoecaria* and *Bruguiera* would increase. This implies that in the future, this ecosystem will not be able to provide the beneficial services such as high carbon sequestration rate (Nandy and Ghose 2001), good timber and a few traditional uses of medicine (which is the speciality of *Heritiera* and *Ceriops*). There is also likely to be a decline in the regulating and supporting services like storm protection, shoreline protection and enhanced primary productivity. On the other hand, with increasing abundance of *Excoecaria* and *Bruguiera*, products from these species such as soft wood for making furniture and ornaments and medicinal values such as anti-tumour promoting drugs (Konoshima et al. 2001) would be better utilised. Increases in *Excoecaria* and *Bruguiera* might also be beneficial for ecosystem services such as soil formation and retention along with organic matter and nutrient enrichment in the forest floor. However, on the whole since the forest extent is predicted to decline by the end of 2100, the magnitude of ecosystem services is expected to decline, especially from those species which are less salt tolerant.

26.4 Management Strategy and Policy Framing

The foremost outcome that is evident from this research is that the mangrove forest cover of Bangladesh Sundarbans is expected to reduce in the future, although not at the alarming rate predicted by previous studies. The most adverse effects of this reduction in mangrove area would be the loss in both quality and quantity of ecosystem services that this ecosystem provides to the local community. In order to compensate for this loss in the main reserve forest area, substantial afforestation is required in the fringe areas and the less vulnerable areas of the forest. Although this kind of afforestation is already underway, local communities could be made more aware of it and their participation in such programmes encouraged. Also, instead of the current mono-species culture, afforestation programmes should be envisaged in such a way that the species diversity and heterogeneity of this ecosystem remain conserved. Special emphasis may be given to the freshwater-loving species like *Heritiera* and *Ceriops* in favourable locations. Mangrove plantation efforts could also be undertaken more in non-forest areas creating a toe line bio-shield for existing embankments.

This re-afforestation would not only strengthen the ecosystem's carbon and biomass stock but also provide easily available forest products which could be utilised by the local poor people for their daily life sustenance and serve as an alternative livelihood option. At present, among the various ecosystem services that this forest provides, collection of honey along with wax are very crucial livelihood options for extremely poor people. In the future, the climate change-driven mangrove forest cover loss could substantially hamper natural production of honey and wax, accentuating the man-animal conflict during honey collection. It may therefore be advisable to promote apiculture in the vicinity of settlements as well as in the core forest areas as a sustainable livelihood option for the local population. This would ensure not only a secure income but the incidences of injury or even casualty during honey collection could be avoided.

A substantial amount of carbon is stored in the forest of Bangladesh Sundarbans at present. This sequestration can play a crucial role in combating the anthropogenic increase of CO_2 in the atmosphere. The natural performance of this mangrove ecosystem which in turn is beneficial to the global climate might be developed as a wealth-creating instrument under a well-functioning, well-informed market mechanism. However, unless a market mechanism, conferring rights to the forest-dependent community to trade carbon credit to the global community, is developed, sustainable development in the delta will remain elusive.

Notes

1. See http://whc.unesco.org/en/list/798
2. 1 Tg C = 1 Teragram carbon = 10^9 kg carbon.

References

Abdullah, A.N.M., N. Stacey, S.T. Garnett, and B. Myers. 2016. Economic dependence on mangrove forest resources for livelihoods in the Sundarbans, Bangladesh. *Forest Policy and Economics* 64: 15–24. https://doi.org/10.1016/j.forpol.2015.12.009.

Ali, S.S. 1998. *Sundarbans: Its resources and eco-system.* Proceedings of the national seminar on Integrated Management of Ganges Flood Plains and Sundarbans Ecosystem, 16–18 July 1994, Khulna University, Khulna.

Alongi, D.M. 2002. Present state and future of the world's mangrove forests. *Environmental Conservation* 29 (3): 331–349. https://doi.org/10.1017/s0376892902000231.

————. 2012. Carbon sequestration in mangrove forests. *Carbon Management* 3 (3): 313–322. https://doi.org/10.4155/cmt.12.20.

Borges, A.V., S. Djenidi, G. Lacroix, J. Theate, B. Delille, and M. Frankignoulle. 2003. Atmospheric CO_2 flux from mangrove surrounding waters. *Geophysical Research Letters* 30 (11). https://doi.org/10.1029/2003gl017143.

Chanda, A., A. Mukhopadhyay, T. Ghosh, A. Akhand, P. Mondal, S. Ghosh, S. Mukherjee, J. Wolf, A.N. Lázár, M.M. Rahman, M. Salehin, S.M. Chowdhury, and S. Hazra. 2016. Blue carbon stock of the Bangladesh Sundarbans mangroves: What could be the scenario after a century? *Wetlands* 36 (6): 1033–1045. https://doi.org/10.1007/s13157-016-0819-7.

Das, S., and J.R. Vincent. 2009. Mangroves protected villages and reduced death toll during Indian super cyclone. *Proceedings of the National Academy of Sciences of the United States of America* 106 (18): 7357–7360. https://doi.org/10.1073/pnas.0810440106.

Donato, D.C., J.B. Kauffman, D. Murdiyarso, S. Kurnianto, M. Stidham, and M. Kanninen. 2011. Mangroves among the most carbon-rich forests in the tropics. *Nature Geoscience* 4 (5): 293–297. https://doi.org/10.1038/ngeo1123.

Duke, N.C., J.-O. Meynecke, S. Dittmann, A.M. Ellison, K. Anger, U. Berger, S. Cannicci, K. Diele, K.C. Ewel, C.D. Field, N. Koedam, S.Y. Lee, C. Marchand, I. Nordhaus, and F. Dahdouh-Guebas. 2007. A world without mangroves? *Science* 317 (5834): 41–42. https://doi.org/10.1126/science.317.5834.41b.

Gani, M.O. 2001. *The giant honeybee (Apis dorsata) and honey hunting in Sundarbans reserved forests of Bangladesh*. Apimondia 2001: Proceedings of the 37th International Apicultural Congress, 28 October–1 November, Durban.

Hoque, M.A., M.S.K.A. Sarkar, A.S.K.U. Khan, M.A.H. Moral, and A.K.M. Khurram. 2006. Present status of salinity rise in Sundarbans area and its effect on Sundari (Heritiera fomes) species. *Research Journal of Agriculture and Biological Sciences* 2 (3): 115–121.

Huq, S., S.I. Ali, and A.A. Rahman. 1995. Sea-level rise and Bangladesh: A preliminary analysis. *Journal of Coastal Research SI* 14: 44–53.

Hutchison, J., A. Manica, R. Swetnam, A. Balmford, and M. Spalding. 2014. Predicting global patterns in mangrove forest biomass. *Conservation Letters* 7 (3): 233–240. https://doi.org/10.1111/conl.12060.

Karim, A. 1988. *Environmental factors and the distribution of mangroves in the Sundarbans with special reference to Heritiera fomes*. Ph.D. thesis, University of Calcutta.

Konoshima, T., T. Konishi, M. Takasaki, K. Yamazoe, and H. Tokuda. 2001. Anti-tumor-promoting activity of the diterpene from Excoecaria agallocha. II. *Biological and Pharmaceutical Bulletin* 24 (12): 1440–1442. https://doi.org/10.1248/bpb.24.1440.

Lang'at, J.K., and J.G. Kairo. 2008. *Conservation and management of mangrove forests in Kenya*. Mangrove Reforestation Program. Mombasa: Kenya Marine and Fisheries Research Institute. www.wrm.org.uy/…on_and_managemen_mangrove_Kenya.pdf. Accessed 5 May 2016.

Loucks, C., S. Barber-Meyer, M.A.A. Hossain, A. Barlow, and R.M. Chowdhury. 2010. Sea level rise and tigers: Predicted impacts to Bangladesh's Sundarbans mangroves. *Climatic Change* 98 (1–2): 291–298. https://doi.org/10.1007/s10584-009-9761-5.

Lovelock, C.E. 2008. Soil respiration and belowground carbon allocation in mangrove forests. *Ecosystems* 11 (2): 342–354. https://doi.org/10.1007/s10021-008-9125-4.

Lovelock, C.E., D.R. Cahoon, D.A. Friess, G.R. Guntenspergen, K.W. Krauss, R. Reef, K. Rogers, M.L. Saunders, F. Sidik, A. Swales, N. Saintilan, L.X. Thuyen, and T. Triet. 2015. The vulnerability of Indo-Pacific mangrove forests to sea-level rise. *Nature* 526 (7574): 559–U217. https://doi.org/10.1038/nature15538.

MEA. 2005. *Ecosystems and human well-being: Synthesis*. Millennium Ecosystem Assessment (MEA). Washington, DC: Island Press. http://www.millenniumassessment.org/documents/document.356.aspx.pdf. Accessed 1 Aug 2016.

Mukhopadhyay, A., P. Mondal, J. Barik, S.M. Chowdhury, T. Ghosh, and S. Hazra. 2015. Changes in mangrove species assemblages and future prediction of the Bangladesh Sundarbans using Markov chain model and cellular automata. *Environmental Science-Processes and Impacts* 17 (6): 1111–1117. https://doi.org/10.1039/c4em00611a.

Nandy, P., and M. Ghose. 2001. Photosynthesis and water-use efficiency of some mangroves from Sundarbans, India. *Journal of Plant Biology* 44 (4): 213–219.

Payo, A., A. Mukhopadhyay, S. Hazra, T. Ghosh, S. Ghosh, S. Brown, R.J. Nicholls, L. Bricheno, J. Wolf, S. Kay, A.N. Lázár, and A. Haque. 2016. Projected changes in area of the Sundarbans mangrove forest in Bangladesh due to SLR by 2100. *Climatic Change*: 1–13. https://doi.org/10.1007/s10584-016-1769-z.

Polidoro, B.A., K.E. Carpenter, L. Collins, N.C. Duke, A.M. Ellison, J.C. Ellison, E.J. Farnsworth, E.S. Fernando, K. Kathiresan, N.E. Koedam, S.R. Livingstone, T. Miyagi, G.E. Moore, N.N. Vien, J.E. Ong, J.H. Primavera, S.G. Salmo, J.C. Sanciangco, S. Sukardjo, Y.M. Wang, and J.W.H. Yong. 2010. The loss of species: Mangrove extinction risk and geographic areas of global concern. *PLoS One* 5 (4). https://doi.org/10.1371/journal.pone.0010095.

Rahman, M.M., M.N.I. Khan, A.K.F. Hoque, and I. Ahmed. 2015. Carbon stock in the Sundarbans mangrove forest: spatial variations in vegetation types and salinity zones. *Wetlands Ecology and Management* 23 (2):269–283. https://doi.org/10.1007/s11273-014-9379-x Licence journal article.

Spencer, T., M. Schuerch, R.J. Nicholls, J. Hinkel, D. Lincke, A.T. Vafeidis, R. Reef, L. McFadden, and S. Brown. 2016. Global coastal wetland change under sea-level rise and related stresses: The DIVA Wetland Change Model. *Global and Planetary Change* 139: 15–30. https://doi.org/10.1016/j.gloplacha.2015.12.018.

27

Hypertension and Malnutrition as Health Outcomes Related to Ecosystem Services

Ali Ahmed, Mahin Al Nahian, Craig W. Hutton, and Attila N. Lázár

27.1 Introduction

This chapter explores the well-being of the coastal population of Bangladesh in relation to their ecosystem services, livelihoods, food consumption and health. This is achieved by exploring the specific nature of the association between human health and well-being with the provision of nutrition

---

A. Ahmed (✉) • M. Al Nahian
Climate Change and Health, International Center for Diarrheal Disease
Research, Bangladesh (icddr,b), Dhaka, Bangladesh

C. W. Hutton
Geodata Institute, Geography and Environment, University of Southampton,
Southampton, UK

A. N. Lázár
Faculty of Engineering and the Environment and Tyndall Centre for Climate
Change Research, University of Southampton, Southampton, UK

© The Author(s) 2018
R. J. Nicholls et al. (eds.), *Ecosystem Services for Well-Being in Deltas*,
https://doi.org/10.1007/978-3-319-71093-8_27

and food diversity by disaggregating elements of this relationship within the seven social-ecological systems (SESs) (see Chap. 22). The specific socio-economic and ecological service components of each SES are identified using data from the sampled household surveys (see, Chap. 23).

Human health and well-being status is closely linked with local ecosystem (see Chap. 7 and MEA 2003). However, the causal relation is complex and often hard to define as they are indirect, displaced in space and time and dependent on a number of modifying forces (Corvalan et al. 2005). The human–ecosystem interaction and how this interaction impacts health is yet to be properly understood or consequently used in policy level decision-making processes.

This work demonstrates how economic conditions, educational levels, environmental conditions and dietary intake differ across SESs and, in turn, is associated with differing health status. A key area of research is the impact of saline water intrusion across the SESs which has critical implications for health, with significant positive associations between hypertension and salinity in drinking water and dietary impact on under-five malnutrition.

27.2 Methods

The analysis in this chapter is based on the primary data collected from three cross-sectional household surveys, as described in Chap. 23. Drinking water samples, anthropometry and blood pressure data are collected from household survey participants. Water samples from primary drinking water sources (except bottled and rain water) are tested with a HACH sensION5 electric conductivity metre and blood pressure measured with a OMRON M2 blood pressure monitor. Measurements of individual height and weight were collected from one eligible male (18–54 years), one female (15–49 years) and the oldest under-five child in the family. Blood pressure is collected for male and female respondents only from the 15–59 age group according to selection criteria based on kinship. Standard definitions of malnutrition and hypertension are used in the study:

- Malnutrition: Malnutrition is a condition which results when a person's diet fails to provide adequate nutrients for growth and maintenance or when a physical condition cannot utilise the food consumed due to illness.

Malnutrition is observed as deficiencies such as thinness, stunting and micronutrient deficiencies, but also as overweight and obesity, known as overnutrition (UNICEF 2012). Under-five child malnutrition is assessed using the World Health Organization (WHO) child growth standard (Z-score) (WHO 2007). Adult health status is assessed using adult body mass index (BMI) category suggested for the Asian communities (Barba et al. 2004).

• Hypertension: Respondent's blood pressure is generated using the average value of last two readings of systolic and diastolic blood pressure. Thresholds for 'normal' are (i) systolic blood pressure (SBP) under 120 mmHg and (ii) the diastolic blood pressure (DBP) under 80 mmHg suggested by the American Heart Association (AHA 2016) and followed in national surveys (NIPORT et al. 2013).

Descriptive statistics are used to show the household characteristics and socio-demographic status of the study area. Health is assessed in terms of under-five child malnutrition and adult health status outcomes by prevalence of underweight, obesity and hypertension. Overall analysis considers explanatory variables of age, sex, educational status, wealth quintiles, seasons, structure of dwellings, occupation, dietary intake, landownership and salinity level in drinking water, and these are mapped across the different SESs. In all the graphs in this chapter, the SESs are listed in order according to the mean raw score of wealth index (poorest (left) to relatively less poor (right)[1]).

27.3   Results and Discussion

Malnutrition and hypertension prevalence was explored for the south-central and south-west coastal Bangladesh. Bivariate analysis is used to map the association between SESs for both under-five child malnutrition and adult malnutrition and hypertension.

27.3.1   Under-Five Child Malnutrition

To assess the malnutrition of under-five children, three main indices are used: (i) height-for-age (stunting), (ii) weight-for-height (wasting) and (iii) weight-for-age (underweight). In Bangladesh, the prevalence of stunting,

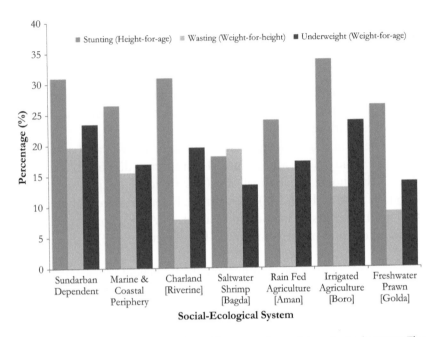

Fig. 27.1 Under-five children malnutrition status by social-ecological system. The SESs are organised left to right order of mean raw score of wealth index (left: poorest to right: relatively least poor)

wasting and underweight for under-five children is roughly 36, 14 and 33 per cent, respectively (NIPORT et al. 2016).

Figure 27.1 shows the under-five child malnutrition status in the SESs. In the study area, the percentage of stunting, wasting and underweight is about 28, 14 and 19, respectively, which is comparable to the national findings. With some noticeable exceptions, there is a weak trend of a decline in all three malnutrition indicators across the SESs as relative wealth increases (i.e. highest in the poorest Sundarbans SES to lowest in the relatively better-off freshwater prawn SES), a trend which may be linked to both access to food diversity including fish, which is discussed in relation to Figs. 27.3 and 27.4.

There are some differences in per capita calorie and protein intake (calculated at household level for all adult and child respondents) among the SESs (Fig. 27.2). The lowest per capita calorie intake is found in Charland

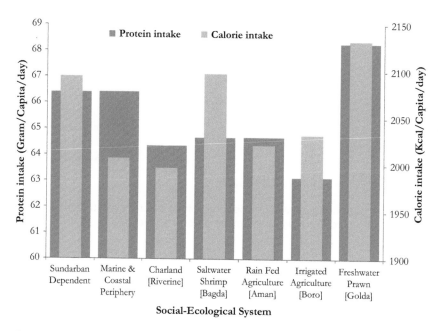

Fig. 27.2 Per capita mean calorie and protein intake per day across the seven social-ecological systems within the delta region

SES and the lowest protein intake in the irrigated agriculture SES. The highest calorie and protein intake is found in the freshwater prawn (*Galda*) SES which supports the comparatively high welfare status of this particular SES.

By investigating the diet patterns within the data, fish consumption might be considered one of the key elements particularly in regards to child malnutrition. Fish consumption and under-five child malnutrition status are found to be inversely associated, with some minor exceptions (Fig. 27.3). Among the total protein source, the proportion of fish consumption is lower among poorer households which consequently suffer more from under-five child malnourishment. The higher the fish consumption with diet, the lower the under-five child malnutrition, suggesting fish provide an important safety net for the poorest households against child malnutrition. The analysis also shows (not plotted) that households consuming domestically produced food and vegetable from homestead gardening have lower likeli-

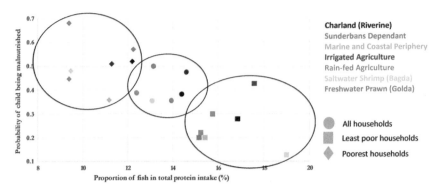

Fig. 27.3 Association between calorie intake, food diversity and under-five child malnutrition by social-ecological system and poverty level

hoods of child malnutrition than households that mainly purchase the food items. Overall findings suggest that, along with poverty, access to nutrition through total dietary diversity is influencing the under-five child malnutrition in the coastal areas. However, model simulations (see Chap. 25) indicate that the total availability of the fish in the Bay of Bengal is likely to decrease under all projected scenarios. This will negatively impact food intake and dietary diversity and may lead to an increase in the prevalence of child malnutrition in coastal Bangladesh (Fernandes et al. 2016).

In this research, food diversity and calorific intake are found to be positively associated for children under five, and the minimum calorie threshold to substantially reduce under-five child malnutrition is 2000 kcal/person/day (Fig. 27.4). Increased calorie intake and food diversity are, in turn, associated with reduced poverty. Figures 27.3 and 27.4 show that, within each SES, the prevalence of under-five child malnutrition increases with the level of poverty, meaning highest malnutrition was prevalent in the lowest wealth quintile or among the poorest group. Both the food diversity index and protein and calorific value is high in the least poor wealth quintile as is the lower probability of child being malnourished; the opposite picture is visible among the poorest wealth quintile.

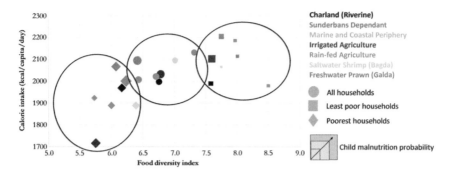

Fig. 27.4 The association between calorie intake, food diversity and under-5 child malnutrition. A marked decline in likelihood of malnutrition is identified at a food diversity index (number of types) of between 6 and 7

27.3.2   Adult Underweight, Overweight and Obesity

BMI is a critical indicator to understand adult nutritional status (Bailey and Ferro-Luzzi 1995). The household survey measured the anthropometry of males aged 18–54 years and females 15–49 years. To understand adult nutritional status, the standard measure, body mass index (BMI), is used to estimate the proportion of adults who are underweight, normal range, overweight or obese according to the health risk categories suggested for Asian communities by Barba et al. (2004): (i) underweight, BMI < 18.5 kg/m^2 (thin or underweight); (ii) normal range, 18.5 kg/m^2 ≤ BMI < 23 kg/m^2 (increasing but acceptable risk); (iii) overweight, 23 kg/m^2 ≤ BMI < 27.5 kg/m^2 (increased risk) and (iv) obese, BMI ≥ 27.5 kg/m^2 (higher risk). Health risks include the occurrence of hypertension, diabetes and cardiovascular disease.

Mean BMI was recently found to be 21.5 kg/m^2 in the coastal population which is slightly below than national average of 22.3 kg/m^2 (NIPORT et al. 2016). Among those surveyed in the study area, for those below the 50 per cent wealth quintile, the mean BMI was found to be 20.7 kg/m^2,

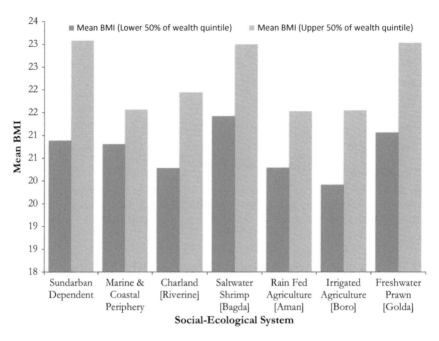

Fig. 27.5 Mean body mass index by wealth quintiles within the social-ecological systems

and for those above this wealth level the BMI reflects the national average. From Fig. 27.5, it is clear that this wealth differential in BMI is also reflected at SES level, a higher mean BMI indicating that the wealthier are more at risk of health issues than those of a poorer wealth status.

Calorie consumption has a direct impact on health and may reasonably be expected to be influenced by poverty and the broader socio-economic status of the community. Generally, the better-off economic situations should relate to good adult nutritional status, but this is not observed in this study. Age and sex disaggregated results (Figs. 27.6 and 27.7) clearly show that adult malnutrition is higher in women for all age groups and in all SESs (except for under 35s in the brackish shrimp (*Bagda*) areas), which may be linked to the established phenomenon of women missing meals to ensure food for other household members or only taking food at the end of a meal. However, women are also considerably more likely than men to be overweight or obese.

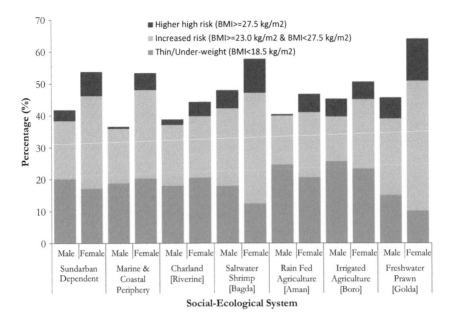

Fig. 27.6 Percentage of adults within each body mass index category by social-ecological system and sex

The overall adult malnutrition (by age and sex) is highest in freshwater prawn (*Golda*), brackish shrimp (*Bagda*) and Sundarban-dependent SESs (Figs. 27.6 and 27.7).

Coastal population from both aquaculture SESs is at higher risk of malnutrition due to overweight and obesity, whereas people in the agriculture SESs are more at risk of being underweight. Along with poverty incidence and dietary intake, local livelihood pattern might have some significant impact on adult nutritional status. However, this will require further investigation to properly understand within the context of the ecosystem-based SES framework.

27.3.3   Adult Hypertension

The World Health Organisation (WHO) identifies hypertension as one of the most important causes of premature death worldwide (WHF 2017). Hypertension is diagnosed through blood pressure measurements.

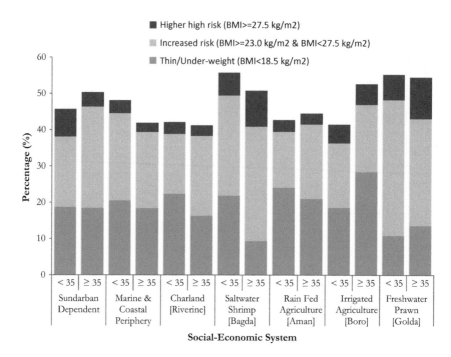

Fig. 27.7 Percentage of adults within each body mass index category by social-ecological system and age

During the household survey, blood pressure of one eligible male and female was collected, and the average value of last two readings from the three readings of SBP and DBP was used to report respondent's blood pressure as well as to maintain comparability with national statistics (NIPORT et al. 2013). The American Heart Association's definition was used for classifying blood pressure: (i) normal (systolic blood pressure < 120 mmHg and diastolic blood pressure < 80 mmHg), (ii) pre-hypertensive (systolic blood pressure 120–139 mmHg or diastolic blood pressure 80–89 mmHg) and (iii) hypertensive (systolic blood pressure ≥ 140 mmHg or diastolic blood pressure ≥ 90 mmHg) (AHA 2016).

Increased blood pressure associated with drinking water salinity has become an immerging public health concern in different parts of the world (Khan et al. 2014; Talukder et al. 2016). This is a concern in coastal Bangladesh, as about 80 per cent of coastal population depends

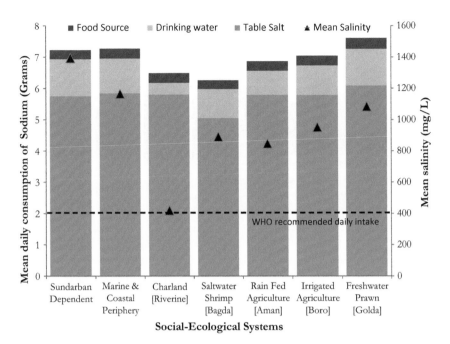

Fig. 27.8 Comparison of per capita daily sodium consumption (grammes) from different sources and mean salinity by social-ecological system (dashed line represents the WHO daily intake recommendation)

on groundwater for drinking purpose (Shamsudduha 2013), a critical ecosystem service in this deltaic country.

Sodium intake from a range of sources was collected: drinking water, food and table salt (Fig. 27.8). From the collected data, the average drinking water salinity across the SESs is found to be 854 mg/l (higher for tube wells where mean salinity is 915 mg/l); although mean salinity level is close or higher than the Bangladesh safe drinking water standard of 1000 mg/l (Ahmed and Rahman 2000) in almost every SESs except Riverine Charland. Furthermore, people's daily per capita sodium intake is much higher than the WHO recommendation (two grammes per capita per day) in coastal Bangladesh (WHO 2012; Rasheed et al. 2014).

For the 35 years and above age group, the gender-specific adult hypertension among the coastal population is shown in Fig. 27.9, compared with national rural statistics (NIPORT et al. 2013). Women are found more

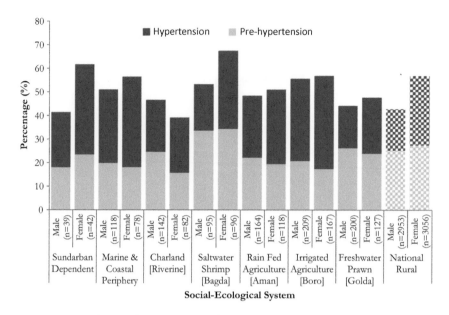

Fig. 27.9 Number and percentage of hypertension and pre-hypertension for males and females 35 and above by social-ecological system (National data from NIPORT et al. 2013)

hypertensive in all SESs compared to men except Charland SES. Both women and men from the irrigated agriculture SES are found to be most hypertensive among the respondents. The prevalence of hypertension is also higher than national rural statistics in most of the SESs for both men and women aged 35 years and above (Fig. 27.9). High blood pressure varies slightly among SESs and about 53 per cent women and 49 per cent male respondents are found to be pre-hypertensive and hypertensive when considering only tube wells as the primary water source.

For further analysis into the association between water salinity, blood pressure and age, drinking water salinity is divided into four categories by source and age into two categories. Firstly, sources of groundwater (tube well, stand post, public tap, etc.) are grouped under 'tube well' and all surface and other sources as 'non-tube well'. As around 97 per cent of the national population depends on underground sources for drinking water (Shamsudduha 2013), the 'tube-well' category was then further divided using (i) the safe water limit (freshwater < 1000 mg/l) following the Bangladesh guideline of safe drinking water (Ahmed and Rahman 2000), (ii) slightly saline (1000–2000 mg/l) and (iii) moderate saline (≥ 2000 mg/l); the latter two categories based on expertise and judgement in the absence of an official standard. A high saline category (≥ 3000 mg/l) was initially created but, due to very limited number of samples, was combined with moderate category. Respondents (only eligible male and female 15–59, no children) were also divided into two age categories namely 'below 35 (15–34 years, N = 2895; 46.6 per cent)' and '35 and above (35–59 years, N = 3319; 53.4 per cent)' to assess the hypertension prevalence in different age groups and also to compare with previous national studies.

The prevalence of high blood pressure (both pre-hypertension and hypertension) steeply increases with drinking water salinity concentration (Fig. 27.10). The results also indicate that the 35 years and above age group are more susceptible to hypertension than below 35 years age group.

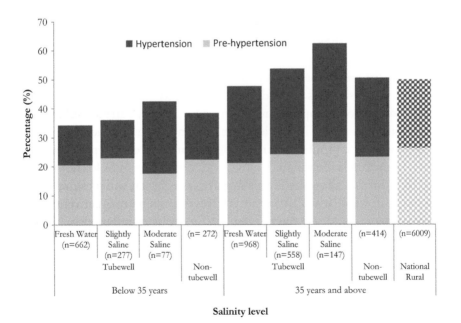

Fig. 27.10 Comparison of drinking water salinity levels with the occurrence of adult pre-hypertension and hypertension. Salinity levels: freshwater (<1000 mg/L), slightly saline (1000–2000 mg/L), moderately saline (>=2000 mg/L). (National data from NIPORT et al. 2013)

The prevalence of hypertension is found higher in the coastal areas than the national rural average in all age group (Fig. 27.10), when only tube-well water is considered. Among people who use non-tube-well water sources, in 35 years and above age group, the prevalence of hypertension is higher and pre-hypertension is lower than the national rural average, which means that there may be some other contributing factors such as dietary practice, livelihood and so on.

Future salinity projections for coastal Bangladesh (see Chaps. 17, 18 and 28) indicate a localised increase in groundwater salinity by mid- and end of the century. This is likely to increase the risk of hypertension prevalence in coastal community unless there is proper policy intervention.

27.4   Conclusion

Overall, in the coastal area of Bangladesh, under-five child malnutrition is lower in some SESs than the national average. However, to some extent, the simple relationship between poverty and malnutrition is disrupted by access to fish and other diverse food groups. Food diversity, sufficient calorific and fish intake and domestically grown food from homestead gardening and consumption are shown to be beneficial against child malnutrition, with a clear indication that there needs to be between six and seven major food groups within the diet of children before a steep decline in malnutrition is identified.

Fish protein is shown to have a negative association with child malnutrition even where it replaces other sources of protein, thus universal fish consumption could have major benefits in terms of population health. While this is already recognised, this work offers the tangible evidence to support policies and interventions to ensure universal access to, and affordability of, protein sources (see Chap. 2).

In terms of adult malnutrition, women are more vulnerable than men in terms of both being under and overweight. Being overweight or obese is more prevalent in higher wealth quintile group and being underweight in the poorer groups.

Hypertension is higher in the delta region than nationally, even when accounting for age and gender. Though table salt is the highest contributor in daily consumption, saline drinking water is another key predictor of adult hypertension. It has been projected that, climate and environmental change will likely exacerbate saline water intrusion into drinking water sources (see Chap. 28). The results here suggest that adaptation interventions and planning for this increased risk could effectively include community sensitisation to salt consumption practices, given the underlying environmental exposure.

Along with poverty incidence, dietary diversity and salinity intake have significant impact on adult nutrition and ultimately on health status, underscoring the direct interaction between the ecosystem services and well-being outcomes.

Note

1. As calculated using STATA 13.0 (http://www.stata.com/).

References

AHA. 2016. *About high blood pressure*. American Heart Association. http://www.heart.org/HEARTORG/Conditions/HighBloodPressure/GettheFactsAboutHighBloodPressure/The-Facts-About-High-Blood-Pressure_UCM_002050_Article.jsp. Accessed 28 Mar 2017.

Ahmed, M.F., and M. Rahman. 2000. *Water supply and sanitation: Rural and low income urban communities*. Dhaka: ITN-Bangladesh, Center for Water Supply and Waste Management, BUET.

Bailey, K., and A. Ferro-Luzzi. 1995. Use of body mass index of adults in assessing individual and community nutritional status. *Bulletin of the World Health Organization* 73 (5): 673–680.

Barba, C., T. Cavalli-Sforza, J. Cutter, I. Darnton-Hill, P. Deurenberg, M. Deurenberg-Yap, T. Gill, P. James, G. Ko, A.H. Miu, V. Kosulwat, S. Kumanyika, A. Kurpad, N. Mascie-Taylor, H.K. Moon, C. Nishida, M.I. Noor, K.S. Reddy, E. Rush, J.T. Schultz, J. Seidell, J. Stevens, B. Swinburn, K. Tan, R. Weisell, Z.S. Wu, C.S. Yajnik, N. Yoshiike, P. Zimmet, and W.H.O. Expert Consultation. 2004. Appropriate body-mass index for Asian populations and its implications for policy and intervention strategies. *The Lancet* 363 (9403): 157–163.

Corvalan, C., S. Hales, and A. McMichael. 2005. *Ecosystems and human well-being: Health synthesis.* Geneva: World Health Organisation (WHO).

Fernandes, J.A., S. Kay, M.A. Hossain, M. Ahmed, W.W. Cheung, A.N. Lazar, and M. Barange. 2016. Projecting marine fish production and catch potential in Bangladesh in the 21st century under long-term environmental change and management scenarios. *ICES Journal of Marine Science: Journal du Conseil* 73 (5): 1357–1369.

Khan, A.E., P.F. Scheelbeek, A.B. Shilpi, Q. Chan, S.K. Mojumder, A. Rahman, A. Haines, and P. Vineis. 2014. Salinity in drinking water and the risk of (pre)eclampsia and gestational hypertension in coastal Bangladesh: A case-control study. *PLoS One* 9 (9): e108715. https://doi.org/10.1371/journal.pone.0108715.

MEA. 2003. *Ecosystems and human well-being: A framework for assessment.* A report of the conceptual framework Working Group of the Millennium Ecosystem Assessment. Washington, DC: Island Press. http://pdf.wri.org/ecosystems_human_wellbeing.pdf. Accessed 01 Aug 2016.

NIPORT, Mitra and Associates, and ICF International. 2013. *Bangladesh Demographic and Health Survey 2011.* Dhaka/Rockville: National Institute of Population Research and Training (NIPORT), Mitra and Associates & ICF International.

———. 2016. *Bangladesh Demographic and Health Survey 2014.* Dhaka/Rockville: National Institute of Population Research and Training (NIPORT), Mitra and Associates, and ICF International.

Rasheed, S., S. Jahan, T. Sharmin, S. Hoque, M.A. Khanam, M.A. Land, M. Iqbal, S.M. Hanifi, F. Khatun, A.K. Siddique, and A. Bhuiya. 2014. How much salt do adults consume in climate vulnerable coastal Bangladesh? *BMC Public Health* 14 (584). https://doi.org/10.1186/1471-2458-14-584.

Shamsudduha, M. 2013. Groundwater-fed irrigation and drinking water supply in Bangladesh: Challenges and opportunities. In *Adaptation to the impact of climate change on socio-economic conditions of Bangladesh,* ed. A. Zahid, M.Q. Hassan, R. Islam, and Q.A. Samad, 150–169. Dhaka: Alumni Association of German Universities in Bangladesh, German Academic Exchange Service (DAAD).

Talukder, M.R.R., S. Rutherford, D. Phung, M.Z. Islam, and C. Chu. 2016. The effect of drinking water salinity on blood pressure in young adults of coastal Bangladesh. *Environmental Pollution* 214: 248–254. https://doi.org/10.1016/j.envpol.2016.03.074.

UNICEF. 2012. *Nutrition Glossary: A resource for communicators.* New York: United Nations Children's Fund (UNICEF). https://www.unicef.org/lac/ Nutrition_Glossary_(3).pdf. Accessed 7 June 2017.

WHF. 2017. *Hypertension: Hypertension and cardiovascular disease.* World Heart Foundation (WHF). http://www.world-heart-federation.org/cardiovascular-health/cardiovascular-disease-risk-factors/hypertension. Accessed 7 Feb 2017.

WHO. 2007. *WHO child growth standards: Head circumference-for-age, arm circumference-for-age, triceps skinfold-for-age and subscapular skinfold-for-age: Methods and development.* WHO child growth standards. Geneva: Multicentre Growth Reference Study Group, World Health Organization (WHO). http://www.who.int/childgrowth/publications/technical_report_2/en/. Accessed 1 Aug 2016.

———. 2012. *Guideline: Sodium intake for adults and children.* Geneva: World Health Organization (WHO). http://www.who.int/nutrition/publications/guidelines/sodium_intake_printversion.pdf. Accessed 1 Aug 2016.

Part 6

Integration and Dissemination

28

Integrative Analysis Applying the Delta Dynamic Integrated Emulator Model in South-West Coastal Bangladesh

Attila N. Lázár, Andres Payo, Helen Adams,
Ali Ahmed, Andrew Allan, Abdur Razzaque Akanda,
Fiifi Amoako Johnson, Emily J. Barbour, Sujit
Kumar Biswas, John Caesar, Alexander Chapman,
Derek Clarke, Jose A. Fernandes, Anisul Haque,
Mostafa A. R. Hossain, Alistair Hunt,
Craig W. Hutton, Susan Kay, Anirban Mukhopadhyay,
Robert J. Nicholls, Abul Fazal M. Saleh,
Mashfiqus Salehin, Sylvia Szabo,
and Paul G. Whitehead

28.1   Introduction

The research described in this book had a vision from its inception of developing a regional integrated assessment model to explore the future of coastal Bangladesh in terms of ecosystem services and human well-being. This requires a systematic framework which brings together all the individual components described in previous chapters, recognising and

A. N. Lázár (✉) • A. Chapman • D. Clarke • R. J. Nicholls
Faculty of Engineering and the Environment and Tyndall Centre for Climate
Change Research, University of Southampton, Southampton, UK

© The Author(s) 2018
R. J. Nicholls et al. (eds.), *Ecosystem Services for Well-Being in Deltas*,
https://doi.org/10.1007/978-3-319-71093-8_28

capturing important cause–effect associations and processes and their relative importance. Integrative assessment enables an analysis of a whole system, promoting the understanding of the importance of individual elements and providing insights into the future across a range of plausible scenarios. Such information enables science and policy processes to better understand current drivers and plausible development trajectories and consider how to steer that development towards favoured future states. In this way, the scientific endeavours described in this book can inform policy.

Achieving a successful, integrated representation of the delta system is highly reliant on the relationship between the integrated modellers and the rest of the project team. A frequent exchange of knowledge and ideas between team members is essential to ensure that system linkages, inputs and outputs are appropriately structured and agreed. This is a challenging

A. Payo
British Geological Survey, Keyworth, Nottingham, UK

H. Adams
Department of Geography, King's College London, London, UK

A. Ahmed
Climate Change and Health, International Center for Diarrheal Disease Research, Bangladesh (icddr,b), Dhaka, Bangladesh

A. Allan
School of Law, University of Dundee, Dundee, UK

A. R. Akanda • S. K. Biswas
Irrigation and Water Management Institute, Bangladesh Agricultural Research Institute, Gazipur, Bangladesh

F. Amoako Johnson
Social Statistics and Demography, Faculty of Social, Human and Mathematical Sciences, University of Southampton, Southampton, UK

E. J. Barbour • P. G. Whitehead
School of Geography and the Environment, University of Oxford, Oxford, UK

and time-consuming process. Researchers need to think about strategic relationships, accounting for factors outside of their specialist areas (domain), evaluating consequences of any changes and fully exchanging ideas with other disciplines. The interaction between the specialists and integrators needs to be iterative not only because the integration method needs to be in line with theories, methods and results of the domain in question but also because the discussions and multiple model representations might prompt domain experts to re-evaluate systems descriptions or do additional analysis. The integrative modellers (or integrators) often need (i) to develop new methods to represent complex, computationally intensive models in a 'rapid' assessment frameworks (i.e. using statistical emulators), (ii) to fill the gaps in the conceptual model by developing new or further developing existing models and (iii) to harmonise scales

J. Caesar
Met Office Hadley Centre for Climate Science and Services, Exeter, Devon, UK

J. A. Fernandes • S. Kay
Plymouth Marine Laboratory, Plymouth, UK

A. Haque • A. F. M. Saleh • M. Salehin
Institute of Water and Flood Management, Bangladesh University of Engineering and Technology, Dhaka, Bangladesh

M. A. R. Hossain
Department of Fisheries Biology and Genetics, Bangladesh Agricultural University, Mymensingh, Bangladesh

A. Hunt
Department of Economics, University of Bath, Bath, UK

C. W. Hutton
Geodata Institute, Geography and Environment, University of Southampton, Southampton, UK

A. Mukhopadhyay
School of Oceanographic Studies, Jadavpur University, Kolkata, India

S. Szabo
Department of Development and Sustainability, Asian Institute of Technology, Bangkok, Thailand

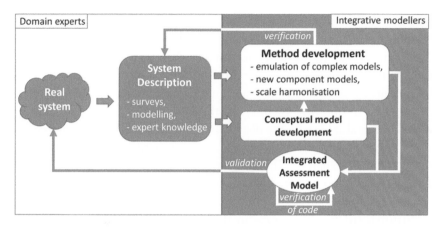

Fig. 28.1 Generic model development process. Domain experts provide the detailed understanding of the system, whereas the integrated modeller(s) develop the methods and create the model environment. The interaction is iterative and results in a more robust system understanding

and units so the components can 'talk' to each other during the model runs. Finally, the integrators always need to ensure that the integrated assessment model is robust and reliable by continuously testing, validating and evaluating the code, the methods and emerging results together in partnership with the domain experts. Ultimately, the process (Fig. 28.1) results in a robust and representative view of the relevant overall system and the consequences of change, as well as generating a highly cohesive research team who share a common understanding expressed in the model framework.

To achieve the aims of this research (i.e. assessment of future environmental change and policy responses relevant to ecosystem services and poverty alleviation in deltas), the quantitative framework combines a variety of approaches in a meta-model to describe the system as simply and efficiently as possible. This flexible model structure is designed to be able to incorporate any type of data, method or other information generated within the project. The strong collaboration with individual specialists ensures the careful selection of model elements and appropriate methods, data and system behaviour. The following sections describe this meta-model, which is called the 'Delta Dynamic Integrated Emulator Model (ΔDIEM)' including the development process, the model elements and illustrative outputs. Validation, testing of the model and the model inputs are described in Appendices 1 and 2 of this chapter.

28.2 Building the Delta Dynamic Integrated Emulator Model (ΔDIEM)

Understanding the system of coastal Bangladesh (Fig. 28.2) is an essential initial step in building the ΔDIEM model. Building upon this system understanding, the skillset requirement and the necessary model elements are identified (Fig. 28.3). Additionally, the types of data required to fulfil

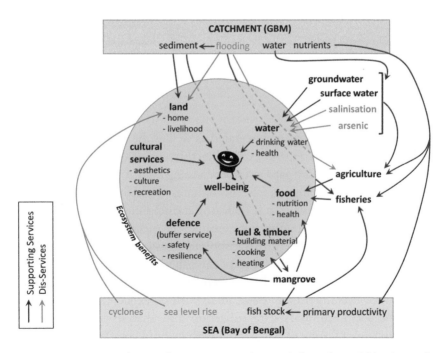

Fig. 28.2 Conceptual map of ecosystem services and disservices within the project domain

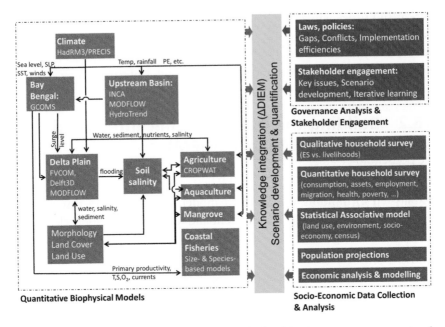

Fig. 28.3 Building on Fig. 28.2, formalised relationships between components of the research

the component models to be integrated as well as the geographical extent of the model (boundaries and external factors) are determined. This process helps to identify any gaps in understanding and/or data/information.

This project mapping exercise illuminates the spatial and temporal scales of the planned analysis. It is clear that disciplines work with different spatial and temporal units. For example, the quantitative bio-physical models are linked for the entire transboundary system with mostly daily/sub-daily calculation routines (except the fisheries and mangrove models that have annual and decadal time steps, respectively). The governance analysis, however, focuses on Bangladesh with no explicit temporal scale, while the socio-economic data collection and analysis focuses on the south-west coastal zone of Bangladesh with multiple spatial (households, Union Parishad, districts, coastal zone) and temporal (seasonal/annual/

five-yearly) scales. To incorporate these multiple scales while keeping the integrated model as efficient as possible with dynamic calculations, the unit of analysis is at the Union Parishad scale and focusses on the southwest coastal zone of Bangladesh. Union Parishads (from now on, called Union) are the smallest planning units of Bangladesh consisting of a few villages and with an average surface area of 26 km^2 and a population of around 20,000 people. To capture seasonality, the temporal scale of ΔDIEM is set as daily for the bio-physical components and monthly for the household-related components.

28.2.1   The Delta Dynamic Integrated Emulator Model (ΔDIEM)

ΔDIEM strongly builds on the frameworks shown in Figs. 28.2 and 28.3. The integrated model (Fig. 28.4) is readily defined into four distinct components: (i) the boundary conditions, (ii) the biophysical calculations, (iii) the process-based household well-being calculations and (iv) the statistical associative well-being calculations.

28.2.1.1   Boundary Conditions

The 'BoundaryConditions' class summarises all input types that ΔDIEM uses. These inputs are the national scenarios (e.g. economics; see Chap. 12), transnational (climate and river flows; see Chaps. 11 and 13 (also Whitehead et al. 2015; Caesar et al. 2015)) and regional (Bay of Bengal sea elevation and fisheries; see Chaps. 14 and 25 (also Fernandes et al. 2016; Kay et al. 2015)) generated for the 1981–2099 period. These inputs are used as look-up tables in ΔDIEM selecting the appropriate time series for each scenario run. There are three exceptions: (i) the population projections which allow the user to change the assumptions of the cohort component population projection method (see Chap. 19 and Szabo et al. 2015), (ii) the Farakka Treaty values (Farakka Treaty 1996) can be adjusted to test plausible governance interventions and (iii) the future economic assumptions can also be

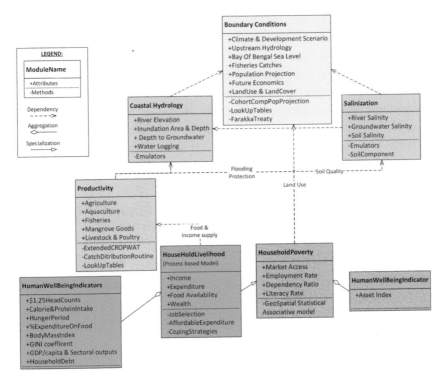

Fig. 28.4 Final conceptual model of ΔDIEM using standard Unified Modelling Language conventions. Colours highlight the four distinct components of ΔDIEM: (1) the boundary conditions (pink), (2) the bio-physical calculations (blue), (3) the process-based household well-being calculations (green) and (4) the statistical associative well-being calculations (brown)

modified by the user. In the future, any of the look-up tables can be replaced or extended by a direct link to the source model.

28.2.1.2   Bio-physical Calculations

The bio-physical calculations include the 'Coastal Hydrology', 'Salinisation' and 'Productivity' classes in Fig. 28.4. These calculations have a daily time step and are carried out on all 653 Unions of the study area.

Coastal hydrology is captured using linear statistical emulators. These distil the full hydrological model outputs (Delft-3D, FVCOM, MODFLOW-SEAWAT) to their core elements (see Payo et al. 2017). Using the Partial Least Square (PLS) regression method, linear emulators of river elevation and salinity are generated for 105 selected river locations and groundwater depth and salinity for each one of the 653 Unions. PLS regression is a technique that combines Principal Component Analysis (PCA) with multiple linear regression (Clark 1975) to predict a set of dependent variables from a set of independent variables or predictors. PLS regression is particularly useful when the prediction of a set of dependent variables from a (very) large set of independent variables (i.e. predictors) is needed. The PCA provides a preprocessing technique to reduce the dimensions of the inputs and outputs, thus making the calculations significantly faster and capturing the spatial autocorrelation simultaneously. The processing time required to emulate all the hydrological variables for all 653 Unions for a 50-year simulation is approximately one minute on a four-core computer (2.70GHz processor, 16GB RAM, 64-bit Win7 operation system). When this one-minute runtime is compared, for example, with the 48 hours Delft-3D computation time of one year for the same area (on an Intel Core i7 processor computer), the benefit of using emulators becomes clear. Inundation area, inundation depth and waterlogging are calculated for: (i) each Union for land behind existing embankments (i.e. protected land) and (ii) the remaining land that is being considered non-protected.

Farming is the dominant ecosystem service-based livelihood in coastal Bangladesh. Thus, the farming component needs to be comprehensive to allow investigation of detailed land use and farm management practices such as new crop varieties, cropping patterns, irrigation water source scenarios and so on. Hence, ΔDIEM utilises the extended CROPWAT model (see Chap. 24 and Lázár et al. 2015) which considers traditional agriculture and pond-based aquaculture together and is fully coupled with a water and soil water balance calculation (Payo et al. 2017). The soil and crop productivity component of ΔDIEM (Fig. 28.5) therefore integrates climate data, the emulated coastal hydrology, crop characteristics (cropping pattern calendar and area cover, crop type, rooting depth and crop coefficient), soil characteristics (soil type, structure and porosity) and ground elevation (Union-specific hypsometric curves). Daily water and salt fluxes due to river and coastal flooding, capillary rise and

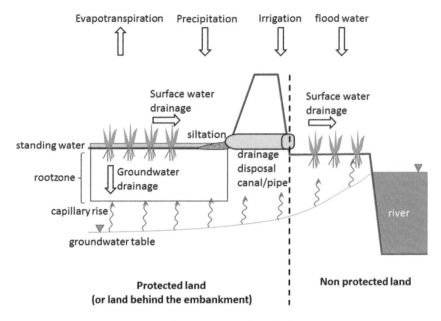

Fig. 28.5 Conceptual model of soil water and salt balance in ΔDIEM

irrigation are automatically calculated in ΔDIEM following the methodology of the FAO CROPWAT model single layer approach (Clarke et al. 1998) and the salt balance model of Clarke et al. (2015). Daily soil salinity for a defined cropping pattern (a sequence of crop types grown on the same field) is calculated as an area averaged value of the soil salinity values beneath the grown crops. These crop-soil salinity time series are calculated both for the protected (i.e. poldered) and non-protected land of the Union with the results area averaged to produce a unique daily, Union-specific soil salinity time series. Thus, the farm productivity calculations are tightly coupled with the water and salt balance calculations and also estimate the agro-economics (i.e. costs and returns to farmers).

28.2.1.3   Process-Based Household Well-Being Calculations

The process-based household well-being calculations are calculated for each Union at monthly time steps. Within each Union, 37 household archetypes are followed (Lázár et al. 2016). This module combines modelled bio-physical outputs (crop yields, required labour, fish catches), input scenarios (population, land cover/land use, economy) and observations from the household survey (see Chap. 23) to approximate household economics, poverty and health of the 37 household archetypes (Fig. 28.6). The household types are developed based on observed seasonal variations of the six occupation types: (i) farming (agriculture/aquaculture/farm animals), (ii) farm labour, (iii) fishing, (iv) forest good collection, (v) manufacturing and (vi) business activities. Farming, farm labour and fishing are dynamically calculated in ΔDIEM, whereas the others are input scenarios. Forest good collection is a static input as the basic ecosystem services of the mangrove forest (timber, fruits, honey, flood protection, etc.) are available in all mangrove species assemblages (see Chap. 26) and, as long as the forest is alive and is present, these services are available for the coastal populations. Finally, as non-ecosystem services-related livelihoods (businesses, services and manufacturing) are outside the scope of research, these use predefined economic scenarios that can be replaced by dynamic components when more detailed understanding and data become available.

At the heart of the household well-being calculations lies an optimisation routine that compares the income and fixed livelihood costs for each household type and approximates the affordable level for food, essential and non-essential house items, education, health and other expenses. This optimisation includes five coping mechanisms that can be activated under stressed economic conditions: (i) use cash savings, (ii) sell assets, (iii) get a formal loan, (iv) get an informal loan and (v) drop expenditure. The 'drop expenditure' option assumes the following order of expenditure reduction:

1. Reduce sporadic house expenses (house improvement), health-related expenses and other expenses (wedding, funeral, etc.)
2. Reduce non-essential house expenses (clothing, furniture, etc.)
3. Reduce education-related expenses
4. Reduce/drop direct livelihood costs
5. Delay loan repayment (if allowed) and reduce essential house expenses (cooking, heating, etc.)
6. Reduce food expenditure and under extreme conditions, rely on friends and family for support (if allowed by the user)

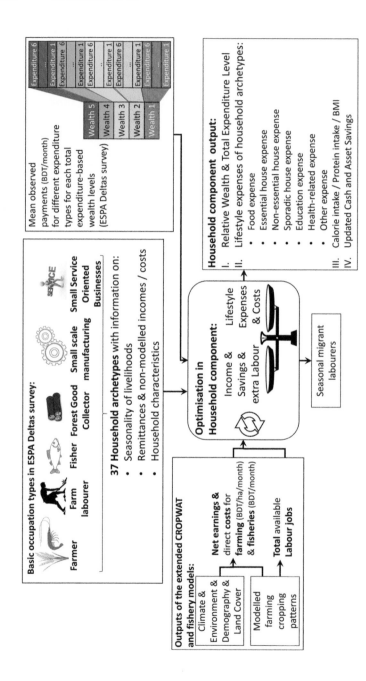

Fig. 28.6 Overview of the process-based household component of ΔDIEM

For household types that engage in farm labour, ΔDIEM also dynamically estimates the number of household members that engage in paid labour jobs, an additional coping strategy in stressed economic situations. The elderly and children are assumed not to be eligible for labour work, unless the household is in the lowest poverty categories (Wealth levels 1 and 2 on Fig. 28.6) and cannot afford the education expenses. The optimisation routine aims to ensure that the household saves a user-defined minimum fraction of the total monthly income (e.g. ten per cent) with the minimum number of household members engaging in labour works.

The outputs of this calculation are household expenditure levels and number of household members engaging in farm labour jobs. Food expenditure levels are assumed to represent different food baskets and thus food quality. Therefore, to estimate health, food expenditure was directly matched with observed levels of calorie- and protein-intake and body mass index (BMI) values based on the household survey dataset (see Chaps. 23 and 27). Finally, by knowing the monthly economics and calorie intake levels of the households, hunger periods of the household and other monetary poverty indicators of the region/community, such as GINI coefficient (i.e. income inequality) and gross domestic product (GDP)/capita, can be easily calculated.

28.2.1.4 Statistical Associative Well-Being Calculation

The statistical associative well-being calculation is completely independent from the process-based well-being calculations and operates at Union scale and annual time steps. Statistical associations were developed among observed land cover, land use, environmental quality, socioeconomy and census-based asset poverty (see Chap. 21 and Amoako Johnson et al. 2016). This association is projected into the future within ΔDIEM, using dynamically modelled bio-physical calculations (soil salinity, waterlogging) and user-defined scenarios (land use, land cover, employment rate, literacy rate, children in school, travel time to cities). ΔDIEM estimates the likelihood of each Union being in the poorest asset poverty class for each year of the simulation.

28.2.1.5   The Software

ΔDIEM is developed in the MATLAB software development environment (MathWorks 2015). MATLAB is a high-level programming language capable of supporting multi-paradigm computations. MATLAB offers not only efficient calculation functions and routines but also allows, among others, plotting, mapping, graphical user interface (GUI) and parallel computing functionalities, thus providing a one-stop-shop model- and application-building environment. The r2015b version of MATLAB and the GUI Layout Toolbox version 2.2 are used to power the ΔDIEM calculations.

ΔDIEM uses a large number of inputs and produces a numerous outputs (Tables 28.1 and 28.2). To support the user in effectively handling these, a GUI was developed that allows the user to change the inputs, run the model and plot the results by using predefined plotting routines.

Table 28.1 Key bio-physical and socio-economic inputs to the integrated model ΔDIEM

Bio-physical inputs	Socio-economic inputs
I. Climate (daily) Precipitation, temperature, evaporation, atmospheric CO_2 concentration **II. Upstream hydrology** (daily) River flow of the Ganges, Brahmaputra and Meghna river system **III. Bay of Bengal** (daily) Mean sea elevation, subsidence, cyclones/storm surges **IV. Levees/polders** Location, height, (horizontal) drainage rate **V. Ecosystem services** Farming patterns (agriculture and aquaculture—pre-2000, present, near future, post-2050), irrigation water use, crop properties, offshore fish biomass (annual), land cover (1991, 2001, 2011, 2050, 2100)	**VI. Demographic indicators** (five-yearly) Life expectancy, Total fertility rate, net migration rate **VII. Economy** (annual) Market price of crops/goods, cost of farm inputs, wages, future changes of incomes/prices by 2030 when compared to the present. Beyond 2030, all economic time series are kept constant to avoid unrealistic market conditions **VIII. Governance** Future water share treaties, fishing intensity, subsidies, loan types/characteristics, land cover/use, infrastructure planning and so on

Table 28.2 Key bio-physical and socio-economic outputs from the integrated model ΔDIEM

Bio-physical outputs (daily results)	Socio-economic outputs (monthly results)
I. Environment Water elevation, inundated area/ depth, waterlogging, soil moisture, river/groundwater/soil salinity **II. Unit area-based productivity** Crop productivity (agriculture and aquaculture), fish catches, economic cost-benefits of farming and fishing (income/costs/net earnings)	**I. Household outputs** (a) Bayesian statistical module: asset-based relative poverty indicator (b) Process-based module: economics (income, costs/ expenses, savings/assets), relative wealth-level, calories/ protein intake/BMI, monetary poverty **II. Regional economic indicators** Sectoral output (tons, BDT), GINI, GDP/capita, potential income tax revenue, household debt level

Due to the detailed daily soil salinity and farming calculations and the large number of simulated Unions ($N = 653$), ΔDIEM is still computationally intensive. Even though the Union-specific calculations are done with parallel computing, a 50-year run of the 653 Unions for one scenario requires ~7 hours computation time on a four-core computer (2.70GHz processor, 16GB RAM, 64-bit Win7 operation system). To make the calculations more effective, ΔDIEM is being run on IRIDIS4, the supercomputer of the University of Southampton. In this way, each scenario takes only four hours to run, and all the scenarios can be initiated at the same time. The user has the option to save all results (17GB per scenario) or just the most important model results (5.8 GB per scenario). Currently, ΔDIEM has a version number of 1.02.

28.2.1.6 Testing and Validating the Model

Model testing/validation and sensitivity/uncertainty analysis are essential parts of model development. Full model validation is not possible for complex, natural system models (Oreskes et al. 1994) and this is especially true for ΔDIEM. Model testing was carried out at three levels: (i) **Code verification** checks the code for bugs to ensure that the model behaves as designed. (ii) **Component testing** checks each individual model element in isolation from the rest of the integrative model to test that the behaviour matches the observations and/or the simulator behaviour (e.g. the goodness of fit of the emulator compared to the Delft-3D model outputs is being tested). Component testing also includes sensitivity tests that aim to explore the behaviour of individual components. Ultimately, when the code is 'bug free' and the individual components worked satisfactorily, the (iii) **global model results** are assessed and validated for their emergent behaviour. The component testing and global analysis results are presented in Appendix 1 of this chapter.

28.3 Overview of the Scenarios Used in ΔDIEM

The model uses harmonised, consistent scenarios and combines three climate scenarios with three development scenarios creating nine distinct and plausible futures for coastal Bangladesh (see Chap. 9). The key assumptions of the ΔDIEM scenarios are described here; see Appendix 2 for the quantified model inputs by scenarios used in ΔDIEM.

28.3.1 Climate

The climate scenarios used the downscaled Bangladesh projections discussed in Chap. 11. When calculating summary statistics for the study area, results are different to regional projections (i.e. South Asia, including the GBM catchments) as shown in Table 28.3. The table shows that projected temperature changes in the study area are anticipated to be slightly lower than the regional projections but they show the same relative pattern with the highest temperature rise under the Q16 scenario. On the other hand, precipitation over the twenty-first century increases until 2050 in all scenarios and slightly decreasing thereafter under Q0.

Table 28.3 Comparison of regional and study area forecasts of the HadRM3/PRECIS Regional Climate Model scenarios (SRES A1B, RCP 6.0–8.5). Significant differences are shown in bold

Scenario name	Annual change by 2041–2060 relative to 1981–2000		Annual change by 2080–2099 relative to 1981–2000	
	Regional	Study area	Regional	Study area
Temperature change				
Q_0	+2.20 °C	+2.17 °C	+3.90 °C	+3.78 °C
Q_8	+2.45 °C	+2.34 °C	+3.98 °C	+3.87 °C
Q_{16}	+2.65 °C	+2.53 °C	+4.75 °C	+4.39 °C
Precipitation change				
Q_0	+8.26%	+10.42%	+11.75%	+9.93%
Q_8	**−1.35%**	**+8.70%**	+13.01%	+10.28%
Q_{16}	+10.28%	+10.22%	**+23.66%**	**+11.36%**

However, these multi-decadal change-indicator values, especially for precipitation, are not useful for analysis. Figure 28.7 shows the mean annual and five-year smoothed precipitation and temperature values. It is clear that the inter-annual variation in precipitation is significant and a clear trend is not obvious. On the other hand, temperature, with slight inter-annual variations, is steadily increasing under all scenarios. Note that the climate scenarios are different for the historical period.

The climate scenarios are important for ΔDIEM's hydrological, coastal flooding, fishery and agriculture productivity calculations. Figure 28.8 explores some of the drivers of change in the study area under three illustrative scenarios: (i) Q16 Less Sustainable (LS), (ii) Q8 Business As Usual (BAU) and (iii) Q0 More Sustainable (MS). The plots show five-year moving averages to make trends visible. As Fig. 28.7 already indicated, the Q16LS scenario is the driest of the three climate scenarios. The mean annual river flow to the study area is much lower than for the other climate scenarios. Similarly, the number of days with potential floods (≥77,000 m³/s) is the lowest, and the number of baseflow days (≤5,500 m³/s) is the highest in Q16LS. This allows more saltwater intrusion along river channels (i.e. high river salinity during the dry season). At the same time, the total dry season precipitation is the lowest for the Q16 scenario with the highest number of dry days and dry consecutive

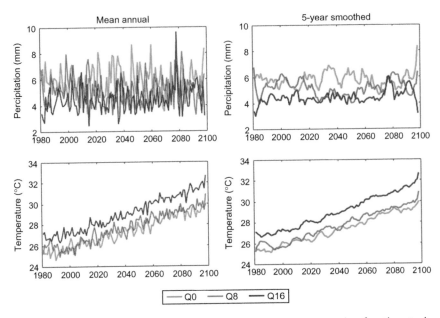

Fig. 28.7 Mean annual precipitation and temperature scenarios for the study area

days. This is the worst case scenario for dry season agriculture where good quality irrigation water is only available from groundwater sources and, without irrigation, no crops can be grown. For the other climate scenarios (Q0, Q8), the river flow seems to slightly increase over time, together with more flood flows and less baseflow. Dry season precipitation has significant inter-annual variability, and thus, the availability depends on which year/period is considered. The 'maximum consecutive dry days' plots have similar values to Q16 (~30 days), indicating that the precipitation will be more intense and comes with a similar frequency.

In summary, the climate scenarios show that the Q0 scenario will increase the variability of available total precipitation with somewhat higher river flows and more intense dry season precipitation (Fig. 28.5). The Q8 scenario shows a decreasing precipitation availability in the coastal zone; however, the upstream river discharges will not be smaller.

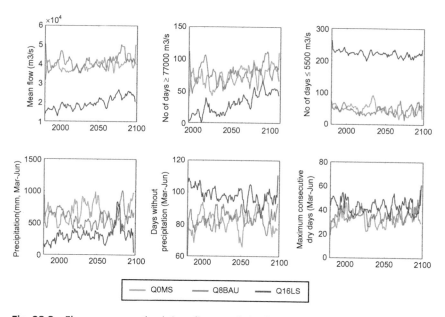

Fig. 28.8 Five-year smoothed river flow statistics (top row) and dry season precipitation trends (bottom row) for coastal Bangladesh under three future scenarios

Q16 is the driest scenario of the three sets but, when analysed by itself, Q16 shows a gently increasing total precipitation trend by 2050 beyond which the total dry season precipitation can greatly increase.

28.3.2   Fisheries

Potential fish catches under climate change and management scenarios in ΔDIEM are directly used from the output projections of SS-DBEM model (see Chap. 25 and Fernandes et al. 2016). The fisheries model projects a decrease of fish availability and catches under all scenarios with a plausible collapse of fisheries (specifically the Hilsha fishery) under the high overfishing scenario if the management does not change. However, if current catches are compared under business-as-usual and sustainable fishing, current catches could be maintained (Fernandes et al. 2016). In ΔDIEM, the fisheries model results are reduced by export quantities (i.e. outside the study area) and then distributed to each of the nine districts of the study area based on observations (i.e. statistical year-book data). Within each district, the caught fish are then distributed to each household archetype based on their fishing intensity and the number of fishers.

28.3.3   Agriculture

The agriculture component of ΔDIEM is built on a review of observed past (1990s) and present (2010) cropping patterns (i.e. sequence of crops on agriculture fields) aiming to identify typical crops, their varieties and their use in coastal Bangladesh. The five most frequent cropping patterns for each *Upazila* (i.e. sub-district) is based on the soil survey reports of Bangladesh. The copping patterns are assumed to be the same in each Union within a specific *Upazila*, and the per cent area used for each cropping pattern is the same as the observed. Table 28.4 shows the agriculture cropping pattern assumptions used in ΔDIEM. The properties of the observed crops (rooting depth, crop coefficient, evaporation depletion factor, salinity tolerance, etc.) are partially collated from field observations (datasets from Bangladesh Agricultural Research Institute [BARI], Bangladesh Rice Research Institute

Table 28.4 Development scenarios for agriculture used in the integrated model

Scenario	Time period (i) Present	(ii) 2041–2060	(iii) 2080–2099
Less Sustainable	Current crop properties 2010 cropping patterns	Current crop properties 1990 cropping patterns	Current crop properties 1990 cropping patterns
Business As Usual	Current crop properties 2010 cropping patterns	Current crop properties 2010 cropping patterns	Increased yield and salt-tolerant varieties 2010 cropping patterns
More Sustainable	Current crop properties 2010 cropping patterns	Increased yield and salt-tolerant varieties 2010 cropping patterns	More increased yield and tolerant varieties (salt, flood tolerant) 2010 cropping patterns

[BRRI] and Department of Agricultural Extension [DAE]) and partially based on a model calibration exercise described in Lázár et al. (2015). To anticipate future crop varieties, properties of future crops (potential yield and salinity tolerance) are modified based on information published in 'Agricultural Technology for Southern Region of Bangladesh' report (BARC 2013). Other basic crop properties that affect water uptake and other tolerances (e.g. temperature) are not changed compared to the existing crops. Over time, cropping patterns change by considering a five-year overlap (i.e. transition period) between the 'old' and 'new' cropping patterns.

28.3.4   Land Use

Historical land cover and land use were identified from remote sensing images for 1989, 2001 and 2011 (see Chap. 21 and Amoako Johnson et al. 2016). Future land cover scenarios, on the other hand, were developed based on a national stakeholder workshop in Bangladesh, where the LS, BAU and MS scenarios were qualitatively described (see Chaps. 10 and 20). These narratives were quantified and then validated during a national technical expert workshop again in Bangladesh (Table 28.5).

28.3.5   Demography

The demographic scenarios are based on the analysis discussed in Chap. 19 (see also Szabo et al. 2015). This work anticipates that the coastal population will decrease from around 14 million in 2011 to about 11–13.5 million in 2050, depending on the development scenario in question. The LS scenario assumes the highest level of out migration, whereas MS assumes the least migration.

28.3.6   Economic Trends

Historical economic time series were collated from different statistical yearbooks and Household Income and Expenditure Survey datasets for coastal Bangladesh. The present-day economic data were collected

Table 28.5 Land cover and land use development scenarios for the study area

Per cent reduction by 2050/by 2100 (compared to present-day land cover)		Gain of agriculture	Gain of aquaculture	Gain of bare land
Less Sustainable				
Loss of Sundarbans (encroaching)	20/25%	0/0%	20/25%	
Loss of other mangroves	10/10%	5/5%	5/5%	
Reduce agriculture if salinity: high	20/20%			20/20%
Reduce agriculture if salinity: moderate	10/10%			10/10%
Reduce agriculture if salinity: low	0/0%			0/0%
Business As Usual				
Loss of Sundarbans (encroaching)	10/15%	5/7.5%	5/7.5%	
Loss of other mangroves	5/10%	2.5/2.5%	2.5/2.5%	
More Sustainable				
Loss of Sundarbans (encroaching)	5/2%	2.5/1%	2.5/1%	
Loss of agriculture	5/0%			5/0%
Loss of agriculture (coastal Unions)	5/0%			5/0%

+ Land zoning scenarios based on FAO and MA (2013) assuming a 30 per cent increase in the promoted sector

through the household survey. Future economic scenarios were developed based on a microeconomic study (see Chap. 12). This analysis estimates expected changes by 2030 (from present-day) for a number of broad economic variables (Table 28.6). At present, no further change is assumed beyond 2030 due to high level of uncertainties in economic trends and to avoid unrealistic 'market' situations. Prices, wages and costs are generally going to increase, with a higher level of increase under the MS future, although there are some exceptions (e.g. aquaculture, services and manufacturing). Incomes are expected to increase for farming and fishing. However, the value of forest goods is expected to decrease in each economic scenario. Note the very high increase in income from services and manufacturing (bold) compared to any other economic increase. The socio-variables (employment rate, literacy rate, children in school, travel time to cities) are all expected to improve by 2030.

Table 28.6 Percentage change in ΔDIEM economic input variables by 2030. Figures in bold highlight the increase in income from services and manufacturing

Economic input variable	Less Sustainable	Business As Usual	More Sustainable
Cost of agriculture (seed, pesticide, fertiliser types)	0	10	20
Cost of aquaculture (feed, post larvae, fishling)	20	10	0
Cost to keep livestock/poultry, fishing, forest collection	0	10	20
Land rent cost (farming)	0	10	20
Cost to undertake services and manufacturing business	20	0	−20
Market (selling) price of agriculture crops	0	10	20
Market (selling) price of fish	30	10	20
Market (selling) price of aquaculture crops (shrimp)	0	10	20
Income from forest goods (honey, fruits, timber, etc.)	−20	−10	0
Income from manufacturing, services and livestock/poultry	**65**	**110**	**165**
Remittances (BDT/month)	20	30	40
Household expenses	0	10	20
Daily wage (without food) (BDT/day)	0	10	30
Cost of diesel (BDT/gallon)	0	10	20
Employment rate (% population)	0	10	30
Literacy rate (% population)	2	4	8
Children in school (% population)	2	5	10
Travel time to major cities	−10	−30	−50
USD/BDT exchange rate and PPP exchange rate	0	0	0

28.4 Illustrative ΔDIEM Outputs

The review of boundary conditions indicates the Q16LS scenario combination is possibly the worst future, due to the importance of agriculture in the study area, while the Q0MS combination represents the best plausible future. To illustrate this, and to limit the number of plots presented, only results for a diagonal transect of the modelled nine plausible futures are presented here, namely, the Q16LS, Q8BAU and Q0MS scenarios.

28.4.1 Composite Indicators Used in ΔDIEM

In order to compare the different scenarios effectively, composite indicators are used to describe the environmental hazards, the provisioning ecosystem services and the resulting socio-economic situation. The elements of these composite indicators are:

1. **Environmental hazards**

    (a) Drought index

    - Number of days when total river inflow is the below the 20 percentile flow
    - Number of days with no precipitation in the dry season (March–June)
    - Maximum number of consecutive dry days in the dry season (March–June)

    (b) Flood index

    - Number of days when the total river inflow exceeds the 90 percentile flow
    - Number of days when the soil is inundated with a minimum of 25 cm deep water. This considers fluvial and coastal floods and water accumulation on the soil surface due to intense precipitation.

    (c) Soil salinity index (area averaged, March–June)

2. **Provisioning ecosystem services**

    (a) Farm income (population averaged)
    (b) Fishing income (population averaged)
    (c) Off-farm, non-ecosystem service related income (population averaged)

3. Socio-economic situation

(a) Food security

- Calorie intake (population averaged)
- Protein intake (population averaged)
- Body mass index (BMI, population averaged)

(b) Income inequality index (GINI coefficient)

All indicators are normalised to the minimum/maximum range for plotting purposes. Figure 28.9 presents the results for the selected three scenarios. The inter-annual variability is the greatest for the environmental hazard, moderate for the provisioning ecosystem services and minimal for the socio-economic indicators. This immediately indicates that the socio-economic conditions have greater influence on the results than the changes in the quality of the environment.

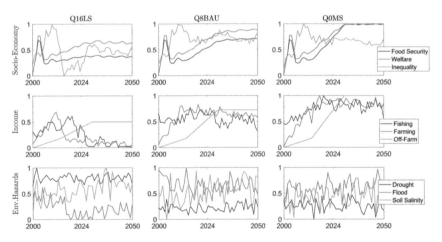

Fig. 28.9 Normalised composite indicator outputs of the ΔDIEM simulations under three future scenarios. Note y-axis values: '0' means low level, '1' means high level

28.4.2   Results for Selected Indicators

The Q16LS scenario indicates a moderate reduction in the prevalence of poverty and food security, but a collapse of rural income (i.e. farming and fishing) and an enhanced off-farm sectors (this is a scenario and is not dynamically modelled). Income inequality (i.e. GINI), after a sudden drop associated with a significant decrease of farming incomes, gently rises after 2025 with some inter-annual variation. Even though soil salinisation intensifies, drought and the return to the traditional 1990 farming practices drive the collapse of farming. Flooding is minimal under this scenario due to low river flows.

The Q8BAU scenario shows a moderate increase in welfare and food security. Inequality is the highest indicating the largest difference between the most poor and the least poor households. Fishing income has a gradual downward trend implying an unsustainable use of fisheries. Farming livelihood similarly declines due to increasing salinisation. Drought also becomes more frequent after 2040. Even though flooding intensity has a high inter-annual variability, it does not have a long-term trend.

The Q0MS scenario has an enhanced agriculture (present-day practices but more resilient crops), sustainable fisheries and intensified off-farm activities. This diversity results in large improvements in poverty and food security. However, income inequality is not lower when compared to the other scenarios. This implies that, while the poorest benefit from favourable economic and environmental conditions, the income gap remains. Due to the economic stability of region, inequality starts to decrease very slightly after 2025. The composite hazard indicators indicate an increase in flood intensity and slight decrease in drought. Soil salinity is highly variable, but this scenario has the lowest soil salinity levels.

Soil salinity in ΔDIEM is the result of climatic (precipitation, temperature), environmental (drainage, capillary rise, flooding) and anthropogenic (irrigation) factors. Considering the three highlighted scenarios, simulations show a slight increase in soil salinity from 2010 to 2050 under all futures. The categories in Fig. 28.9 correspond to critical soil salinity thresholds. For example, above four deciSiemens per metre (dS/m), vegetables generally do not grow, and above 8 dS/m, most rice varieties stop growing. This makes farming livelihoods very challenging during the dry

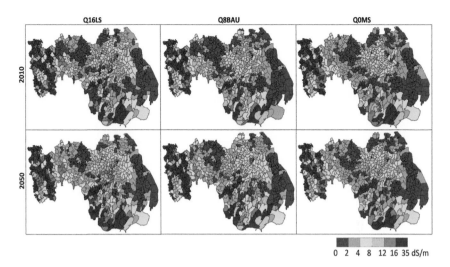

0 2 4 8 12 16 35 dS/m

Fig. 28.10 ΔDIEM soil salinity results for 2010 and 2050 (Union average)

season and might force the farmers into off-farm livelihoods during the driest months. Figure 28.10 indicates that the already high soil salinity levels found in the coastal fringe and Khulna and Satkhira districts (the western part of the study area) in 2010 become slightly worse. In the central Barisal division (the north-east), soil salinity increases by one category in many Unions between 2010 and 2050. This means that this region, which traditionally has three agriculture crops per year, will require careful management to be able to produce *rabi* (dry season) crops. Finally, the less sustainable scenarios result in the largest soil salinisation (spatial extent-wise). The main driver of soil salinisation in the model simulations is the salinity of the irrigation water.

Soil salinity is just one factor affecting farm productivity, but it is often thought to be the most important. Figure 28.11 shows the crop potential for the study area in general (i.e. over the year) and for the *rabi* (dry) season. When the 'all crop' maps are compared to the soil salinity maps, the crop potential results are somewhat similar. However, the dry season situation is strikingly different. This is because drought becomes more limiting than soil salinity in the simulations. This is especially true under the Q16 climate scenario, where drought is significantly worse than in

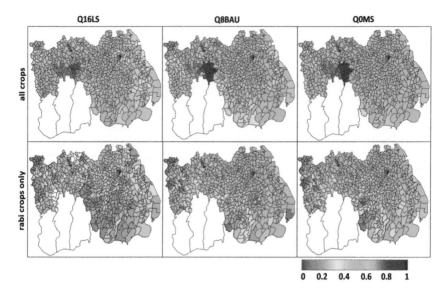

Fig. 28.11 Simulated crop potential in 2050 for the overall year (top panels) and for the *rabi* (dry) season (lower panels) (Note: '1' indicates maximum potential, '0' indicates no potential)

the other climate scenarios. The western part of the study area remains close to the maximum crop potential (i.e. dark blue), even though the soil salinity values are high. This is a consequence of the cropping pattern scenarios that assume that shrimp farming is significant there.

Whether a household can make a decent living from farming depends on the crop yield, the market price of the crop, the direct costs associated with the crop production and the land size. For example, the farmers that have homestead-size farm (0.01–0.5 acre) earn on average 500 Bangladesh taka (BDT) per month. A small land owner (0.5–2.5 acre) earns about 2,000 BDT per month and a large land owner (>2.5 acre) 15,000 BDT per month, when selling crops on the open market. This is however not the actual net earnings of the household from farming as direct agriculture costs can consume almost the entire farm income. Net earnings (i.e. income minus costs), and thus the household income, increases with the land area and cultivation efficiency. Thus, the larger the land area, the more profit remains for the farming household.

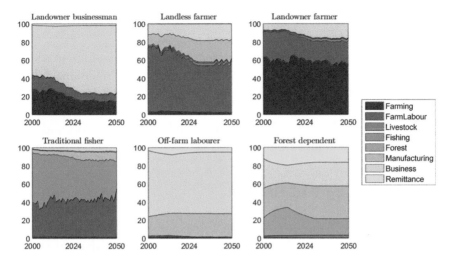

Fig. 28.12 Trajectory of simulated livelihood income composition (as percentage of total income) for six selected household archetypes under the Q8BAU scenario

ΔDIEM also allows the dominance of income type for the defined household archetypes (Lázár et al. 2016) to be investigated over time. Figure 28.12 shows the importance of livelihoods for six selected household archetypes which illustrate the spread of livelihood strategies and associated land sizes within the study area. The total number of household archetypes found in the survey (37) means that the population percentage identified in each category is generally low: business-business-business (large land owner, two per cent of population), farming-farming-farming (landless, nine per cent of population), farming-farming-farming (large land owner, one per cent of population), fishing-fishing-fishing (landless, three per cent of population), business-manufacturing-business (landless, ten per cent of population) and forest dependent (less than one per cent of population). However, it is clear that ecosystem services-based livelihoods decline in importance over time for all household types. Wealthy land owner businessmen start with a fairly balanced income distribution, but by the end of the simulation, they predominantly rely on off-farm business activities (i.e. not ecosystem service-based). The landless farmer archetype has marginal land, but pond-based aquaculture and farm labour provide the bulk of their income. The land owner farmer archetype has enough land to sustain their well-being, with occasional farm labour and some business activities providing an extra income. Fishing and livestock provide the livelihood for the traditional fisher archetype with increasingly limited fishing income. The off-farm labourer archetype only undertakes off-farm activities, whereas the forest-dependent archetype does both ecosystem service-based and off-farm activities with no significant change in income dominance.

28.4.3   Summary of Findings

The study area in Bangladesh gradually develops under all scenarios, but the level of improvement depends on the future population dynamics and the economic situation. Present-day agriculture-based livelihoods are constrained by land availability, dry season salinity and agro-economics. Monsoon rains supply adequate water to grow a main season rice crop but farmers' incomes are constrained by the market price for rice compared with the direct agriculture costs (i.e. allows no or only a modest profit). However, second and third crops grown in the drier months require irrigation even today. Low irrigation water quality contributes to soil salinisation reducing farm production potential. Future crop production, however, is also likely to be constrained by drought (i.e. lack of water and high air temperature). The development of drought and salt-tolerant varieties of rice and other crops is in progress and will be essential to sustain and improve production under the future scenarios considered here. Fisheries also require careful management to avoid a collapse of fish stocks. In addition to the provisioning ecosystem services (agriculture, fisheries), inter-annual variability is also significant for environmental hazards (i.e. floods).

To 2050, when strong cyclone and storm surge activities are not considered, household socio-economics appear largely decoupled from climate and environmental change although strongly related to future population dynamics. This means that non-ecosystem services-based income sources can maintain well-being even though the traditional fishing and farming-based livelihoods become less and less profitable and potentially unsustainable under the LS and BAU socio-economic scenarios. Therefore, governance related to market conditions, job opportunities and access to resources and markets has a critical impact on the future welfare of the study area. However, a word of caution needs to be raised: even though the economy and well-being of the population can increase under appropriate governance and social policies, environmental change can still have a detrimental effect on the food security for Bangladesh. For example, soil salinity, together with household socio-economic characteristics, is one of the main explanatory variables of food security for coastal households (negative effect) regardless of wealth levels (Szabo et al. 2015). In the modelling discussed here, soil salinisation is likely to continue and even accelerate after 2050 due to sea-level rise and subsidence. Furthermore, increasing future cyclone frequency and magnitude together with increasing sea-level rise can be catastrophic for agriculture and, through damage, can impact all sectors in the coastal zone. Therefore, appropriate adaptation will be required to sustain an ecosystem service rich coastal zone and food-secure coastal population. This might include some combination of enhanced polders, tidal river management, water diversions and nature-based approaches. Such adaptation requires additional assessment.

28.5   Conclusions

This chapter presents the vision, design and summary results of a novel integrated assessment model, the Delta Dynamic Integrated Emulator Model (ΔDIEM). ΔDIEM is a holistic integrated assessment framework coupling bio-physical and socio-economic changes with governance to assess livelihoods, poverty and health of the rural coastal population in Bangladesh. The model strongly builds on the expertise and results of the research undertaken and utilises high fidelity models, statistical associations and observations including a bespoke household survey dataset. ΔDIEM was successfully tested and validated against both the high fidelity models and observations, and thus the results represent the observed system of coastal Bangladesh well.

   ΔDIEM is being used to integrate consistent bio-physical and socio-economic scenarios developed across the project, initially up to 2050.

The presented results show the economy, and thus the generic well-being is expected to steadily increase across all scenarios, although the rate of change will depend on governance. Ecosystem services are, however, expected to decline by 2050 across all three scenarios. In the least desirable world, both agriculture and fisheries collapse, but even under the more sustainable scenario, both fish catches and farm productivity decline slightly due to environmental stress such as soil salinisation and heat stress on crops. Therefore, even though economic development can decouple from ecosystem service trends, the future food security of Bangladesh and the sustainability and habitability of the coastal zone will depend on how the extreme events and persistent environmental processes are managed, and what adaptation and development is implemented.

The integrated analysis capacity that ΔDIEM provides was previously non-existent in the literature. It is crucial to integrate the available knowledge to understand and analyse the big picture (environment, socio-economy, welfare together) to be well prepared for the challenges that climate and environmental change and a shift in macroeconomics might bring in the future. The ΔDIEM approach is the first of its kind to combine top-down and bottom-up approaches while tightly coupling climate environmental, demographic, social behaviour, economics and governance at the relevant scale of decision making. Decision makers have to be aware of cause–effect relationships and long-term trends to make more informed and thus better decisions. ΔDIEM is designed to investigate such tough questions and has demonstrated itself to be capable of providing the necessary insights and trends. Future work is planned, in the context of the Bangladesh Delta Plan 2100, to use ΔDIEM to evaluate some potential interventions and demonstrate its detailed application to policy.

Appendix 1: Testing and Validating the ΔDIEM Model

This appendix reviews the outcomes of the key testing and sensitivity analysis of the Delta Dynamic Integrated Emulator Model (ΔDIEM) as follows: (i) uncertainty of the ΔDIEM in the emulators, (ii) behaviour of the soil salinity calculations compared to observations, (iii) fit to observations and sensitivity of the agriculture model (i.e. the extended CROPWAT model) and (iv) the goodness of fit and sensitivity of the household component model.

Testing of the Emulators

An emulator statistically represents the input-output relationship of the simulator (i.e. the real, process-based expert model) to achieve an efficient calculation speed that enables a tight-coupled integrative assessment model. Therefore, the accuracy of the emulated results is judged based on the outputs of the simulator model (e.g. inundation depth from the Delft-3D), and thus the emulated accuracy cannot be better than the accuracy of the simulator model.

Emulators' prediction accuracy (compared to the simulator outputs) can be evaluated using several standard metrics. In this research, the root-mean-square error (RMSE) was used. To assess the performance of the emulators, different percentages of available simulation datasets (e.g. Delft-3D outputs) are used to train the emulators (described as the 'training dataset'), and the remaining dataset is used to validate the prediction accuracy. The training dataset is randomly selected and the goodness of fit (i.e. RMSE) is calculated for the remaining dataset that is not used for training. This sampling-predicting-validation process was repeated 30 times to obtain a robust, mean RMSE for each percentage of data used for training.

Table 28.7 summarises the accuracies of the trained emulators compared to the Delft-3D and other model results. Figure 28.13 shows an example for the accuracy test results of the emulators. The magnitude of errors is acceptable for the purposes of the integrated model (i.e. assess trends and cause–effect relationships). Larger scatter occurs for smaller values (e.g. the largest uncertainties are when groundwater salinity <~0.5 ppt). From hydrological point of view, the higher values are much more important than the low values (e.g. higher river elevation are the ones that cause flooding) and the emulators are considered adequate. During the normal simulation of ΔDIEM, the emulators are trained by using all training data, thus having the lowest possible error.

Table 28.7 Accuracy of the emulators in ∆DIEM (see also Payo et al. 2017)

Emulator	Simulator	RMSE (min-max): all randomly trained emulators	Additional notes
Groundwater salinity (top and irrigation layer of unconfined aquifer)	Modflow-Sewat	0.072–0.13 ppt	Subtle trends at the beginning of the simulation period are not captured well. Largest uncertainty when < ~0.5 ppt
Depth to groundwater	Modflow-Sewat	0.1–0.4 m	While the daily variability is captured by the emulator, the amplitude of this variability is under predicted. Largest uncertainty when <1 m
River elevation	Delft-3D	0.35–0.4 m	Largest uncertainty when <2 m
Inundation depth	Delft-3D	0.012–0.13 m	Largest uncertainty when <1 m
River salinity	FVCOM	1.36–2.7 ppt	Largest uncertainty when <1 ppt

Testing of the Soil Salinity Calculations

The soil salinity calculations of ∆DIEM are not based on existing expert models, rather a new process-based soil simulation model that is fully coupled in the integrated assessment model is developed (Payo et al. 2017).

Figure 28.14 shows the averaged May 2009 simulated soil salinity of ∆DIEM for all three climate scenarios to check the sensitivity of the results to the different climate scenarios. The western Unions are most severely affected by soil salinity under Q0 scenario and less under Q8 scenario. Differences between the three scenarios and the north-east and south-east regions are visually less evident but the summary table of the areal extent (Fig. 28.14) suggests that soil salinity has the minimum effect in the Q8 scenario (i.e. largest nonsaline soil area) and has the maximum effect under the Q0 climate scenario. The extent of soil salinisation is sensitive to the inter-annual variability of the climate, thus can

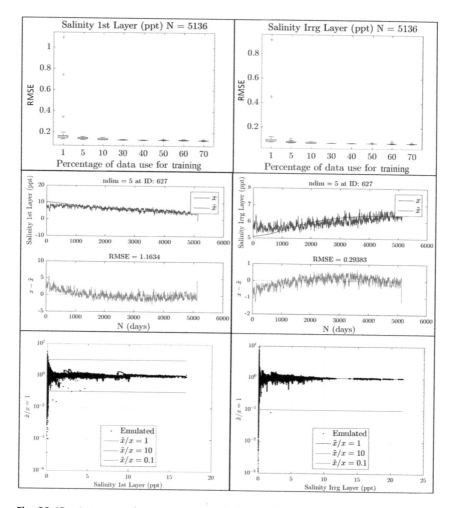

Fig. 28.13 Accuracy of groundwater salinity emulator in ΔDIEM at ~10 m depth (left panels) and ~100 m depth (right panels). (RMSE, root-mean-square error; N, sample size; ndim number of principal components used to reduce the dimensionality of the inputs; ID, Union ID; x, simulated value; x^, emulated value; N, simulation days)

be significantly different in different years of the simulation. For example, if the period 2001–2009 is used to characterise the monthly mean soil salinity, the total area affected by soil salinity decreases. The non-affected area for the period 2001–2009 is 2.1, 1.6 and 1.5 times larger than the

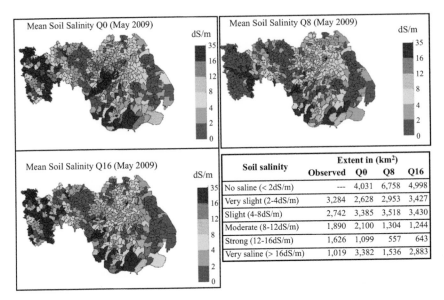

Soil salinity	Extent in (km²)			
	Observed	Q0	Q8	Q16
No saline (< 2dS/m)	---	4,031	6,758	4,998
Very slight (2-4dS/m)	3,284	2,628	2,953	3,427
Slight (4-8dS/m)	2,742	3,385	3,518	3,430
Moderate (8-12dS/m)	1,890	2,100	1,304	1,244
Strong (12-16dS/m)	1,626	1,099	557	643
Very saline (> 16dS/m)	1,019	3,382	1,536	2,883

Fig. 28.14 ΔDIEM soil salinity results for May 2009 (Source of observed area, SRDI (2012))

extent simulated for May 2009 for the scenarios Q0, Q8 and Q16, respectively. An inspection of the reported values at 41 stations by Dasgupta et al. (2014) and the simulated values at Union level suggest that the main spatial variability is well captured for all three scenarios, with Q8 being the closest to the observations (Payo et al. 2017).

Soil salinity is affected by the crops grown on the land through evaporation and irrigation water requirement. ΔDIEM simulates soil salinity for each crop within each cropping pattern, and the soil salinity of the broader area is calculated as the area averaged mean. Thus, the model allows comparison of the soil salinity results for each simulated crop. Figure 28.15 contrasts the results with one of the available observations (Mondal et al. 2001) for sesame. Simulated (Q0, Q8, Q16) soil salinity of non-protected areas (i.e. without the protection of an embankment) under sesame farming is in better agreement with observed values than the protected simulations. Unfortunately, the exact location of the experimental field is not known limiting firm conclusions. Furthermore, the Q8 climate scenario-driven simulations produce the best agreement with the observations.

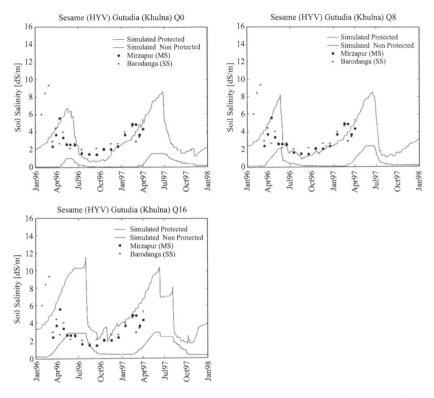

Fig. 28.15 Comparison of Protected and Non-protected simulated soil salinity under Sesame farming with observed values by Mondal et al. (2001) for the Q0, Q8 and Q16 climate scenarios

Testing of the Extended CROPWAT Model

Similarly to the soil salinity calculations, ΔDIEM uses its own process-based model to capture the potential of agriculture in the simulations. This section reports the sensitivity and performance of this module.

A preliminary sensitivity analysis revealed that the crop simulations are most sensitive to five out of 23 parameters (Lázár et al. 2015): yield response factor (Ky), the optimum temperature range of maximum growth ($Topt_1$ and $Topt_2$), the crop coefficient of the middle growth period (Kc,mid) and the salinity related yield reduction parameter (Ece,b). The sensitivity of the parameters changes with crops and season. The yield response factor is the single most important variable, resulting in the largest changes in the

Table 28.8 Crop simulation results of ΔDIEM at *Upazila* and district levels (Lázár et al. 2015)

Crops	Goodness-of-fit results					
	RMSE (2000, Upazila, %)	RMSE (2005, Upazila, %)	RMSE (2010, Upazila, %)	RMSE (1990, district, %)	RMSE (2000, district, %)	RMSE (2010, district, %)
T. Aman (local)	14.1	9.7	4.1	8.2		2.3
T Aman (HYV)	8.8	8.8	8.6	60.3	28.1	6.9
T.Aus (local)	38.8	5.2	5.9	6.8		2.7
T Aus (HYV)	16.0	18.3	10.1	47.4	17.9	9.6
Boro (HYV)	8.7	12.0	11.0	50.3	13.6	11.9
Chilli (local, rabi)				15.4		8.4
Chilli (hybrid, rabi)	18.7	14.4	10.4	58.3	32.9	7.4
Grass pea (HYV)	48	26	29	123	44	24
Potato (HYV)	76	49	65	315	207	70
Wheat (HYV)	28.4	28.8	24.6	144.1	99.8	22.0

Notes: green, very good agreement (<15 per cent); blue, acceptable agreement
(15–30 per cent); orange, fair agreement (30–50 per cent); red, poor agreement
(>50 per cent); white cells, no observation

results (up to 55 per cent difference compared to the baseline results). The four other parameters have smaller but still significant impact resulting in a 15–35 per cent difference in the outputs.

Observations on crop productivity were available from the Department of Agricultural Extension (DAE) at *Upazila* and district levels. The calibrated crop parameters resulted in a good fit in most cases for 2010 at both levels (Table 28.8). Representation of the year 2000 conditions was mostly acceptable, but district-level simulation results for the year 1990 almost always greatly deviated from the district-level observations. Conversely, the *Upazila*-level simulations showed good correlation with the observations for 2000, 2005 and 2010. The deviation from the observed values for 1990 due to a mixture of four issues:

1. Model structural error and parameter uncertainty
2. Uncertainty around the observed farmers' yield

   (a) The way it was collected and entered into databases might have changed over time.
   (b) Data for different varieties are mixed up in one average yield value (e.g. *T.Aus* HYV), and the proportion of these varieties in the statistics has changed over time.

3. The management of the crops could have drastically improved since 1990, but the CROPWAT model does not any include management-related equations/parameters apart from irrigation.

4. Soil salinity is highly spatially and temporally variable in the coastal zone of Bangladesh, and observed, homogenous soil salinity time series are not available. The present study used average *Upazila*-level, yearly salinity values for 1971, 2000 and 2009 and carried out a linear interpolation in between the observed values. Finally, the seasonality of soil salinity was assumed to be the same as for river salinity. This approach holds considerable uncertainties for the model results.

The *Upazila*-level simulated yields are generally representing the observations well. The fit to observations was not so good for some minor crops, such as potato and grass pea, but they are generally accounted for less than ten per cent of the total agriculture area in coastal Bangladesh. The recalibration of these crops would require further data that is currently not available.

Testing of the Process-Based Socio-economic Component

The ΔDIEM uses its own simple agent-based-type model to capture the behaviour and monetary well-being of the coastal population. Validation of socio-economic results is very difficult, because monthly observations for millions of people with such detail are not available. However, national survey results are accessible for comparison with ΔDIEM aggregated results; the summary statistics of the HIES surveys (BBS 2011) for five variables (Fig. 28.16) were used. Mean total household expenditure matches well the observations, both in terms of trend and magnitude. However, the range (grey area) shows calculated values from almost nothing to significant amounts. Calculated mean calorie- and protein-intake values agree well with the rural-specific observations. In the same way as the total household expenditure, these figures also show that for some households the food intake is at very low levels (food poverty line of Bangladesh is 2,122 kcal/capita/day; BBS (2011), page 59). The GINI coefficient measures the income inequality of the population.

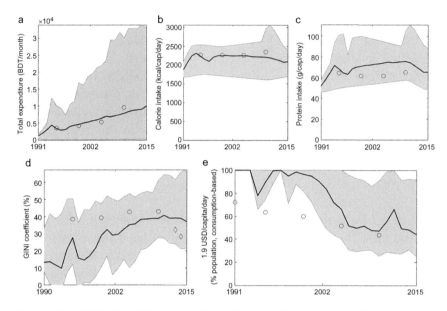

Fig. 28.16 Validation of the process-based household component of ΔDIEM. Black lines show the simulated mean study area values, shaded area shows the min-max simulated range within the study area, and grey dots and diamonds are observations. Observations: (a) BBS (2011) Table 4.4; (b) BBS (2011) Table 5.3; (c) BBS (2011) Table 5.4; (d) dots: rural inequality (Ferdousi and Dehai (2014)), diamonds: national inequality (UNDP (http://hdr.undp.org/en/content/income-gini-coefficient accessed on 08/07/2016)); (e) World Bank: People living on less than $1.90 a day (http://povertydata.worldbank.org/poverty/country/BGD accessed on 08/07/2016)

The larger the inequality index, the larger the income difference between the poorest and richest households. The calculated mean inequality is lower than the observed, meaning that most of the population have more similar income levels. The model trend is increasing inequality over time, and the calculated value reaches the observed level by 2010, beyond which the inequality stabilises and decreases slightly. ΔDIEM results are closer to the rural observations than the national average. Finally, the World Banks's national 'People living on less than $1.90 a day' indicator is also used to assess the calculated results. To calculate this, household consumption is used, as opposed to household income, because consumption can be more reliably measured. The calculated total household consumption levels are adjusted with the World Bank's purchase power parity (PPP) conversion factor. In the 1990s, this calculated headcount indicator is hugely overestimated compared to the 'observed' national average. All households are below this $1.90 poverty threshold. As the simulation progresses, however, the poverty prevalence decreases substantially and, by 2005, it reaches the magnitude of the observations. This model behaviour can be attributed partially to the uncertainty of the input data for ΔDIEM and partially that the model needs a few years at the beginning of the simulation before an equilibrium of the simulated household finances is reached. Overall, the model behaves as expected and reproduces the national trends well.

A sensitivity analysis, similar methodology to the CROPWAT study, investigated the importance of all 17 model parameters of the process-based household component. This exercise only uses observations (BBS 2011 and the ESPA household survey dataset) and does not use any inputs from the full ΔDIEM model to avoid introducing additional errors. Five model parameters showed no sensitivity (Fig. 28.17; initial cash savings, initial asset savings, loan grace period, minimum saved income and coping strategies). Therefore, they could be excluded from the model if a more rigorous analysis also finds them insensitive. When the mean sensitivity is assessed, it is clear that only the 'target affordability', 'official loan APR' (i.e. annual percentage rate charged for borrowing) and 'waiting time to increase expenditure level' are the important parameters. However, when the minimum-maximum range of sensitivity is assessed, a few parameters are very important for all household archetypes: 'target affordability', 'APR of official loans', 'waiting time to increase expenditure level', 'savings that cannot be spent', and 'income drop' (i.e. how much the income is reduced if not all the livelihood costs are paid). Other parameters have moderate or no sensitivity depending on the household type in question. The most sensitive household types are landless or small land owners with lower income levels who try to balance farm-related activities with other off-farm occupations.

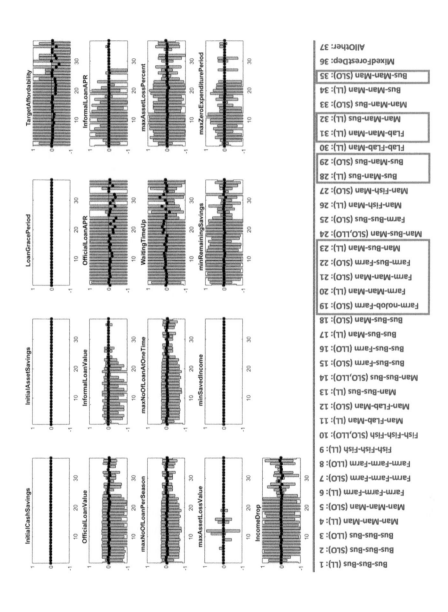

Fig. 28.17 Sensitivity of household model parameters in ΔDIEM. The 'x' axis shows the 37 household archetypes, whereas the 'y' axis shows the simulated sensitivity values (grey bars, minimum-maximum range; black dots, mean value) by considering all Unions and simulated months (1990–2015). Description of each household ID is at the bottom of figure: the seasonally dominant occupation type and land size (**Farm** farming, **Flab** farm labour, **Fish** fishing, **MixedForestDep** mixed forest dependent occupations, **Man** small-scale manufacturing, **Bus** small service-oriented business, **LL** landless/homestead, **SLO** small land owner, **LLO** large land owner). The highlighted household types (i.e. red box) are the most sensitive based on the mean sensitivity value

References

BBS. 2011. *Report of the household income and expenditure survey 2010.* Dhaka: Bangladesh Bureau of Statistics (BBS).

Dasgupta, S., M. Hossain, M. Huq, and D. Wheeler. 2014. *Climate change, soil salinity, and the economics of high-yield riceproduction in coastal Bangladesh.* Policy Research Working Paper 7140. Washington, DC: Environment and EnergyTeam, Development Research Group, World Bank. http://documents.worldbank.org/curated/en/131161468004833954/pdf/WPS7140.pdf. Accessed 9 Aug 2016.

Ferdousi, S., and W. Dehai. 2014. Economic growth, poverty and inequality trend in Bangladeah. *Asian Journal of Social Sciences and Humanities* 3 (1): 1–11.

Lázár, A.N., D. Clarke, H. Adams, A.R. Akanda, S. Szabo, R.J. Nicholls, Z. Matthews, D. Begum, A.F.M. Saleh, M.A. Abedin, A. Payo, P.K. Streatfield, C. Hutton, M.S. Mondal, and A.Z.M. Moslehuddin. 2015. Agricultural livelihoods in coastal Bangladesh under climate and environmental change – A model framework. *Environmental Science-Processes and Impacts* 17 (6):1018–1031. https://doi.org/10.1039/c4em00600c.

Mondal, M.K., S.I. Bhuiyan, and D.T. Franco. 2001. Soil salinity reduction and prediction of salt dynamics in the coastal ricelands of Bangladesh. *Agricultural Water Management* 47 (1): 9–23. https://doi.org/10.1016/s0378-3774(00)00098-6.

Payo, A., A.N. Lázár, D. Clarke, R.J. Nicholls, L. Bricheno, S. Mashfiqus, and A. Haque. 2017. Modeling daily soil salinity dynamics in response to agricultural and environmental changes in coastal Bangladesh. *Earth's Future.* https://doi.org/10.1002/2016EF000530.

SRDI. 2012. *Analytical methods: Soil, water, plant material and fertilizer.* 2nd ed. Dhaka: Soil Resource Development Institute (SRDI), Ministry of Agriculture, Government of the People's Republic of Bangladesh.

Appendix 2: ΔDIEM Model Scenario Inputs

Summary tables of indicator changes within each ΔDIEM forcing scenario. Climate change scenarios Q0, Q8, Q16 (see Chap. 11) and socio-economic scenarios LS (Less Sustainable), BAU (Business As Usual), MS (More Sustainable) (see Appendix to Chap. 10 for scenario narratives).

Indicator	Units	Scenario	Changes from 2015		Notes
Temperature			2050	2098	
Average	°C	Q0	2.4	3.7	
temperature		Q8	2.5	3.7	
maximum		Q16	2.8	4.1	

Indicator	Units	Scenario	Changes from 2000			Notes
Sea level			2030	2050	2098	
Sea level	Cm	Low	12	24	54	Subsidence: 2.5 mm/year (reference to year 2000)
		High	20	45	148	

Indicator	Units	Scenario	Changes from 2015			Notes
Water			2030	2050	2098	
Total monsoon season rainfall (June–Sep.)	%	Q0	2	5	0	
		Q8	−17	−21	−9	
		Q16	3	−4	14	
Total dry season rainfall (Dec.–Feb.)	%	Q0	−37	25	0	
		Q8	−57	−56	−38	
		Q16	30	−3	29	
Total number of dry days per year	%	Q0	1	−2	−1	Dry day =precipitation below 1 mm
		Q8	8	5	6	
		Q16	−1	1	0	
Length of longest consecutive dry day period	%	Q0	−23	−24	−1	
		Q8	−4	−3	28	
		Q16	7	20	1	
Total annual rainfall	%	Q0	6	10	11	
		Q8	−23	−21	−13	
		Q16	2	−3	14	

(continued)

(continued)

Indicator	Units	Scenario	Changes from 2015			Notes
Water			2030	2050	2098	
Ganges annual maximum discharge	%	Q0	5	10	19	
		Q8	13	18	31	
		Q16	24	35	67	
Brahmaputra annual maximum discharge	%	Q0	5	13	17	
		Q8	3	−5	2	
		Q16	12	39	82	
Meghna annual maximum discharge	%	Q0	7	11	20	
		Q8	11	17	31	
		Q16	25	33	64	
Water sharing	Farraka treaty arrange-ments	LS	Constant: 30,000 cfs guaranteed			
		BAU	Constant: 35,000 cfs guaranteed			
		MS	Constant: 40,000 cfs guaranteed			
Water transfers	% flow reduction	No change				
		Small transfer (5% of Brahmaputra flow)			Negligible	
		Large transfer (20% of Ganges, 30% of Brahmaputra flow)			22 (monsoon), 48 (dry season)	
Cyclone frequency	Per year	Set manually using illustrative cyclones				

Indicator		Scenario	Changes from 2015			Notes
Infrastructure			2030	2050	2098	
Embankment height	Metres	All scenarios	Unchanged from design height			
Travel time to main settlements	%	LS	−10	−10	−10	
		BAU	−30	−30	−30	
		MS	−50	−50	−50	

Indicator	Units	Scenario	Changes with 2015		Notes
Land use			2050	2098	
Agricultural land cover	%	LS	−8	−26	
		BAU	3	7	
		MS	−3	−3	
Urban land cover	%	LS	No change		
		BAU	No change		
		MS	92	92	
Rural settlement land cover	%	LS	No change		
		BAU	No change		
		MS	−1	−1	

(continued)

(continued)

Indicator Land use	Units	Scenario	Changes with 2015 2050	2098	Notes
Mangrove land cover within the Unions	%	LS	−10	−19	
		BAU	−5	−15	
		MS	92	92	
Crops grown	Crop name	LS	1990s crops		
		BAU	Present day crops		
		MS	Present day crops		
Crop varieties	% yield improve- ment	LS	Present day (0)		
		BAU	Present day (0)	15	
		MS	15	50	

Indicator Socioeconomics	Units	Scenario	Changes from 2015 2030	2050	2098	Notes
Population	%	LS	1.1	−0.7	−25.1	
		BAU	0.5	−7.7	−32.9	
		MS	−3.8	−7.2	−36.7	
Literacy rate	%	LS	2	2	2	
		BAU	4	4	4	
		MS	8	8	8	
Children in school	%	LS	2	2	2	
		BAU	5	5	5	
		MS	10	10	10	
Employment rate	%	LS	0	0	0	
		BAU	10	10	10	
		MS	30	30	30	
Total migration	%	LS	3	6	6.5	
		BAU	2.2	4.8	6	
		MS	3	5	5	
Migration rate	%	LS	0	−50	−100	No further migration assumed beyond 2050
		BAU	−37	−37	−100	
		MS	−25	−75	−100	
Food variety	Kcal/day	All scenarios	Unchanged (1738–2422 with 48–88 protein)			

References

Amoako Johnson, F., C.W. Hutton, D. Hornby, A.N. Lázár, and A. Mukhopadhyay. 2016. Is shrimp farming a successful adaptation to salinity intrusion? A geospatial associative analysis of poverty in the populous Ganges–Brahmaputra–Meghna Delta of Bangladesh. *Sustainability Science* 11 (3): 423–439. https://doi.org/10.1007/s11625-016-0356-6.

BARC. 2013. *Agricultural technology for Southern Region of Bangladesh* (in Bangla). Dhaka: Bangladesh Agricultural Research Council (BARC). https://archive.org/details/AgriculturalTechnologySouthernRegion. Accessed 26 Oct 2016.

Caesar, J., T. Janes, A. Lindsay, and B. Bhaskaran. 2015. Temperature and precipitation projections over Bangladesh and the upstream Ganges, Brahmaputra and Meghna systems. *Environmental Science-Processes and Impacts* 17 (6): 1047–1056. https://doi.org/10.1039/c4em00650j.

Clark, D. 1975. *Understanding canonical correlation analysis, Concepts and techniques in modern geography No. 3.* Norwich: Geo Abstracts Ltd.

Clarke, D., M. Smith, and K. El-Askari. 1998. New software for crop water requirements and irrigation scheduling. *Irrigation and Drainage* 47 (2): 45–58.

Clarke, D., S. Williams, M. Jahiruddin, K. Parks, and M. Salehin. 2015. Projections of on-farm salinity in coastal Bangladesh. *Environmental Science-Processes and Impacts* 17 (6): 1127–1136. https://doi.org/10.1039/c4em00682h.

FAO, and MA. 2013. *Master plan for agricultural development in the Southern Region of Bangladesh.* Food and Agriculture Organization of the United Nations (FAO) and Government of the People's Republic of Bangladesh, Ministry of Agriculture (MA).

Farakka Treaty. 1996. *Treaty between the Government of the People's Republic of Bangladesh and the Government of the Republic of India on sharing of the Ganga/Ganges water at Farakka.* 36 I.L.M 523 (1997).

Fernandes, J.A., S. Kay, M.A.R. Hossain, M. Ahmed, W.W.L. Cheung, A.N. Lázár, and M. Barange. 2016. Projecting marine fish production and catch potential in Bangladesh in the 21st century under long-term environmental change and management scenarios. *ICES Journal of Marine Science* 73 (5): 1357–1369. https://doi.org/10.1093/icesjms/fsv217.

Kay, S., J. Caesar, J. Wolf, L. Bricheno, R.J. Nicholls, A.K.M.S. Islam, A. Haque, A. Pardaens, and J.A. Lowe. 2015. Modelling the increased frequency of extreme sea levels in the Ganges-Brahmaputra-Meghna delta due to sea level rise and other effects of climate change. *Environmental Science-Processes and Impacts* 17 (7): 1311–1322. https://doi.org/10.1039/c4em00683f.

Lázár, A.N., D. Clarke, H. Adams, A.R. Akanda, S. Szabo, R.J. Nicholls, Z. Matthews, D. Begum, A.F.M. Saleh, M.A. Abedin, A. Payo, P.K. Streatfield, C. Hutton, M.S. Mondal, and A.Z.M. Moslehuddin. 2015. Agricultural livelihoods in coastal Bangladesh under climate and environmental change – A model framework. *Environmental Science-Processes and Impacts* 17 (6): 1018–1031. https://doi.org/10.1039/c4em00600c.

Lázár, A.N., H. Adams, W.N. Adger, and R.J. Nicholls. 2016. *Characterising and modelling households in coastal Bangladesh for model development.* Project working paper. ESPA Deltas.

MathWorks. 2015. *MATLAB and statistics toolbox release 2015b.* Natick: The MathWorks, Inc.

Oreskes, N., K. Shraderfrechette, and K. Belitz. 1994. Verification, validation and confirmation on numerical-models in the earth sciences. *Science* 263 (5147): 641–646. https://doi.org/10.1126/science.263.5147.641.

Payo, A., A.N. Lázár, D. Clarke, R.J. Nicholls, L. Bricheno, S. Mashfiqus, and A. Haque. 2017. Modeling daily soil salinity dynamics in response to agricultural and environmental changes in coastal Bangladesh. *Earth's Future.* https://doi.org/10.1002/2016EF000530.

Szabo, S., D. Begum, S. Ahmad, Z. Matthews, and P.K. Streatfield. 2015. Scenarios of population change in the coastal Ganges Brahmaputra Delta (2011–2051). *Asia Pacific Population Journal* 30 (2): 51–72.

Whitehead, P.G., E. Barbour, M.N. Futter, S. Sarkar, H. Rodda, J. Caesar, D. Butterfield, L. Jin, R. Sinha, R. Nicholls, and M. Salehin. 2015. Impacts of climate change and socio-economic scenarios on flow and water quality of the Ganges, Brahmaputra and Meghna (GBM) river systems: Low flow and flood statistics. *Environmental Science-Processes and Impacts* 17 (6): 1057–1069. https://doi.org/10.1039/c4em00619d.

29

Communicating Integrated Analysis Research Findings

Mashrekur Rahman and Md. Munsur Rahman

29.1 Introduction

A key component of any research is to ensure that results and implications are communicated to those most affected and those in a position to successfully implement any response as well as those who have committed time and effort to the research; in this case it is the relationship between ecosystem services and poverty at a community and social-ecological system level to the local communities who provided survey information (Chap. 23) and decision-makers. This helps the configuration and adoption of adaptive responses to a changing environment, which, with the historical and projected changes discussed earlier in this book, necessitates the modes of survival and livelihoods found within the delta to adapt and evolve. Understanding the context, having situational awareness and executing plans through partnership form an indispensable part of effective research dissemination at a community level. It is therefore

M. Rahman (✉) • Md. Munsur Rahman
Institute of Water and Flood Management, Bangladesh University of Engineering and Technology, Dhaka, Bangladesh

© The Author(s) 2018
R. J. Nicholls et al. (eds.), *Ecosystem Services for Well-Being in Deltas*,
https://doi.org/10.1007/978-3-319-71093-8_29

important to provide information and perspectives of practical value to communities and associated Non-Governmental Organisations (NGOs) as well as local government.

There are diverse ways in which research can be disseminated at a community and local governance level. To reach and be understood by the remotest of rural population in Bangladesh, knowledge is best translated into a non-scientific, jargon-free and practical package. This may consist of a variety of tailored dissemination tools, such as indigenous cultural performances or press releases. There are pros and cons to each of these tools in reaching audiences, and therefore a combination of these tools constitutes a stronger dissemination package and increases the chances of meaningful impact.

29.2   Engaging News Media

Bangladesh in recent years has seen an impressive rise in the number of local and national newspapers, news channels, online news portals and radio stations. The number of internet users in Bangladesh surpassed 60 million in 2016 according to Bangladesh Telecommunication Regulatory Commission[1], allowing news media to deliver news of relevant interest to millions of rural people, directly to their homes. Translating knowledge from research findings to newsworthy segments requires certain strategies and ethical considerations before dissemination. Indeed, contextual awareness forms a major part of the strategy for research dissemination in Bangladesh; cultural differences, understanding what the public are willing to accept and political interests are some of the aspects to be aware of while engaging with the media. Critically, language should be easy to understand for the general population. For example, in case of newspaper articles and blogs, it is imperative to publish in both native Bengali and English languages to encourage uptake. Bengali newspapers have overwhelmingly larger reader numbers when compared to English language newspapers particularly in rural Bangladesh. Social media can be an effective tool to further proliferate impact.

A significant portion of the rural population in Bangladesh is either illiterate or has limited literary skills. Printed media cannot therefore ensure inclusiveness. Television and radio, as audio-visual media, have the capacity to improve this although there are still millions of people in the coastal region without access to electricity. For television and radio appearances, it is vital to plan ahead on the questions and answers to ensure the findings are not misrepresented or misinterpreted. Planning ahead also averts the possibility of revealing unconfirmed findings, which may cause confusion or ill-founded responses. Overtly political statements should be avoided or, where strictly necessary, made indirectly to avoid conflict. Proper acknowledgement of contributors is also a substantial part of research ethics (Sula 2016).

Research described in this book has been published in various national newspapers and blogs. A brief presentation on the significance of ecosystem services and the project was delivered at the First Climate Leaders Convention 2015 in Dhaka, the Fourth Asia Youth Summit in Japan and various other national and international forums. An interview with Prof. Dr. Md. Munsur Rahman was published by the Bengali news daily *Kaler Kantho* in August 2016. In the interview, a variety of questions were answered ranging from the goals and vision of research to the development of Bangladesh Delta Plan 2100. In the same month, an article titled 'Children of the Bengal Delta' was published in the *Daily Observer Bangladesh*, followed by three consecutive articles in the month of September 2016: 'Bangladesh's rich ecosystem and its role in poverty alleviation', 'Exploring the changes in ecosystem resources' and 'Ecosystem science, policy and people'. These articles were aimed at raising awareness on ecosystem protection in the wake of climate change and summarised the research in a manner understandable by the reader. After publication, these materials were shared in various social media platforms to increase their reach and to generate feedback from readers. Feedback was mostly positive; the general consensus from those who responded on ecosystem protection was overwhelmingly supportive of the work. Readers also emphasised the value of the larger and more comprehensive circulated news pieces in national Bangladesh papers to build support for protecting ecosystems against inappropriate development and industrialisation.

There are obvious challenges while engaging the national media. One of the major challenges is persuading the newspapers of the significance of this information, with some papers preferring to prioritise political stories over environmental issues. Building a relationship with media entities in Bangladesh is often time consuming. A dedicated media communication programme is therefore beneficial to expedite dissemination of research in the country.

29.3   Dissemination to Stakeholders

Disseminating research findings at a local and regional level is aimed at ensuring inclusiveness, and a regional to local strategy should be adopted. This approach advocates that the most practical and enduring method for passing information back to communities is to engage with the regional network of NGO's who in turn have sustained links into the local population. Within the study area, a series of stakeholder workshops (Fig. 29.1) with the most remote of the local coastal population of Bangladesh were found to be particularly beneficial. The specific objectives of these workshops were (i) to get feedback and input from regional stakeholders on research findings and appropriate dissemination techniques, (ii) to review the current status of local issues through communication with local participants and (iii) to introduce indigenous 'Pot Song' as a research dissemination

Nalchity (02/08/2016)   Tala (04/08/2016)   Dumuria (04/08/2016)

Sharankhola (05/08/2016)   Koyra (06/08/2016)   Bhandaria (07/08/2016)

Fig. 29.1 Regional level stakeholder workshops. The total number of people involved in the process was 360, representing local and regional institutions (Photographs: Mashrekur Rahman)

tool for rural populations. These workshops were held in six locations across the coastal belt of Bangladesh: (i) Nalchity, Jhalokathi; (ii) Tala, Satkhira; (iii) Dumuria, Khulna; (iv) Sharankhola, Bagerhat; (v) Koyra, Khulna; and (vi) Bhandaria, Pirojpur. Among these locations, Sharankhola and Koyra are in close proximity to the Sundarbans, and often serve as entry-points into the mangrove forest. Tala and Dumuria are located inside the Western Estuarine System of Bangladesh and are subject to a wide range of environmental issues. Nalchity and Bhandaria face their own unique set of challenges.

29.3.1   Pot Songs: Talking with Communities in Their Own Language

Pot songs (Bengali: পটসঙ্গীত) are south-west Bangladesh's indigenous cultural performances which entertain hundreds of thousands of people at a community level. These songs are traditionally performed by local entertainment groups at village congregations and events where they depict certain social situations through rolling paintings, improvised lyrics and traditional dancing. Low-income people in remote coastal regions, who rarely have access to print or electronic media, draw much pleasure from these performances, while the songs can subtly convey vital social messages. Although pot songs have previously been used to raise awareness on social responsibility and raising disaster awareness, they are rarely used to disseminate specific research findings.

Pot songs were performed at the six regional level workshops (Fig. 29.2). The song used paintings depicting ecosystem degradation scenarios and the ways in which rural population can partake in improving them. The bespoke song written for these workshops attempted to cover all the research findings (Nicholls et al. 2015, 2016) in a very non-technical language and using local examples. It begins with a warm welcome to the listeners and continues by describing visible and perceptible changes in the weather and environment, gradually linking to global climate change, potential impacts on ecosystems and ultimately consequences for lives and livelihoods. The song also makes the listeners aware of the consequences of sea-level rise, land subsidence, salinity intrusion, the inter-relationship between ecosystems and human

Fig. 29.2 Pot Song based on ecosystem research being performed at Nalchity, Jhalokathi (Photograph: Mashrekur Rahman)

well-being (Hossain et al. 2017) and why a healthy ecosystem has important benefits. Agricultural malpractices such as use of chemical fertilisers and pesticides, overexploitation of land resources, deforestation and deliberately allowing saline water into land for shrimp aquaculture are discouraged in the song. The song lyrics then include warnings on depleting groundwater, dwindling river flow and predicted salinity intrusion, reiterating the importance of reducing water pollution and conserving freshwater. As the song nears its ending, the protective role of the Sundarbans is highlighted: that people living in the study area are shielded from cyclones and storm surges by the mangroves (Sakib et al. 2016) and therefore afforestation is encouraged. At the end of the song, unity among people, cooperation with the government and non-government bodies, and taking a stance against corruption are emphasised. The last line translates into 'survival is not possible without saving the environment'.

This effort was widely appreciated by participants taking part in the workshop, including the participants from remote places who came there to make their voices heard. *"Not everybody can understand scientific terms; particularly the underprivileged population, who have not been exposed to quality education amenities"*, said one participant who came from an area damaged by Cyclone Sidr. *"This song is an extremely entertaining way to know about our ecosystem services and how they are quintessential to our survival and what we can do to protect it"*, he added.

Attendees in general wanted to see this song played in rural centres, bazaars, schools and other places of congregation for wider circulation of knowledge through entertainment. Audiences suggested that these messages may also be delivered using staged dramas, storytelling and other local forms of entertainment. Therefore, it may be recommended that these types of performances are staged regularly at the most remote coastal regions. Documentaries on such performances may be broadcast by television channels to augment their popularity. More tailored dissemination tools and other versions of the song may be introduced to further bolster the research dissemination process.

29.3.2   Stakeholder Workshops Outcomes

Among the guests and participants at the stakeholder workshops were *Upazila* Chairman, vice-chairman, government officers, local administrative staffs, teachers, community leaders, people's representatives, professionals, farmers, fishermen, journalists and people from various other walks of life. The mix of participants was equally diverse for all of the workshops.[2] Following the programme, presentations and the Pot Song performance, an interactive open discussion session was held with the participants, the key points from which are summarised in Table 29.1 and Fig. 29.3.

29.3.3   Stakeholder Feedback

These stakeholder workshops and pot songs enabled participants to engage in meaningful discussions about the issues regarding ecosystem degradation and its impact on their lives and livelihoods. Dissemination of scientific research, through indigenous entertainment performances such as this pot song, allows the most laid-back of remote population to understand the perils facing them. It may even empower these people with the

Table 29.1 Summary of key issues discussed during the stakeholder workshops

Location	Issues raised	Action required
Nalchity, Jhalokathi	Overfishing in rivers and canals	Awareness raising
	Fishing during restricted months	Better governance
	Use of poisonous substances to catch fish	Institutional reforms for proper implementation of future policies
	River bank erosion	Greater engagement of government engineering
	River depth maintenance	Alternate employment opportunities
	Persistent floods	Better water management
	Potable water crisis	
Tala, Satkhira	Riverbed siltation	Riverbed dredging and bank protection
	Inundation and waterlogging inside polders	Repair and maintenance of existent water control infrastructures
	Decreasing rice cultivation and proliferation of aquaculture	Rehabilitation of waterlogged population inside polders
	Rise of salinity levels leading to freshwater scarcity	Area-specific strategies
Dumuria, Khulna	Frequent structural breach of polders causing damage to people and property, dislocating people (Fig. 29.3)	Better monitoring of polder maintenance by government authorities
	Substandard construction and repair of polders	Good governance
	River and canal grabbing by influential people	
Sharankhola, Bagerhat	Fishermen are being exploited by 'Mahajon' (fishermen lords) and pirates	Increased policing from government law enforcement authorities
	Increasing number of fishers leading to lower catches per fisher	Alternate employment for fishermen
	Decreasing fish diversity	Raising awareness on the protection of ecosystem
	Rise of multiple sand bars at the Bay of Bengal	Better water management and maintenance of existent water management infrastructures
	Use of poisonous chemicals to catch fish also led to some cases of Tiger death	
	'Current nets' are being used to trap fish hatchlings	
	Honey collectors are not adhering to the conventional collection times	
	Rampant deforestation inside Sundarbans	
	Drinking water crisis	

Location	Issues	Recommendations
Koyra, Khulna	Riverbank erosion Shrimp farmers deliberately damage polders to allow saline water inland Decreasing catches of fish and shrimp hatchlings Malnutrition of mother and child due to degradation of ecosystem services Water scarcity despite the use of deep tube-well and rainwater harvesting Organic fertilisers are not available in bulk quantities Increased water scarcity post-Aila due to rising freshwater salinity levels Substandard and inadequate polder repair and maintenance Sundarbans dependent people are facing extortion from pirates Government authorities are not proactive about protecting the ecosystem of Sundarbans	Repair and maintenance of embankments Policing and monitoring of polders Shift towards organic farming Better water management strategies Greater policing and monitoring of polder repair and maintenance by concerned authorities Transparency and accountability in the use of government funds Good governance and institutional reform More proactive stance on protecting Sundarbans from government bodies, especially the department of forestry
Bhandaria, Pirojpur	Repeated inundation and bank erosion lead to salt water infiltration during high tides, causing damage to seed and plant beds Arsenic contamination is present in some areas Lack of tube-wells and sanitation problems have exacerbated the freshwater crisis The effect of dry season low flow, delayed monsoon and rapid flood flow during monsoon can be felt Crop diseases, insecticide-resistant crop insects, depleting productivity of soil General disregard towards organic fertilisers	Embankment for protection from flooding Better freshwater water management strategies Greater engagement of the government agriculture authorities with the farmers Promotion of organic fertilisers

Fig. 29.3 Locals living in shanties on polder 29 after it was breached by the Bhadra River (Photograph: Mashrekur Rahman)

knowledge to cope with future scenarios. Pot songs in the six stakeholder workshops discussed in this chapter were appreciated by the audiences, particularly for the innovative path it took to reach them at a very basic level. Such performances, if used more widely for proliferating research findings, have the potential to generate significant impact among the vulnerable coastal population of Bangladesh, an opinion seconded by many audiences. It is however important to realise that different communities respond to diverse tailored dissemination tools. NGOs and government institutions attending these workshops were intrigued by the level of penetration these performances were able to achieve. They expressed their willingness to use such research dissemination techniques in their future endeavours to reach communities. It is important to ensure expert consultation while developing these dissemination tools, and also take into account political and religious sensitivity, especially considering the conservative nature of audiences in remote regions. Audience feedbacks should be recorded for future reference and may be used to improve on the performances. Local languages and cultures differ quite noticeably from place to place, opening opportunities of area-specific dissemination tools to further bolster their reach. Government and NGOs can collaborate in the future to use such Pot Song performances to better disseminate research findings, as well as deliver important social messages which may eventually result in elevated socio-economic stature and reduced vulnerabilities of the coastal population.

29.4   Conclusions

Plenty of scientific research conducted in Bangladesh ends up on bookshelves and in scientific archives, inaccessible for the mass population—depriving people from knowledge and benefits. Disseminating research findings effectively to people who need it most is probably as important as conducting the research itself. The true value of science is only realised when it benefits humankind and protects the planet's precious natural resource systems. Fresh new ideas and innovative approaches to such dissemination such as the pot songs can be the way forward.

Notes

1. www.btrc.gov.bd
2. For participants, see Participating organisation list at front of this book.

References

Hossain, M.S., F. Eigenbrod, F. Amoako Johnson, and J.A. Dearing. 2017. Unravelling the interrelationships between ecosystem services and human wellbeing in the Bangladesh delta. *International Journal of Sustainable Development & World Ecology* 24 (2): 120–134. https://doi.org/10.1080/135 04509.2016.1182087.

Nicholls, R.J., P. Whitehead, J. Wolf, M. Rahman, and M. Salehin. 2015. The Ganges-Brahmaputra-Meghna delta system: Biophysical models to support analysis of ecosystem services and poverty alleviation. *Environmental Science-Processes and Impacts* 17 (6): 1016–1017. https://doi.org/10.1039/c5em90022k.

Nicholls, R.J., C.W. Hutton, A.N. Lázár, A. Allan, W.N. Adger, H. Adams, J. Wolf, M. Rahman, and M. Salehin. 2016. Integrated assessment of social and environmental sustainability dynamics in the Ganges-Brahmaputra-Meghna delta, Bangladesh. *Estuarine, Coastal and Shelf Science* 183, Part B: 370–381. https://doi.org/10.1016/j.ecss.2016.08.017.

Sakib, M., F. Nihal, A. Haque, M. Rahman, R. Akter, M. Maruf, M. Akter, S. Noor, and R.A. Rimi. 2016. *Afforestation as a buffer against storm surge flooding along the Bangladesh coast.* 12th International Conference on Hydroscience Engineering Hydro-Science; Engineering for Environmental Resilience, November 6–10, Tainan.

Sula, C.A. 2016. Research ethics in an age of big data. *Bulletin of the Association for Information Science and Technology* 42 (2): 17–21. https://doi.org/10.1002/bul2.2016.1720420207.

Erratum to: Ecosystem Services for Well-Being in Deltas: Integrated Assessment for Policy Analysis

Robert J. Nicholls, Craig W. Hutton, W. Neil Adger, Susan E. Hanson, Md. Munsur Rahman, and Mashfiqus Salehin

Erratum to:

The Author(s) 2018
R. J. Nicholls et al. (eds.), *Ecosystem Services for Well-Being in Deltas*,
https://doi.org/10.1007/978-3-319-71093-8

The book was inadvertently published with an incorrect spelling of the author's name in Chapter 5 as Sarwar Hossain whereas it should be Md. Sarwar Hossain.

In addition to this, the name of the author in Chapter 18 should be Md. Mahabub Arefin Chowdhury instead of Shahad Mahabub Chowdhury and the affiliation should be Institute of Water and Flood Management, Bangladesh University of Engineering and Technology, Dhaka, Bangladesh.

The updated online version of these chapters can be found at
https://doi.org/10.1007/978-3-319-71093-8_5
https://doi.org/10.1007/978-3-319-71093-8_18

© The Author(s) 2018
R. J. Nicholls et al. (eds.), *Ecosystem Services for Well-Being in Deltas*,
https://doi.org/10.1007/978-3-319-71093-8_30

Glossary of Bengali Words

Aman A type of rice crop which is grown in the *kharif* season. There are two types of *Aman* (transplanted and broadcasted); broadcasted tend to be used in the delta area.

Aus A rice crop which is planted during the *kharif* season in the delta

Bagda Giant tiger shrimp (*Penaeus* sp.); also used to refer to socio-ecological system with brackish shrimp production as its primary ecosystem service.

Beel A wetland area covered by static rather than running water.

Boro A rice crop usually planted November–December and harvested in April–May. High-yielding *Boro* requires irrigation.

Charland Vegetated sandbar island within a river or estuary; also used to refer to socio-ecological system found on the islands.

Division Administrative unit within Bangladesh. At the time of this research there were seven divisions: Chittagong, Dhaka, Khulna, Rajshahi, Barisal, Sylhet and Rangpur. Mymensingh Division, outside the study area, was created in 2015.

Galda Giant freshwater prawn (*Macrobrachium rosenbergii*); also used to refer to socio-ecological system with freshwater prawn production as its primary ecosystem service.

Ghers Ponds used for aquaculture (shrimp, prawn, white fish or rice or a combination) in Bangladesh.

Ilish/ilisha Locally important fish species Hilsa shad (*Tenualosa ilisha*).

© The Author(s) 2018
R. J. Nicholls et al. (eds.), *Ecosystem Services for Well-Being in Deltas*,
https://doi.org/10.1007/978-3-319-71093-8

Jhupris Basic type of housing in Bangladesh which have rooves made of inexpensive materials (e.g. straw, bamboo, grass, leaves, polythene, gunny bags).

Kacha Type of housing in Bangladesh, often temporary, usually made from organic materials. Usually bamboo framed.

Kharif Cropping season (April–November).

Khas land Government-owned fallow land where nobody has property rights.

Mouza A Bengali district that approximates a village.

Pucka and *semi-pucka* Housing structure in Bangladesh. Pucka houses are constructed of inorganic materials being mainly concrete or brick. *Semi-pucka* houses are generally timber framed with foundations made of earth, brick and/or concrete. The wall cladding is most often bamboo mats or wood panelling.

Rabi Cropping season (November–May). Usually vegetables, pulses and *Boro* rice are grown during *Rabi* season.

Samaj Refers to the broader community or society that convenes a 'shalish'.

Shalish Refers to a small-scale local council, distinct from the formal village court, which is convened for the purposes of civil and criminal conflict resolution. It dates back to pre-British times. While it is formally recognised as a mediation body, it has no legal authority in relation to criminal cases or marriage and dowry disputes. It is commonly used as a forum for the adjudication of community disputes, and although rulings are required to be formally registered with the police station, this is not common in practice. Views differ as to the impartiality and quality of decision-making.

Taka/BDT Bangladesh unit of currency.

Upazila/Union Administrative unit in Bangladesh, sub-division of a zila. In the mid-1980s thanas were named upazilas (smaller than zila); renamed thanas in 1991 but still commonly used.

Zila Administration district within Bangladesh below Division. A zila is made up of upazilas (Unions).

Index

© The Author(s) 2018
R. J. Nicholls et al. (eds.), *Ecosystem Services for Well-Being in Deltas*,
https://doi.org/10.1007/978-3-319-71093-8